The Melanotropic Peptides

Volume I: Source, Synthesis, Chemistry, Secretion, Circulation, and Metabolism

Editor

Mac E. Hadley, Ph.D.
Department of Anatomy
University of Arizona
Tucson, Arizona

CRC Press, Inc.
Boca Raton, Florida

Library of Congress Cataloging-in-Publication Data

The Melanotropic peptides / editor. Mac E. Hadley.
   p.   cm.
   Contents: v. 1. source, synthesis, chemistry, secretion,
circulation, and metabolism — v. 2. Biological roles — v.
3. Mechanisms of action and biomedical applications.
   Includes bibliographies and index.
   1. MSH (Hormone)   I. Hadley, Mac E.
QP572.M75M45 1988
[DNLM: 1. MSH--physiology.   WK 515 M517]
596'.0142--dc19
DNLM/DLC
   ISBN 0-8493-5277-0 (v. 1)
   ISBN 0-8493-5278-9 (v. 2)   87-29903
   ISBN 0-8493-5279-7 (v. 3)   CIP

This book represents information obtained from authentic and highly regarded sources. Reprinted material is quoted with permission, and sources are indicated. A wide variety of references are listed. Every reasonable effort has been made to give reliable data and information, but the author and the publisher cannot assume responsibility for the validity of all materials or for the consequences of their use.

All rights reserved. This book, or any parts thereof, may not be reproduced in any form without written consent from the publisher.

Direct all inquiries to CRC Press, Inc., 2000 Corporate Blvd., N.W., Boca Raton, Florida, 33431.

© 1988 by CRC Press, Inc.

International Standard Book Number 0-8493-5277-0 (Volume I)
International Standard Book Number 0-8493-5278-9 (Volume II)
International Standard Book Number 0-8493-5279-7 (Volume III)

Library of Congress Card Number 87-29903
Printed in the United States

PREFACE

The three volumes on "*The Melanotropic Peptides*" are the outcome of a conference of the same name that was held in Tucson, Arizona, from October 11-12, 1986.

The melanotropic peptides are the hormones that are generally recognized as playing a major role in the control of vertebrate integumental pigmentation. The two physiologically relevant melanotropins are α-melanocyte-stimulating hormone (α-MSH, α-melanotropin) and melanin-concentrating hormone (MCH). Because α-MSH and corticotropin (adrenal cortical-stimulating hormone, ACTH) share some structural similarity, these other melanocorticotropins possess MSH-like activity when used at higher (unphysiological) concentrations. A quick glance at the topics that are included within these volumes makes it clear that these peptidergic hormones probably also play important roles in the control of a number of diverse physiological processes.

The format of the three volumes provides a complete coverage of what is known about the Melanotropic Peptides. Volume I provides information on the source, synthesis, chemistry, mechanism of secretion, control of secretion, and the circulation and metabolism of the melanotropic peptides. In addition, a chapter on the pharmacokinetics of secreted MSH as determined by RIA, is provided.

Volume II discusses the diverse putative physiological roles of MSH and MCH. A concluding chapter of this volume provides a discussion of the possible evolutionary relationships between MSH and MCH in the control of color change. Volume III discusses the known mechanisms of action of the two melanotropic peptides, and concludes with a discussion of the possible biomedical applications of the melanotropins.

As editor, I am indebted to each of the authors who have contributed their time, effort, and expertise to their particular chapter. Although we can, of course, expect advancements in our knowledge of the melanotropins, these volumes will provide the primary source of information on the melanotropic peptides for many years to come.

THE EDITOR

Mac Eugene Hadley, Ph.D., is a professor in the Department of Anatomy at the University of Arizona. He also holds a joint appointment with the Department of Molecular and Cellular Biology.

Dr. Hadley received his doctorate degree from Brown University in 1966. He then did his postdoctoral work at the University of Arizona after which he became an assistant professor, associate professor, and a full professor at the same institution.

Dr. Hadley was trained as a biologist and developed a particular interest in endocrinology. He has written a popular textbook of *Endocrinology* which is used in many college classes.

Dr. Hadley's major research interest has been the endocrine control of vertebrate integumental pigmentation. He has published numerous manuscripts and reviews in leading scientific journals. He is a member of the International Pigment Cell Society which presented him the Myron Gordon Award in 1986 for his research in the area of the neuroendocrine control of pigmentation.

The broad scope of Dr. Hadley's research interests is reflected in the range of topics that comprise the three volumes on *"The Melanotropic Peptides."*

CONTRIBUTORS

Volume I

Saida Adjeroud, Ph.D.
INSERM
Faculté des Sciences
Université de Rouen
Mort-St.-Aignan, France

Ann C. Andersen, Ph.D.
INSERM
Faculté des Sciences
Université de Rouen
Mont-St.-Aignan, France

Arlette Burlet, Ph.D.
Laboratoire d'Histologie A
INSERM
Faculté de Medecine
Vandoeuvre-les-Nancy
France

Ana Maria de Lauro Castrucci, Ph.D.
Assistant Professor
Department of Physiology
University of São Paulo
São Paulo, Brazil

Bibie M. Chronwall, Ph.D.
Associate Professor
University of Missouri-Kansas City
School of Basic Life Sciences
Division of Structural and Systems
  Biology
Kansas City, Missouri

Jean-Michel Danger, Ph.D.
INSERM
Faculté des Sciences
Université de Rouen
Mont-St.-Aignan, France

Robert M. Dores, Ph.D.
Assistant Professor
Department of Biological Sciences
University of Denver
Denver, Colorado

Stela Elkabes, Ph.D.
Visiting Fellow
National Institutes of Health
Bethesda, Maryland

Mac E. Hadley, Ph.D.
Professor
Department of Anatomy
University of Arizona
Tucson, Arizona

Victor J. Hruby, Ph.D.
Professor
Department of Chemistry
University of Arizona
Tucson, Arizona

Bruce G. Jenks, Ph.D.
Department of Animal Physiology
University of Nijmegen
Nijmegen, The Netherlands

Hiroshi Kawauchi, Ph.D.
Professor
Laboratory of Molecular Endocrinology
School of Fisheries Sciences
Kitasato University
Sanriku, Iwate, Japan

Gotfryd Kupryszewski, Ph.D.
Laboratory of Bioorganic Chemistry
Institute of Chemistry
University of Gdansk
Gdansk, Poland

Marek Lamacz, Ph.D.
INSERM
Faculté des Sciences
Université de Rouen
Mont-St.-Aignan, France

Philippe Leroux, Ph.D.
INSERM
Faculté des Sciences
Université de Rouen
Mont-St.-Aignan, France

Y. Peng Loh, Ph.D.
National Institutes of Health
Bethesda, Maryland

Philip J. Lowry, Ph.D., D.Sc., F.I. Biol.
Professor
Department of Physiology and Biochemistry
University of Reading
Reading, England

Gerard J. M. Martens, Ph.D.
Research Associate
Departments of Animal Physiology and Molecular Biology
University of Nijmegen
Nijmegen, The Netherlands

Brenda Myers, B.A.
Student
Department of Biology
Johns Hopkins University
Baltimore, Maryland

Tom O'Donohue, Ph.D. (Deceased)
Director
CNS Research
Monsanto Company
St. Louis, Missouri

Georges Pelletier
MRC Group for Molecular Endocrinology
Centre Hospitalier de L'Université Laval
Laval, Quebec, Canada

Elizabeth K. Perryman, Ph.D.
Professor
Department of Biological Sciences
California Polytechnic State University
San Luis Obispo, California

Lise Stoeckel, Ph.D.
CNRS
Laboratoire de Physiologie Générale
Université Louis Pasteur
Strasbourg, France

Marie-Christine Tonon, Ph.D.
INSERM
Faculté des Sciences
Université de Rouen
Mont-St.-Aignan, France

Hubert Vaudry, D.Sc.
Directeur de Recherche
INSERM
Départment d'Endocrinologie Moleculaire
Université de Rouen
Mont-St.-Aignan, France

B. M. Lidy Verburg-van Kemenade, Ph.D.
Department of Animal Physiology
University of Nijmegen
Nijmegen, The Netherlands

John F. Wilson, Ph.D.
Department of Pharmacology and Therapeutics
University of Wales College of Medicine
Cardiff, Wales

CONTRIBUTORS

Volume II

Zalfa A. Abdel-Malek, Ph.D.
Assistant Professor
Department of Dermatology
College of Medicine
University of Cincinnati
Cincinnati, Ohio

Joseph T. Bagnara, Ph.D.
Professor
Department of Anatomy
University of Arizona
Tucson, Arizona

Bridget I. Baker, Ph.D.
Doctor
School of Biological Sciences
University of Bath
Bath, England

C. Wayne Bardin, M.D.
Director of Biomedical Research
Center for Biomedical Research
The Population Council
New York, New York

Bill E. Beckwith, Ph.D.
Associate Professor
Department of Psychology
University of North Dakota
Grand Forks, North Dakota

Carla Boitani, Ph.D.
Research Associate
Institute of Histology and General
  Embryology
University of Rome, "La Sapienza"
Rome, Italy

Ana Maria de Lauro Castrucci, Ph.D.
Assistant Professor
Department of Physiology
University of São Paulo
São Paulo, Brazil

Ching-Ling Chen, Ph.D.
Scientist
Center for Biomedical Research
The Population Council
New York, New York

P. De Koning, Ph.D.
Institute of Molecular Biology and
  Medical Biotechnology
University of Utrecht
Utrecht, The Netherlands

A. Dell, Ph.D.
Department of Biochemistry
Imperial College London
London, England

P. M. Edwards, Ph.D.
Institute of Molecular Biology and
  Medical Biotechnology
University of Utrecht
Utrecht, The Netherlands

W. H. Gispen, Ph.D.
Professor
Institute of Molecular Biology and
  Medical Biotechnology
University of Utrecht
Utrecht, The Netherlands

Mac E. Hadley, Ph.D.
Professor
Department of Anatomy
University of Arizona
Tucson, Arizona

K. L. Henville, Ph.D.
Department of Biochemistry
St. Bartholomew's Hospital Medical
  College
London, England

J. P. Hinson, Ph.D.
Department of Biochemistry
St. Bartholomew's Hospital Medical
 College
London, England

Omid Khorram, Ph.D.
Medical Student IV
School of Medicine
Texas Tech University
Lubbock, Texas

James M. Lipton, Ph.D.
Professor
Department of Physiology
University of Texas
Health Science Center at Dallas
Dallas, Texas

Samuel M. McCann, Ph.D.
Professor
Department of Physiology
University of Texas
Health Science Center at Dallas
Dallas, Texas

H. R. Morris, Ph.D.
Department of Biochemistry
Imperial College London
London, England

Patricia L. Morris, Ph.D.
Research Investigator
Center for Biomedical Research
The Population Council
New York, New York

Chandrima Shaha, Ph.D.
National Institute of Immunology
JNU Complex
New Delhi, India

Wade C. Sherbrooke, Ph.D.
Resident Director
Southwestern Research Station
American Museum of Natural History
Portal, Arizona

Anthony J. Thody, Ph.D.
Reader in Experimental Dermatology
Department of Dermatology
University of Newcastle-upon-Tyne
Newcastle-upon-Tyne, England

C. E. E. M. Van der Zee, Ph.D.
Institute of Molecular Biology and
 Medical Biotechnology
University of Utrecht
Utrecht, The Netherlands

R. Gerritsen van der Hoop, Ph.D.
Institute of Molecular Biology and
 Medical Biotechnology
University of Utrecht
Utrecht, The Netherlands

J. Verhaagen, Ph.D.
Institute of Molecular Biology and
 Medical Biotechnology
University of Utrecht
Utrecht, The Netherlands

Gavin P. Vinson, D.Sc.
Professor
Department of Biochemistry
St. Bartholomew's Hospital Medical
 College
London, England

**Brian Weatherhead, M.A., Ph.D.,
C.Biol., F.I. Biol.**
Professor and Head of Department
Department of Anatomy
University of Hong Kong
Hong Kong

B. J. Whitehouse, Ph.D.
Department of Physiology
King's College London (KQC)
London, England

Catherine A. Wilson, Ph.D.
Doctor
Department of Obstetrics and
 Gynaecology
St. George's Hospital Medical School
London, England

CONTRIBUTORS

Volume III

Fahad Al-Obeidi, Ph.D.
Department of Chemistry
University of Arizona
Tucson, Arizona

Ana Maria de Lauro Castrucci, Ph.D.
Assistant Professor
Department of Physiology
University of São Paulo
São Paulo, Brazil

Dhirendra Chaturvedi, Ph.D.
Senior Chemist
Research and Development Laboratories
Vega Biotechnologies, Inc.
Tucson, Arizona

Wayne L. Cody, Ph.D.
Research Laboratory
Eastman Chemicals Company
Kingsport, Tennessee

Pierre N. E. de Graan, Ph.D.
Doctor
Department of Molecular Neurobiology
Rudolf Magnus Institute for
  Pharmacology and
Institute of Molecular Biology and
  Biotechnology
University of Utrecht
Utrecht, The Netherlands

Alex N. Eberle, Ph.D.
Head, Laboratory of Endocrinology
University Hospital
Basel, Switzerland

Jürg Girard, Ph.D.
Professor
Department of Pediatric Endocrinology
University Hospital
Basel, Switzerland

Mac E. Hadley, Ph.D.
Professor
Department of Anatomy
University of Arizona
Tucson, Arizona

Victor J. Hruby, Ph.D.
Professor
Department of Chemistry
University of Arizona
Tucson, Arizona

Aaron B. Lerner, Ph.D.
Department of Dermatology
Yale University School of Medicine
New Haven, Connecticut

Gisela E. Moellmann, Ph.D.
Associate Professor
Department of Dermatology
Yale University School of Medicine
New Haven, Connecticut

John M. Pawelek, Ph.D.
Senior Research Scientist
Department of Dermatology
Yale University School of Medicine
New Haven, Connecticut

Tomi K. Sawyer, Ph.D.
Senior Research Scientist III
Biotechnology Division
The Upjohn Company
Kalamazoo, Michigan

Walter Siegrist, Ph.D.
Professor
Department of Pediatric Endocrinology
University Hospital
Basel, Switzerland

J. W. Sam Stevenson, Ph.D.
Research Chemist
Department of Physical and Analytical
  Chemistry Research
Tennessee Eastman Company
Kingsport, Tennessee

Elizabeth Sugg, Ph.D.
Department of Chemistry
University of Arizona
Tucson, Arizona

TABLE OF CONTENTS

Volume I

Chapter 1
The Melanotropic Peptides: An Introduction 1
Philip J. Lowry

Chapter 2
Melanotropins: Pars Intermedia Structure and Secretion 5
Elizabeth K. Perryman

Chapter 3
Pituitary Melanotropin Biosynthesis 25
Robert M. Dores

Chapter 4
The Melanotropic Peptides: Structure and Chemistry 39
Hiroshi Kawauchi

Chapter 5
Melanocyte-Stimulating Hormone in the Central Nervous System 55
Bibie Chronwall and Thomas L. O'Donohue

Chapter 6
The Pro-Opiomelanocortin Gene in *Xenopus Laevis:* Structure, Expression, and
Evolutionary Aspects 67
Gerard J. M. Martens

Chapter 7
Regulation of Pro-Opiomelanocortin Biosynthesis in the Amphibian and Mouse Pituitary
Intermediate Lobe 85
Y. Peng Loh, Stella Elkabes, and Brenda Myers

Chapter 8
Pro-Opiomelanocortin in the Amphibian Pars Intermedia: a Neuroendocrine Model
System 103
Bruce G. Jenks, B. M. L. Verburg-van Kemenade, and Gerard J. M. Martens

Chapter 9
Multihormonal Control of Melanotropin Secretion in Cold-Blooded Vertebrates 127
**M. C. Tonon, J. M. Danger, M. Lamacz, P. Leroux, S. Adjeroud, A. Andersen,
B.M.L. Verburg-van Kemenade, B. G. Jenks, G. Pelletier, L. Stoekel, A. Burlet,
G. Kupryszewski, and H. Vaudry**

Chapter 10
Melanotropin Enzymology 171
Ana Maria de L. Castrucci, Mac E. Hadley, and Victor J. Hruby

Chapter 11
Peripheral and Central Pharmacokinetics of the Melanotropins 183
John F. Wilson

Index 211

TABLE OF CONTENTS

Volume II

Chapter 1
Melanotropins and Melanin Pigmentation of the Skin of Mammals 1
Brian Weatherhead

Chapter 2
Melanotropins, Chromatophores, and Color Change 21
Joseph T. Bagnara

Chapter 3
Melanotropin Effects on Pigment Cell Proliferation 29
Zalfa A. Abdel-Malek

Chapter 4
The Melanotropins, Learning and Memory 43
Bill E. Beckwith

Chapter 5
Melanotropins and Peripheral Nerve Regeneration 73
C. E. E. M. Van der Zee, P. M. Edwards, R. Gerritsen van der Hoop, P. De Koning, J. Verhaagen, and W. H. Gispen

Chapter 6
The Actions of α-MSH on the Adrenal Cortex 87
Gavin P. Vinson, B. J. Whitehouse, K. L. Henville, J. P. Hinson, A. Dell, and H. R. Morris

Chapter 7
MSH in CNS Control of Fever and its Influence on Inflammation/Immune Responses.. 97
James M. Lipton

Chapter 8
The Presence and Possible Function of α-MSH and Other POMC-Derived Peptides in the Reproductive Tract 115
Carla Boitani, Patricia L. Morris, Ching-Ling Chen, Chandrima Shaha, and C. Wayne Bardin

Chapter 9
The Role of Melanotropins in Sexual Behavior 131
Anthony J. Thody and Catherine A. Wilson

Chapter 10
Neuroendocrine Effects of Melanotropins 145
Omid Khorram and Samuel M. McCann

Chapter 11
Melanin-Concentrating Hormone 159
Bridget I. Baker

Chapter 12
Melanotropic Peptides and Receptors: An Evolutionary Perspective in Vertebrates..... 175
Wade C. Sherbrooke, Mac E. Hadley, and Ana Maria de L. Castrucci

Index 191

TABLE OF CONTENTS

Volume III

Chapter 1
Melanotropin Bioassays 1
Mac E. Hadley and Ana Maria de L. Castrucci

Chapter 2
Melanotropin Mechanisms of Action: Melanosome Movements 15
Mac E. Hadley and Ana Maria de L. Castrucci

Chapter 3
The Role of Protein Phosphorylation in the Mechanism of Action of α-MSH 27
Pierre N. E. De Graan and Alex N. Eberle

Chapter 4
Melanotropin Mechanisms of Action: Melanogenesis 47
John M. Pawelek, John McLane, and Michael Osber

Chapter 5
Melanotropins and Receptor Signal Transduction 59
Tomi K. Sawyer, Mac E. Hadley, and Victor J. Hruby

Chapter 6
Cyclic Conformationally Constrained Melanotropin Analogues: Structure-Function and
Conformational Relationships 75
Wayne L. Cody, Mac E. Hadley, and Victor J. Hruby

Chapter 7
Melanotropin Three-Dimensional Structural Studies by Physical Methods and Computer-
Assisted Molecular Modeling...
**Wayne L. Cody, J. W. Sam Stevenson, Fahad Al-Obeidi, Elizabeth Sugg,
and Victor J. Hruby**

Chapter 8
Melanotropin Receptors: Studies with Labeled Melanotropins 111
Alex N. Eberle, Pierre N. E. De Graan, Walter Siegrist, and Jürg Girard

Chapter 9
Melanotropins: Biomedical Applications 129
Dhirendra Chaturvedi and Mac E. Hadley

Chapter 10
The Melanotropic Peptides: A Summary 151
Aaron B. Lerner and Gisela E. Moellmann

Index 153

Chapter 1

THE MELANOTROPIC PEPTIDES: AN INTRODUCTION

Philip J. Lowry

It was early in this century that the involvement of the pituitary gland in the control of color change was first recognized. Naturally, experiments were performed on amphibia since by simple observation they could be seen to change their pigmentation in adapting to background color. By surgical removal of the pituitary rudiments it could be shown that larval amphibians developed into pale individuals,[1,2] and that either implanting pituitaries into pale tadpoles or using pituitary extracts it was concluded that a humoral factor was responsible in causing the dispersion of melanin in the melanophores of the skin, this movement of pigment being seen as an overall darkening. Thereafter, it was generally agreed that where an anatomically distinct pars intermedia existed then it was there that most of the melanotropic bioactivity resided.

Although it was the amphibians which provided the means of measuring melanotropic activity, it was the domestic mammals which were to provide a rich source of the material responsible for their final characterization. To add to the then already confused state of the knowledge of the endocrinology of these hormones in mammals, during the 1950s, two distinct peptides were purified and characterized and thus termed α- and β-MSH.

After sequence analysis, both peptides were found to be related structurally, possessing the common heptapeptide core sequence Met-Glu-His-Phe-Arg-Trp-Gly which was shown to possess inherent melanotropic activity (for review see Reference 3). Since it was recognized that α-MSH had the same amino-acid sequence as the N-terminal portion of corticotropin and the β-MSH sequence occupied the mid-portion of β-lipotropin (β-LPH), it seemed natural to postulate that these larger peptides acted as precursors for the smaller melanotropins. The discovery of the C-terminal fragment of ACTH (CLIP) and its occurrence in the intermediate lobe[4] and later on the C-terminal fragment from β-LPH the opiate peptide β-endorphin provided further evidence for this hypothesis. Since the melanotropic activity had been observed in early studies to reside mainly in the intermediate lobe, it was proposed that ACTH and β-LPH were synthesized in both the corticotrophs of the pars distalis and all the cells of the pars intermedia. In the former, the parent peptides were secreted intact, whereas in the pars intermedia they were fragmented by specific enzymic mechanisms and modified (e.g., acelytation) to form the smaller fragments, particularly the melanotropins.[5] One peptide which did not fit into this pattern was human β-MSH, but this was later found to be an extraction artifact rather than a true processed peptide.[6]

It was the elegant work of Mains and Eipper (for review see Reference 7) who originally used the rat ATt20 pituitary tumor-cell line and later extended their studies to the pars intermedia, who, by using a variety of antisera and pulse-chase experiments, demonstrated the existence of a large precursor for ACTH and LPH of a molecular weight of some 30,000 daltons. A similar conclusion was reached by Roberts and Herbert[8] who immunologically characterized the translated product in vitro of m-RNA isolated from the neurointermediate lobe.

The cloning and sequencing of the c-DNA encoding the whole of the ACTH/LPH precursor, finally resolved this fascinating phenomenon.[9] More importantly, they had used as their starting material mRNA isolated from the pars intermedia not normally thought of as a source of ACTH and β-LPH. They demonstrated, however, from this pars-intermedia-derived material that ACTH occupied the mid-portion and β-LPH the C-terminal of a large precursor. These peptides were separated from each other by pairs of basic amino acids and from a large N-terminal peptide which contained a third melanotropic sequence (γ-MSH).

This γ-MSH precursor had been identified earlier by Mains and Eipper[7] to be glycosylated and had been termed the 16K fragment, although they did not realize at that time that it contained a melanotropic sequence. Thus, it would appear that three melanotropic sequences occur within the intermediate lobe (and pars distalis) precursor for these peptides. While the cloning of this precursor has vastly increased our knowledge of the way in which these peptides are biosynthesized, it has done little to help in our understanding of their biological significance. The situation in the lower vertebrates is not so much a problem since here it can be clearly demonstrated that they do at least change color and do respond to physiological changes in blood melanotropin levels by changing color. In mammals, the situation is difficult to fully understand. First of all, there are a number of mammals which have little or no intermedia tissue at all, this includes: man, the higher apes, elephants, manatees, whales, dolphins, and the armadillo. When attempts have been made, little evidence has been found for the existence of melanotropins being manufactured in the pars distalis corticotrophs and certainly in man, the only time when α-MSH can be identified has been in the fetal pituitary where a band of cells resembling intermedia tissue can be observed. Even then, the peptide was found to be lacking N-terminal acetylation,[10] and a peptide resembling β-MSH has never been identified.[11] If the pars intermedia in the other mammals where the melanotropins are found displayed little biosynthetic activity, then one could assign the tissue to join the ranks of the vestigial organs. Unfortunately, intense biosynthetic activity can be seen, particularly if the dopaminergic tonic inhibition is removed either surgically or pharmacologically. Despite the hypersecretion of several other peptides along with the melanotropins under these conditions, most scientists agree that in mammals at least, it is extremely difficult to see any physiological change in the animals. At the 1987 Ciba Foundation Symposium, we suggested that species with a highly evolved brain, indeed, the species listed above represent some of the most intelligent in the animal kingdom, have lost the need for a pars intermedia in adult life and lent credence for a role for the pars intermedia peptides in learning and behavior.[11] Despite the ensuing 5 years of intense scientific activity, little evidence has come to light of a convincing function for the pars intermedia and the melanotropins in mammals. The problem is compounded by the fact that when a weak, but unique biological action can be assigned to for example α-MSH, because at least six other peptides with other unique biological actions are co-secreted with it, the physiological consequences are difficult to explain. Even in the lower vertebrates, the co-secretion of three melanotropins in an animal such as the dogfish appears a little paradoxical. Immunoneutralizing α-MSH with an antiserum which only recognizes the heptapeptide core of α-MSH caused dark animals to pallor.[12] The antiserum, because of its specificity, however, would not be expected to immunoneutralize γ-MSH and β-MSH where the core sequence is changed despite their melanotropic activity. What other biological actions the other POMC peptides are expressing in these animals is a matter for further research.

Let us hope that at the culmination of this meeting on melanotropins held in Tucson, Arizona, new aspects of this tantalizing subject will come to light and we will be able to at last assign a function to the mammalian pars intermedia and the peptides it elaborates, especially the melanotropins.

REFERENCES

1. **Smith, D. C.**, The effect of hypophysectomy in the early embryo upon the growth and development of the frog, *Anat. Rec.*, 11, 57, 1916.
2. **Atwell, W. J.**, On the nature of the pigmentation changes following hypophysectomy in the frog larva, *Science*, 49, 48, 1919.

3. **Hofman, K.,** Chemistry and function in polypeptide hormones, *Ann. Rev. Biochem.,* 31, 213, 1962.
4. **Scott, A. P., Ratcliffe, J. G., Rees, L. H., Landon, J., Bennett, H. P. J., Lowry, P. J., and McMartin, C.,** A new pituitary peptide, *Nature, (London) New Biol.,* 224, 65, 1973.
5. **Lowry, P. J. and Scott, A. P.,** The evolution of ACTH and MSH, *Gen. Comp. Endocrinol.,* 26, 16, 1975.
6. **Bloomfield, G. A., Scott, A. P., Silkes, J. J. H., Lowry, P. J., and Rees, L. H.,** Human beta-MSH, a reappraisal. *Nature,* 252, 492, 1974.
7. **Mains, R. E. and Eipper, B. A.,** Biosynthetic studies on ACTH, β-endorphin and α-melanotropin in the rat, *Ann. N.Y. Acad. Sci.,* 343, 94, 1980.
8. **Roberts, J. L. and Herbert, E.,** Characterization of a common precursor to corticotropin and β-lipotropin, Identification of β-lipotropin peptides and their arrangement relative to corticotropin in the precursor synthesized in a cell free system, *Proc. Natl. Acad. Sci. U.S.A.* 74, 5300, 1977.
9. **Nakanishi, S., Inoue, A., Kita, T., Nakamura, M., Chang, A. C. Y., Cohen, S. N., and Numa, S.,** Nucleotide sequence of cloned c-DNA for bovine corticotropin-β-lipotropin precursor, *Nature,* 278, 423, 1978.
10. **Tilders, F. J. H., Parker, C. R., Barnea, A. and Porter, J. C.,** The major immunoreactive melanocyte stimulating hormone (γ-MSH)-like substance found in human fetal pituitary is not α-MSH but may be desacetyl α-MSH (adrenocorticotropin 1-13-NH$_2$), *J. Clin. Endocrinol., Metab.,* 52, 319, 1981.
11. **Jackson, S., Hope, J., Estivariz, F. E., and Lowry, P. J.,** Nature and control of peptide release from the pars intermedia, in *Peptides of the Pars Intermedia, Ciba Foundation Symposium 81,* Evered, D. and Lawrenson, G., Eds., Pitman Medical, London, 1981, 141.
12. **Sumpter, J. P., Denning-Kendall, P. A., and Lowry, P. J.,** The involvement of melanotropins in physiological colour change in the dogfish, *Scyliorhinus canicula, Gen. Comp. Endocrinol.,* 56, 360, 1984.

Chapter 2

MELANOTROPINS: PARS INTERMEDIA STRUCTURE AND FUNCTION

Elizabeth K. Perryman

TABLE OF CONTENTS

I. Introduction 6

II. Morphology of the Pars Intermedia 6
    A. Overview of Development 6
    B. Comparative Aspects of Gross Anatomy 6
    C. Vascular Supply 7

III. Granular Cells 8
    A. Cytology 8
    B. Innervation 10

IV. Nongranular Cells 11
    A. Cytology 11
        1. Folliculo-Stellate System 11
        2. Follicle and Cleft 14
    B. Suggested Roles for Nongranular Cells 15
    C. Extracellular Channel System 17

V. Concluding Remarks 19

Acknowledgments 19

References 19

I. INTRODUCTION

While this review of the structure of the pars intermedia (PI) and the morphological aspects of secretion will emphasize the recent scientific advances, it is important to consider the past literature which has led to these discoveries. Thus, an effort will be made to relate the data of the last 10 years to important concepts already established by the more classical anatomical techniques. Also, at times, a comparison of the PI of various vertebrates will be used in order to provide a more comprehensive picture. Because of the present limitation of space and previously published volumes, it is suggested that the reader be aware of the following publications. An excellent comprehensive volume with a comparative approach is *The Pituitary Gland*[1] published in 1974. More recent data of the 1980s are included in a review of the PI by Weatherhead,[2] relating structure with function, and also the book, *Ultrastructure of Endocrine Cells and Tissues*,[3] which contains chapters on fine structure of various portions of the pituitary gland. Within the present review, the morphology of the components (e.g., cell types, follicles, vascularity) will be related to melanotropin secretion.

II. MORPHOLOGY OF THE PARS INTERMEDIA

A. Overview of Development

The pituitary gland has two embryonic origins. The neurohypophysis differentiates from the neural ectoderm of the infundibular floor of the third ventricle of the brain; the adenohypophysis forms from Rathke's pouch, an oral ectodermal invagination from the stomodeum. The neurohypophysis in tetrapods and lungfish is divided into two neurohemal areas, an anterior median eminence and a posterior neural lobe or pars nervosa (PN). The form and degree of anatomical association of the neurohypophysis with the adenohypophysis vary in the different vertebrate classes. Evidently, the neural ectoderm induces the formation of the pars intermedia (PI) where this ectoderm makes contact with the hindmost part of Rathke's pouch. No PI forms in vertebrates, such as birds, in which no contact is made. Artificial barriers to the passage of the inducers can experimentally block the formation of the PI, as shown especially in amphibians. Thus, in most vertebrates, there is a close contact between the neurohypophysis and the PI. With the development of the pars distalis (PD) of the adenohypophysis, some investigators now support the possibility of a neural contribution. Until recently, it was believed that Rathke's pouch was formed entirely by oral ectoderm. Now with the Amine or Amine Precursor Uptake and Decarboxylation (APUD) hypothesis proposed by Pearse,[4,5] there is the concept of adenohypophysial secretory cells derived from neural components that migrate to the site at some very early stage. Or, alternatively, possibly both cells of the nervous system and pituitary gland may retain the ability to independently produce various enzymes and regulators — the evidence upon which this hypothesis is based. The acronym APUD is applied to groups of cells which contain similar peptides and/or amines along with enzymes that catalyze conversion of amino acids to amines; thus they display similar immunocytochemical staining properties. This controversial hypothesis, with further supporting data, could therefore be used to explain the presence of these cells in such diverse sites as the brain, pituitary, digestive tract, etc. In support, neuron-specific enolase (NSE), which had been thought to be a marker confined to neurons, has been found in the peptide-secreting cells of the PI and PD and their tumors.[6,7,8]

B. Comparative Aspects of Gross Anatomy

The PI varies considerably in mammals; it may exist as only an epithelium of only two or three cells lining the hypophysial cleft or even be completely absent as in cetaceans (whales and dolphins). To the other extreme, it may be so large as to be lobulated. In many species of adult mammals, the PI is separated, to varying extents, from the PD by an intraglandular (hypophysial) cleft which is the remains of the lumen within Rathke's pouch. Even though the PI and PD are separated by the intraglandular cleft, the PI can be directly continuous with the PD superiorly and laterally. The continuity of the PI with the PD means that it is difficult to determine the exact boundary between the two. As in the mouse and ferret, there is an extension into the PI of cells characteristic of cells of the PD. In humans, the PI is present in the fetus, but is represented chiefly as a group of cysts or enlarged follicles in the adult. Also, follicles have been described as a regular feature in the PI of several mammals, including the ferret,[9] rat,[10] and in particular the jird,[11] *Meriones*, a desert gerbil, where they were found both within the PI and partially extending in the PN. The size of the PI may be correlated with the habitat of the animal. For example, in the jird, the PI represents 27% of the pituitary; this is compared to 0.2% in the garden doormouse. Thus, mammals living in a dehydrating environment are said to have a larger PI.[1]

The term "fish" covers several groups of vertebrates; within this group is the lamprey, a cyclostome, whose PI is posterior to the PD and fused with the overlying neural lobe. Only a thin connective tissue septum with capillary plexus separates the PI from the neural lobe. In the elasmobranchs, there is a very large posterior PI which is actually invaded by the neurohypophysis to form a fused neuro-intermedia (neuro-intermediate lobe). The degree of intermingling of the two varies with the species as does the arrangement of the PI cells themselves which may be arranged into a parenchymal mass or into lobes separated by a vascular connective tissue. Within the teleosts, the dominant fishes alive today, the distinctive feature is the interdigitation of the central neurohypophysis with the entire adenohypophysis, composed of the posterior PI and an anterior PD. These two are more closely fused than in tetrapods to form a compact adenohypophysis, with the interdigitations of the neurohypophysis with the PI usually more deep and elaborate than with the PD.

With the anuran adenohypophysis, the PD is flattened, elongated, and dorsally continuous with the PI and there is no hypophysial cleft between the two areas. The PI, together with the neural lobe, forms a so-described "dumbbell-shaped organ" which can be easily separated mechanically from the PD. Even though the plexus intermedius capillary network is between the PI and PN, there is some interdigitation of the two. Along the ascending scale of vertebrates, the amphibians are the first to have a pars tuberalis which, however, may be lacking in some groups of reptiles. In general, for the reptilian pituitary, the intermedia and distalis are not so separate as in amphibians. The PI can be unusually large in comparison to other vetebrates or, to the other extreme, it can be only a couple of cell layers, as in the lizard, *Klauberina riversiana*.[12] A hypophysial cleft may be present in the adult. Although attempts have been made to corrleate the size of the PI with background adaptation, the degree of development may even be related to the amount of direct sunlight experienced.[1]

C. Vascular Supply

When describing the vascularity of the PI parenchyma in relation to other areas of the pituitary, the phrase "relatively avascular" is commonly used. Thus, the peptide secretion must travel some distance from the site of production to the nearest capillaries. There is a capillary network, the plexus intermedius of Benda between the PI and the PN. For the toad, *Bufo*, Rodriguez and Piezzi[13] determined that this capillary network receives blood from three sources; the median eminence, the hypophysial artery, and the encephalo-neurohypophysial portal system. From the plexus intermedius are smaller capillaries to the periphery of the PI. As stressed by Holmes and Ball,[1] in the amphibians, fishes, reptiles, and mammals, blood within the intermedius has first passed through the PN; therefore, it may contain high concentrations of neurohypophysial hormones. Also, it should be emphasized that the intermedius receives blood from wide areas of the hypothalamus and mesencephalon. With a very thin reptilian PI; the plexus intermedius seems to be a sufficient blood supply, but in specimens with a massive PI (e.g., *Anolis*), the intermedius branches to continue within the intermedia tissue.[14,15] In those mammalian specimens with a larger parenchyma or even a lobulated mass, vessels are in the connective tissue at the periphery of the parenchyma or in the septa at the periphery of the individual lobules. In those species with an intraglandular cleft, the direct passage of vessels between the PD and PI is prevented, but vessels can cross into the PI by way of the rostral and caudal zones of continuity. Recently, an "intra-adenohypophysial portal system" has been described for circulation of blood from the PI to the PD; its functional significance is yet unknown.[16] As investigators have emphasized, the "low degree of vascularity of the PI seems hardly in accord with an endocrine function."[1] The intimate relationship of the secretory cells of the mammalian PD to capillaries is documented.[17] Lametschwandtner and co-workers[18,19] described in the toad PD, but not for the PI, a network of capillaries as forming polygonal meshes which enable individual secretory cells to have direct vascular contact.

III. GRANULAR CELLS

A. Cytology

The PI contains two basic cell types, granular and nongranular. The more numerous granular ones are classified as producing proteins and/or peptides, while a number of roles are suggested for the nongranular. Of the granular, the melanotroph is the most common. Its fine structure is clearly demonstrated in an amphibian responding to a black background (i.e., synthesizing and secreting melanophore-stimulating hormone, MSH).[20,21,22] The well-developed rough endoplasmic reticulum (rER), distributed in the greater part of the cytoplasm, is in parallel stacks or whorls (Figure 1A). Secondly, a filamentous precipitate is found in the dilated rER cisternae and there are 150 to 200 nm, electron-dense, membrane-bound granules, mostly near the Golgi regions. Near the plexus intermedius capillaries, the rER cisternae of active melanotrophs are likely to contain 4 to 0.5 μm electron-dense droplets[20,22] (Figure 1B). With the distalisectomized toad, Rodríguez and Cuello[23] described melanotrophs containing these so-named "colloid vesicles."[24] They were additionally present near vessels then seen penetrating the PI, indicating that the formation of the intracisternal droplets is related to the proximity of blood vessels. There has been confusion about the intracellular location and source of these vesicles. Because of their name, they were confused with the extracellular colloid of the hypophysial cleft. We now know that both the filamentous precipitate and intracisternal droplets of the rER are morphological characters of protein production.

Secretory granules can be found throughout the cytoplasm as illustrated with a frog responding by adaptation to a white background (i.e., storing not releasing MSH) (Figure 1C). These granules, differing from those associated with a Golgi complex, are larger in diameter, less electron-dense, and enclosed by a membrane not always continuous. These changes appear to be a result of retention within the cell or granule maturation.[25] Various ultrastructural changes associated with secretion can be quantified by stereological analysis. For example, when *Xenopus laevis* were transferred from a white to a black background, the melanotrophs showed over a three-day period a decrease of these larger granules, from approximately 37% of total cell volume occupied to approximately 10%. Also, indicating synthesis, the percentage of rER increased from approximately 5 to approximately 30%.[26] It was determined that these larger, mature, secretory granules contain various peptides (MSH, ACTH, β-endorphin) derived from the prohormone, pro-opiomelanocortin (POMC).[27] It is now acknowledged that newly synthesized POMC is intracellularly channeled from the rER through the Golgi complex and then packaged into secretory granules. Isolation studies indicate that the processing of the POMC occurs intragranularly and that the POMC-converting enzymes were found in both the membrane and soluble fractions of the secretory granule.[28,29] It is thought that the enzymes are not an integral protein of the granule membrane.

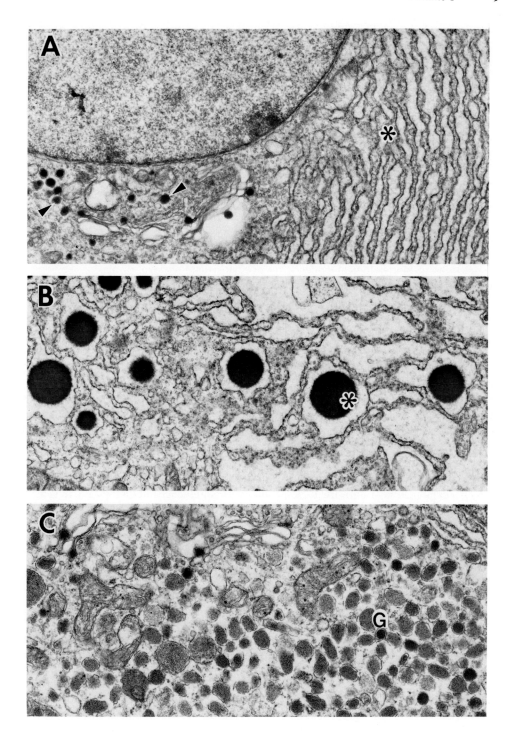

FIGURE 1. Melanotrophs of anuran PI. (A) Cell of animal adapted to black background; note well-developed rER (*) and secretory granules (arrowheads) at Golgi complex. Magnification × 12,400. (B) Dilated rER containing "colloid vesicles" (*). Magnification × 13,200. (C) Cell of animal adapted to white background; note secretory granules (G) filling cytoplasm. Magnification × 21,900.

Furthermore, Loh and Tam[30] reported that in the frog and the mouse, newly synthesized POMC is associated with the membrane of the secretory granule prior to processing, suggesting that such membrane association may be important in the intracellular sorting and transporting of prohormones to their destination, the secretory granule. Immunocytochemical data support these biochemical experiments, indicating processing at the level of secretory granules.[31] Additionally, Stoeckel and co-workers,[32] using pyroantimonate and electron microprobe techniques, localized calcium mainly in the Golgi saccules and suggested them as a storage site for calcium, which is essential for secretion.

Although it is accepted, as emphasized by Weatherhead,[2] that secretion in melanotrophs occurs by exocytosis, the classical, obvious interaction of granule and plasma membranes is not routinely observed. Secretory granules are often seen in close contact with the plasma membrane, and electron-dense material is seen in adjacent extracellular spaces.[33] With thin sections of sheep PI and freeze-fracture replicas of rat PI, Perry and co-workers[34] and Saland,[35] respectively, observed evidence of exocytosis. In the sheep fetus, the first appearance of fenestrated endothelium in capillaries at 100 days of gestation coincided with the detection of morphological evidence for exocytosis and retrieval of plasma membrane. Overall, these events were considerably more common in the late-term and newborn than in the adult.[34,36] Data exist for an alternative mode of secretion within the PI of the lizard, *Anolis*.[37,38] To summarize, secretory granules within the melanotroph fuse with large pale vacuoles which in turn connect to one another and then with the extracellular space by canaliculi. This mode could account for a rapid and large release of MSH, regardless of the distance to a capillary. Concerning released hormone, it is suggested that high levels of circulating α-MSH can act as a direct feedback control mechanism in teleosts.[39] Synthetic α-MSH released from an Alzet osmotic pump implanted into the abdomen of black-background-adapted fish decreased the metabolic activity of the melanotrophs so that they appeared comparable to those from an untreated fish on a white background. With background adaptation and intraocular transplants of neurointermediate lobes in frogs, the fine structure of the melanotrophs suggest that MSH can act directly at the pituitary gland to inhibit further release.[22,40] With data from in vitro incubations, others fail to support this concept.[41]

With a combination of ultrastructural and immunofluorescence microscopy, there is much documentation for an additional cell type, the corticotroph, within the mammalian PI. For the details describing this cell in the mouse, rat, and cat, the reader is referred to the review of the mammalian PI.[25] Generally, these corticotrophs were seen within the PI where it adjoins the PD or PN and also beneath the marginal cells lining the hypophysial cleft. In the sheep PI, these PD-like cells, comprising less than 5% of the glandular cells, were found throughout the width, but were frequently near the cleft.[34] Additionally, the PI of teleosts shows two distinct granular cell types when stained with periodic acid Schiff (PAS) and lead hematoxylin (PbH).[42] With electron microscopy and immunocytochemistry, the more predominant PbH-positive cell has been identified as producing MSH. The role of the PAS-positive one remains debatable, having responded to such stimuli as ionic changes and background adaptation.[43-46]

B. Innervation

Within the classes of vertebrates there is variation in the innervation of the PI, even a lack thereof in some reptiles.[15] Type I fibers (i.e., peptidergic, neurosecretory) are present in amphibians and localized in that region nearest the PN. Few immunocytochemical identifications of the peptides have been done, but in *Rana*, mesotocin and arginine-vasotocin fibers were confirmed.[47] In the mammalian PI, peptidergic fibers are more common in some members of the class but, for example, uncommon and restricted to areas of the PI near the neural lobe in the cat, rat, and mouse.[25] The PI of the rabbit and hare is abundantly innervated throughout with oxytocin-immunoreactive fibers.[48] In the rabbit, the absence of the dopamine fibers coincides with the lack of inhibitory effect of dopamine on in vitro superfused intermedia and with the absence of specific receptors known to mediate dopamine inhibition of α-MSH release in other species.[49] The presence of the oxytocin-containing fibers again raises the controversial question of the melanophore-inhibiting factor (MIF) characterized as the terminal three amino acids of oxytocin.

Type II fibers which predominate, especially in most mammalian PI, are accepted to be aminergic with most containing the inhibitory dopamine. Hopkins,[50] injecting the neurotoxic 6-hydroxydopamine (6-OHDA) into *Xenopus,* destroyed fibers not of the Type I category. With destruction of these fibers in *Xenopus,* there is, as measured by sterological analysis of the melanotrophs, increased hormone release, loss of secretory granules and increase in volume of rER and Golgi apparatus.[51] Similarly, in rats, the degeneration of Type II endings is coupled with enlargement of the Golgi areas within the melanotrophs.[52] Thus, the tonic inhibitor, dopamine, is inhibiting synthesis. As recently demonstrated, the enzymes involved in processing POMC are under dopaminergic control.[53,54] In light of the data indicating that 6-OHDA does not necessarily destroy all the dopaminergic fibers of the mammalian PI, it may be unsuitable for ultrastructurally distinguishing these from the proposed cholinergic fibers.[25] Also, possibly some of these Type II fibers are the γ-aminobutyric acid (GABA)-containing fibers demonstrated in the rat PI.[55] Verburg-Van Kemenade and co-workers[56] immunocytochemically demonstrated a rich GABAergic network in the PI of *Xenopus;* administration of GABA to superfused neurointermediate lobes caused inhibition of release of MSH. Recently, serotonin-immunoreactive fibers were demonstrated in the PI of the rat.[57] It was supposed previously that the presence of serotonin was related to the mast cells rather than to axons.[25] With the immunocytochemical identification of fibers, there is renewed interest in neural control by way of a network between the melanotrophs. But, as Weatherhead[2] cautioned, "in any consideration of the regulation of the PI it is essential not to restrict the discussion solely to the nerve fibers which directly innervate the secretory cells . . .". Because of the apparent lack of nerves in some species and the close proximity of the PN and PI, separated by the plexus intermedius, there may be a humoral route for modulators of secretion. With the absence of direct innervation in *Anolis* PI, this humoral route provides access for monoamine control.[58,59] Schimchowitsch and colleagues[48] affirmed in the rabbit the flow of neurohypophysial hormones from the PN to the PI by way of the perivascular space associated with the plexus intermedius. In addition, there are now data to support speculation for the nongranular folliculo-stellate cell of the PI to act in this control scheme (*cf.* Section IV.B. and C).

IV. NONGRANULAR CELLS

A. Cytology
1. Folliculo-Stellate System

Rinehart and Farquhar,[60,61] studying the ultrastructure of the rat anterior pituitary, first revealed the presence of certain star-shaped (stellate) nongranular cells with processes extending between the other parenchymal cells. Other studies followed describing the nongranular cells as stellate or follicular (forming follicles) in the adenohypophysis of various vertebrate classes, for example in PI of amphibians,[20,33,62] reptiles,[63-65] and mammals.[10,11,66] The term "folliculo-stellate" (FS)[66] has been accepted by most researchers as naming the same nongranular cell which has two aspects. Moreover, from developmental stages of the reptilian pituitary, Pearson and Licht[65] followed the extension of a FS system composed of marginal (lining the cleft), follicular, and stellate cells; it is composed of a single class of cell derived from the lining of Rathke's pouch. Because much remains to be learned about the role(s) of the FS cells, they will be discussed at some length.

The amphibian FS cell will be used as the model to describe this nongranular cell type with comparative references to the other vertebrate groups. Within the amphibian PI, they

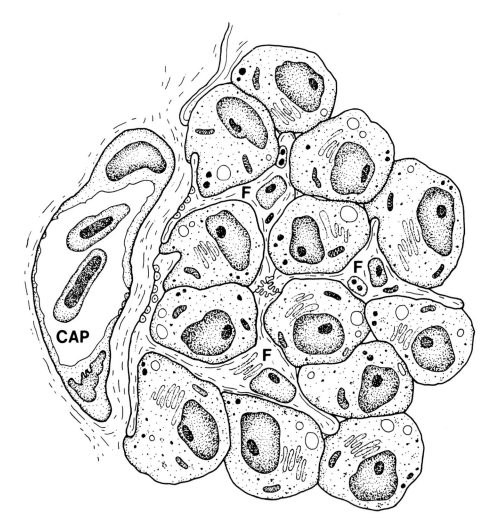

FIGURE 2. Diagram of area of anuran PI as it borders a capillary (CAP) of the plexus intermedius network located between the PI and PN. Note arrangement of folliculo-stellate cells (F) as their "end-feet", some with pinocytotic vesicles, abut the pericapillary space. Processes of the FS cells extend between the secretory cells and meet to form a small follicle at their point of interdigitation. (Note: pinocytotic vesicles have been enlarged to stress their presence.)

form a thin capsule that almost encloses the secretory cells. At the surface of the gland and beneath the basement membrane of the capillaries at the junction of the PI and the PN is a narrow envelope of FS cells. Usually their cell bodies are found between the secretory cells in the interior of the gland. Their long processes extend through the intercellular spaces between the secretory cells to terminate at the pericapillary space. Some of these terminations, referred to as "end-feet" are expanded. Figure 2 is a diagram summarizing the relationship between the FS cells and the other components near the PI and PN junction. Note also that these nongranular cells form small follicle-like structures where their slender extensions interdigitate with one another. These spaces are sometimes very small and usually are not as large as those in the PD of mammals or even amphibians. As emphasized,[67] FS cells have elongated nuclei; lack characteristic secretory granules and prominent rER and Golgi complexes; lysosomes are usually few. They do show numerous mitochondria as well as microfilaments and microtubules in both their cytoplasmic processes (Figure 3A) and "end-

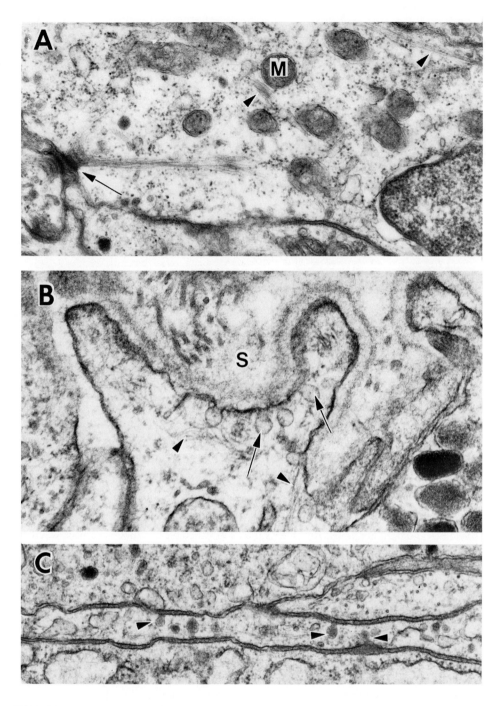

FIGURE 3. Processes and "end-feet" of folliculo-stellate cells in the frog. (A) Process extending from cell body; note mitochondria (M), microfilaments (arrowheads), and desmosome (arrows). Magnification × 46,300. (B) "End-foot" abutting on pericapillary space (S); note microfilaments (arrowheads) and pinocytotic vesicles (arrows). Magnification × 59,500. (C) Process extending between two melanotrophs; note vesicles (arrowheads) attached to plasma membrane of FS cell. Magnification × 31,000. (B and C from Perryman, E. K., *Cell Tiss. Res.*, 164, 387, 1975. With permission.)

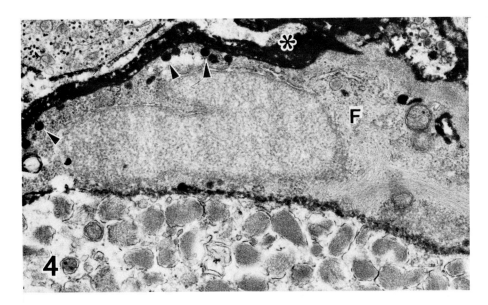

FIGURE 4. Folliculo-stellate cell (F) after injection of HRP. The tracer is largely accumulated in the pericapillary space (*) upon which the cell abuts. Also the HRP is in the pinocytotic vesicles (arrowheads) within the cell. Magnification × 34,300. (From Perryman, E. K. and Bagnara, J. T., *Cell Tiss. Res.*, 193, 297, 1978. With permission.)

feet'' (Figure 3B). As in the expanded ''end-feet'' in the lizard PI,[63] pinocytotic vesicles are frequently seen scattered along their plasma membranes (Figure 3B). Also, vesicles can be seen associated with the plasma membranes of the processes in the interior of the gland (Figure 3C).[33] Using horseradish peroxidase (HRP) injected intracardially into the frog, we demonstrated that this electron-dense marker moves from the capillaries into the pericapillary space, and into these vesicles of the nongranular cells (Figure 4). In addition, this marker freely moves by way of the intercellular spaces into the interior of the PI within 2 min after injection.[62,68] Cell junctions can be seen linking FS cells to one another and, at times, to secretory cells (Figure 3A). In earlier studies of the adenohypophysis, authors proposed desmosomes and tight junctions. But, later studies, in which we and others[68-70] used electron-dense tracers, do not support the presence of this impermeable tight junction in the frog or rat PI. Recently, with both HRP- and freeze-etch techniques, Krisch and Buchheim[71] described in the rat PD maculae adhaerentes (desmosomes) and small maculae occludentes which are not barriers to intercellular transport. Vila-Porcile and Oliver,[67] in a review of nongranular cells, also affirm the presence of both desmosomes and a modified tight junction.

2. Follicle and Cleft

Although most of the follicle-like structures in the amphibian PI are small, transparent intercellular spaces (Figure 2), enlarged follicles containing electron-dense amorphous material are observed. Nongranular cells were recognized as forming their boundaries. These follicles were seen, at times, to be associated with an extension of the pericapillary space from the plexus intermedius. Thus, it appears that the apex of these cells can contact with the lumen of the follicle itself while the nonluminal portions abut onto the extensions of the pericapillary space. Perry and co-workers[34] frequently observed similar, enlarged follicles in the sheep PI; particularly noteworthy were vesicles clustered beneath the luminal plasma membrane. With the rat PD, Soji[72] recently demonstrated alkaline phosphatase on the luminal rather than the lateral plasma membranes of the FS cells; thus he theorizes that this surface, with its microvilli, is functioning in absorption from the follicle. Weatherhead[15] proposed that the presence of the nongranular cells is related to the degree of evolutionary development of the PI because in reptiles they are seen only where the lobe is more than a few cells thick. He comments on the relationship of their extended processes with the extensions of the plexus intermedius. Enlarged follicles were frequently observed in the Djungarian hamster at the junction of the PI and PN (Perryman and Weatherhead,[101] unpublished observations). The pericapillary space was seen to underlie these structures bounded by nongranular cells which exhibited microvilli and pinocytotic vesicles at the luminal membranes. In light of these observations, it is tempting to speculate that substances are exchanged between the follicles and the pericapillary space. Experimentation has shown changes in the nongranular cells themselves as well as their follicles. Gracia-Navarro and co-workers[73] examining the PD of *Rana ridibunda*, after injections of thyrotrophic-releasing hormone (TRH), found large, opaque cavities surrounded by hyperactive FS cells. Also, Perry and co-workers[36] examined sheep PI at 120 days of gestation and correlated the appearance of amorphous colloid within the follicle with the onset of hormone release from the secretory cells.

Since the hypophysial cleft is absent in amphibians, the mammal, especially the rat, will be the model for the nongranular cells lining the cleft. These cells sometimes referred to as "marginal" or even "ependymal" are usually restricted to a single layer of epithelial cells varying from squamous to columnar. As summarized by Barberini and Correr,[74] these cells forming both the anterior (i.e., PD) and posterior (e.g., PI) marginal layers in the rat have a varying number of microvilli, cilia, and blebs over their apical surfaces and, in addition, irregular basal processes penetrate spaces between secretory cells of the PD. In the wallaby, both the anterior and posterior layers of the cleft exhibit gaps (pits) between the lining cells.[75] Ciocca and Gonzalez[76] noted within and below the anterior layer of the rat, dilated intercellular spaces filled with electron-dense material. They supported Vila-Porcile's[66] postulation of a three-dimensional network of cavities connecting the follicular luminae and the cleft. On the posterior side of the rat cleft, the marginal cells are separated from the glandular parenchyma by basal laminae within the pericapillary space; other vessels are present between the large lobules. Even with this vascular pattern, intravenously injected HRP diffuses rapidly into the PI and is present in the cleft within 3 min of starting the injections.[25] Therefore, there is also a lacunar system here that appears to interconnect the cleft with the vascular system.

B. Suggested Roles for Nongranular Cells

Several roles, based mainly on studies utilizing electron microscopy, have been theorized. These can be grouped into the following: support, endocytosis (pinocytosis and phagocytosis), and regulatory (transport and modification of molecules). It is probably unrealistic to divide the roles as if they are so distinct and separate. The support role is one proposed earlier than the others. In amphibians, Cardell[77] proposed this sustentacular role for the nongranular cells of the salamander. He emphasized their position (contacting all the secretory cells in addition to having the arrangement of end-feet at the capillaries), the long processes (containing microfilaments and microtubules), and finally the linking desmosomes. Based on these characteristics, researchers still continue to list this as a possible role.

In many vertebrate groups, phagocytosis has been observed, both in vivo and *in vitro*. As early as 1972, Castel[78] mentioned that the stellate cells of the P.I. of *Rana* tadpoles had phagocytic properties *in situ* after sectioning the pituitary stalk, and when the gland was transplanted into, and subsequently cultured in, a piece of isolated tailfin. Båge and Fernholm[79] suggested that the nongranular cells in the pro-adenohypophysis of the river lamprey transform into phagocytes during the last month before spawning. These authors showed the cells phagocytosing large, protruding portions of the gonadotrophs. In the PD of *Rana*, the nongranular cells were observed to contain lysosome-like bodies and to be phagocytosing extracellular debris, mostly granules from the secretory cells.[80] This study was only one of

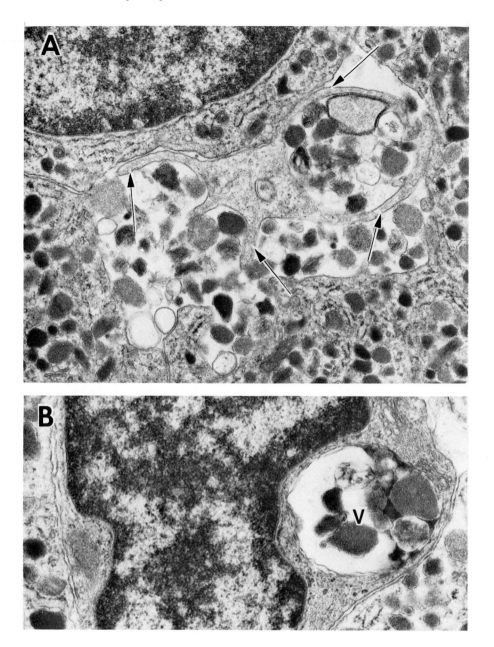

FIGURE 5. Anuran neurointermediate lobe incubated $3^{1}/_{2}$ hours in organ culture. (A) Processes (arrows) from the terminal portion of this FS cell cut in cross section. Note macrophage-like activity of cell toward extracellular debris. Magnification × 20,280. (B) Phagocytic vacuole (V) contained within the FS cell; note secretory granules within the vacuole. Magnification × 25,150. (A and B from Perryman, E. K., deVellis, J., and Bagnara, J. T., Cell Tiss. Res., 208, 85, 1980. With permission.)

ours showing phagocytic activity by nongranular cells of the PI and PD, using *in situ* glands, ectopic transplants, and both primary cell and explant organ cultures[81] (Figures 5A and 5B). What are the "triggers" for this phagocytic activity and the subsequent degradation by lysosomal bodies? One may be an extra "surge" of exocytosed gonadotropin from the PD secretory cells of the gravid female frogs used. Recently Stokreef and co-workers[82] observed the FS cells as disposing of residual lipid bodies released from adjacent mammotropes which were induced to dispose of secretory granules, thereby leading to excess lipid bodies being formed within the mammotropes. They stimulated this activity by withdrawl of the estrogen from rats previously implanted with capsules containing estradiol. The hypertrophy, including an increase in the number of cell junctions and lysosomes, of the FS cells was associated with altered estrogen-dependent prolactin secretion in mammotropes.

Much has been written about the formation of pinocytotic vesicles along the membranes of the nongranular cells and the filling of these vesicles with HRP previously injected into the vascular system.[66,64,62,68,59] Within the PI of *Rana*, these cells pinocytose this tracer all along their processes, including those that abut a perivascular space or its extensions as well as those adjacent to secretory cells of the PI. But, these observations have not supplied answers for the question of what triggers this pinocytotic activity or, needless to say, what is carried within the vesicles. Stoeckel and co-workers[32] indicated that the cells of the FS system, including the marginal cells, accumulate intravesicular calcium in their cytoplasm, "suggesting their intervention in ionic regulation".[25] Beyond this, Semoff and Hadley[83] demonstrated the localization of an ionic ATPase at the plasmalemma of the stellate cells of the anuran PI, and theorized that these cells act to depolarize the secretory cell, thereby allowing for the release of MSH from the adjacent secretory cells. In light of this localization, they also proposed that these cells are similar to glial cells of the central nervous system. Recently, Bambauer and co-workers[84] demonstrated Ca^{++}-ATPase activity restricted to the extracellular aspect of the plasmalemma of guinea pig pituicytes of the neurohypophysis as well as stellate cells of the adenohypophysis.[85] They concluded that these cells regulate extracellular Ca^{++} concentration, found to be necessary for the secretory process, by acting as pumps to decrease Ca^{++} levels in the extracellular spaces. Cultured, purified bovine FS cells, showing dome-forming behavior, provide more evidence for this local control of ion transport.[86] It is known that secretion by melanotrophs is Ca^{++} dependent, relying upon the movement of Ca^{++} through channels which are associated with the plasma membrane.[87-89] The Ca^{++}-binding protein, S-100, initially used as a neural marker, has been immunocytochemically localized in the FS system[90,91] as well as various extra-pituitary sites and even lately within somatotrophs of the PD.[92] In light of the fact that calcium has been localized within both the FS and secretory cells of the PI,[25,32] perhaps the previous significance of S-100 as a marker of nervous tissue is overshadowed by its importance as a Ca^{++}-binding agent. But, data supporting a neural origin for FS cells are provided by Stoeckel and co-workers[25] who observed strong reactions of these cells with antiserum to glial fibrillary acidic protein (GFAP), a structural protein of nervous tissue. Using monoclonal antibody U J13A, a reported neuroectodermal marker, Morris and Hitchcock[8] labeled cultured FS cells and thus reaffirmed their neuroectodermal origin. Moreover, FS cells are believed to increase their synthesis and accumulation of GFAP in response to events such as injury or metabolic changes in adjacent secretory cells.[93] Finally, in light of data showing these cells capable of modulating growth hormone release from perifused anterior pituitary cell aggregates, FS cells may communicate with hormone-secreting cells through an inhibitory paracrine factor they themselves release.[94]

C. Extracellular Channel System

As indicated in the preceding section, the cells of the FS system may constitute a pathway for the transport and exchange of molecules into and throughout the pituitary in various vertebrates. From scanning electron microscopy reports on mammals, the pathway would allow transfer between capillaries, the follicular spaces, and the cleft.[95,74,96] Blood proteins have been localized within the rat cleft,[97] as have blood cells. These findings support the thought of continuity between the cleft and the vascular system. Using HRP injected into the heart of frogs, we described an "extravascular system" of channels in the PI[62,68] Also, we could recognize, without the HRP, extensions of the perivascular space which were

18 *The Melanotropic Peptides*

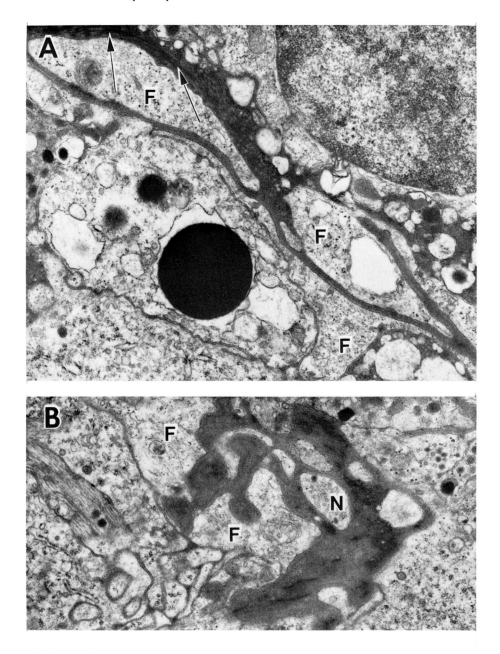

FIGURE 6. Ramifications of the pericapillary space penetrate into the parenchyma of the anuran PI. (A) Collagen fibers (arrows) are visible in these extensions which parallel the FS cell (F) processes. Magnification × 21,300. (B) Enlarged extracellular space which has characteristics of the extensions from the pericapillary space; note collagen fibers within the space; note interdigitating FS cell (F) processes and nerves (N). Magnification × 22,100. (A and B from Perryman, E. K., and Bagnara, J. T., *Cell Tiss. Res.*, 193, 297, 1978. With permission.)

identified on the basis of collagen fibers and basal laminae (Figures 6A and 6B). Frequently, these extensions paralleled the processes of the FS cells and pinocytotic vesicles were seen within these processes of FS cells. A few fortuitous observations showed these spaces to be directly continuous with the perivascular spaces of the plexus intermedius. These perivascular spaces form a link to the pericellular spaces of the parenchyma. The question remains as to what molecules are thus transferred into and from the PI parenchyma. Concerning modulators of secretion, in addition to molecules from the vascular system, neurotransmitters may possibly use the space to diffuse and thus influence several secretory cells. This has been proposed since not every secretory cell in the PI appears directly innervated.[70,98] The inhibitory dopamine released from these Type II fibers may possibly diffuse and then be taken up for degradation by the FS cells. Supporting data show that the biogenic amine degrader, monoamine oxidase B, has been localized in both pituicytes and some cells of the FS system.[99] Considering the absence of nerves in the PI of the lizard, *Anolis carolinensis*, Larsson[59] used HRP to outline transport channels which could be used by modulators flowing from the neural lobe. With this proposal of regulation of MSH release by utilization of the extracellular space, there is the interrelationship between the FS cells and the extracellular spaces of the PI.

V. CONCLUDING REMARKS

Even with continued research about the production of the melanotropins, there are obviously many questions remaining. To hasten the answering, it will be to the benefit of all if there is greater communication between those trained initially as anatomists and those primarily interested in the biochemical events or electrical activity in the melanotrophs. Also, with a comparative approach, we may develop more than the somewhat sketchy picture of the relationship of all the PI parenchymal elements. We now have alternative hypotheses for the embryological origin of the parenchymal cells as well as continued debate over nervous control of the PI. In addition, as stated by Nunez and Gershon,[100] the FS cell is still "in search of a function;" it awaits the quantity of investigations performed on the melanotroph.

ACKNOWLEDGMENTS

The author thanks Dr. George E. Snow not only for his invaluable discussions during the writing of this chapter, but also for his encouragement. Thanks to Dean Crawford for the stylized drawing shown in Figure 2.

REFERENCES

1. **Holmes, R. L. and Ball, J. N.,** *The Pituitary Gland,* Cambridge University Press, London, 1974.
2. **Weatherhead, B.,** The pars intermedia of the pituitary gland, in *Progress in Anatomy* Vol. 3, Navaratnam, V. and Harrison, R. J., Eds., Cambridge University Press, London, 1983, 1.
3. **Motta, P. M.,** *Ultrastructure of Endocrine Cells and Tissues.,* Martinus Nijhoff Publishers, Boston, 1984.
4. **Pearse, A. G. E.,** The cytochemistry and ultrastructure of polypeptide hormone producing cells of the APUD series, and the embryonic, physiologic, and pathologic implications of the concept, *J. Histochem. Cytochem.,* 17, 303, 1969.
5. **Pearse, A. G. E.,** The diffuse neuroendocrine system and the APUD concept. Related "endocrine" peptides in brain, intestine, pituitary, placenta, and anuran cutaneous glands., *Med. Biol.,* 55, 115, 1977.
6. **Schmechel, D., Marangos, P. J., and Brightman, M.,** Neurone-specific enolase is a molecular marker for peripheral and central neuroendocrine cells, *Nature (London),* 276, 834, 1978.
7. **Van Noorden, S., Polak, J. M., Robinson, M., Pearse, A. G. E., and Marangos, P. J.,** Neuron-specific enolase in the pituitary gland, *Neuroendocrinology,* 38, 309, 1984.
8. **Morris, C. S. and Hitchcock, E.,** Immunocytochemistry of folliculo-stellate cells of normal and neoplastic human pituitary gland, *J. Clin. Pathol.,* 38, 481, 1985.
9. **Holmes, R. L.,** The pituitary gland of the female ferret, *J. Endocrinol.* 20, 48, 1960.
10. **Vanha-Perttula, T. and Arstila, A.,** On the epithelium of the rat residual lumen, *Z. Zellforsch.,* 108, 487, 1970.

11. **Bhattacharjee, D. K., Chatterjee, P., and Holmes, R. L.,** Follicles and related structures in the pars intermedia of the adenohypophysis of the jird *(Meriones unguiculatus), J. Anat.,* 130, 63, 1980.
12. **Rodríguez, E. M. and LaPointe, J.,** Light and electron microscopic study of the pars intermedia of the lizard, *Klauberina riversiana, Z. Zellforsch.,* 104, 1, 1970.
13. **Rodríguez, E. M. and Piezzi, R. S.,** Vascularization of the hypophysial region of the normal and adenohypophysectomized toad, *Z. Zellforsch.* 83, 207, 1967.
14. **Meurling, P. and Willstedt, A.,** Vascular connections in the pituitary of *Anolis carolinensis* with special reference to the pars intermedia, *Acta Zool.,* 51, 211, 1970.
15. **Weatherhead, B.,** Comparative cytology of the neuro-intermediate lobe of the reptilian pituitary, *Zbl. Vet. Med. C. Anat. Histol. Embryol.,* 7, 84, 1978.
16. **Murakami, T., Ohtsuka, A., Taguchi, T., Kikuta, A., and Ohtani, O.,** Blood vascular bed of the rat pituitary intermediate lobe, with special reference to its development and portal drainage into the anterior lobe. A scanning electron microscope study of vascular casts, *Arch. Histol. Jap.* 48, 69, 1985.
17. **Farquhar, M. G.,** Fine structure and function in capillaries of the anterior pituitary gland, *Angiology,* 12, 270, 1961.
18. **Lametschwandtner, A., Simonsberger, P., and Adam, H.,** Vascularization of the pars distalis of the hypophysis in the toad, *Bufo bufo* (L.) (Amphibia, Anura). A comparative light microscopical and scanning electron microscopical study. I, *Cell Tiss. Res.,* 179, 1, 1977.
19. **Lametschwandtner, A., Simonsberger, P., and Adam, H.,** Vascularization of the pars intermedia of the hypophysis in the toad, *Bufo bufo* (L.) (Amphibia, Anura). A comparative light microscopical and scanning electron microscopical study. II, *Cell Tiss. Res.,* 179, 11, 1977.
20. **Saland, L. C.,** Ultrastructure of the frog pars intermedia in relation to hypothalamic control of hormone release, *Neuroendocrinology,* 3, 72, 1968.
21. **Hopkins, C. R.,** Studies on secretory activity in the pars intermedia of *Xenopus laevis.* 1. Fine structural changes related to the onset of secretory activity *in vivo, Tissue and Cell,* 2, 59, 1970.
22. **Perryman, E. K.,** Fine structure of the secretory activity of the pars intermedia of *Rana pipiens, Gen. Comp. Endocrinol.,* 23, 94, 1974.
23. **Rodríguez, E. M. and Cuello, C. A.,** Ultrastructure of the toad pars intermedia after the extirpation of the pars distalis, *J. Ultrastruct. Res.,* 79, 207, 1982.
24. **Iturriza, F. C.,** Electron-microscopic study of the pars intermedia of the pituitary of the toad, *Bufo arenarum, Gen. Comp. Endocrinol.,* 4, 492, 1964.
25. **Stoeckel, M. E., Schmitt, G., and Porte, A.,** Fine structure and cytochemistry of the mammalian pars intermedia, in *Peptides of the Pars Intermedia,* Evered, D. and Lawrenson, G., Eds., Pitman Press, London, 1981, 101.
26. **de Volcanes, B. and Weatherhead, B.,** Early changes in the ultrastructure of the pars intermedia of the pituitary of *Xenopus laevis* after change of background colour, *Neuroendocrinology,* 22, 127, 1976.
27. **Martin, R., Weber, E., and Voigt, K. H.,** Localisation of corticotropin- and endorphin-related peptides in the intermediate lobe of the rat pituitary, *Cell Tiss. Res.,* 196, 307, 1979.
28. **Loh, Y. P. and Gainer, H.,** Characterization of pro-opiocortin converting activity in purified secretory granules from rat pituitary neurointermediate lobe, *Proc. Natl. Acad. Sci. U.S.A.,* 79, 108, 1982.
29. **Chang, T.-L. and Loh, Y. P.,** *In vitro* processing of proopiocortin by membrane-associated and soluble-converting enzyme activities from rat intermediate lobe secretory granules, *Endocrinology,* 114, 2094, 1984.
30. **Loh, Y. P. and Tam, W. W. H.,** Association of newly synthesized pro-opiomelanocortin with secretory granule membranes in pituitary pars intermedia cells, *FEBS Lett.,* 184, 40, 1985.
31. **Stoeckel, M. E., Schimchowitsch, S., Garaud, J. C., Schmitt, G., Vandry, H., Klein, M. J., and Porte, A.,** Immunocytochemical evidence for intragranular processing of pro-opiomelanocortin in the melanotropic cells of the rabbit, *Cell Tiss. Res.,* 242, 365, 1985.
32. **Stoeckel, M. E., Hindelang-Gertner, C., Dellmann, H.-D., Porte, A., and Stutinsky, F.,** Subcellular localization of calcium in the mouse hypophysis. I. Calcium distribution in the adeno- and neurohypophysis under normal conditions, *Cell Tiss. Res.,* 157, 307, 1975.
33. **Perryman, E. K.,** Ultrastructure of the stellate cell in the pars intermedia of the frog, *Rana pipiens, Cell Tiss. Res.,* 164, 387, 1975.
34. **Perry, R. A., Robinson, P. M., and Ryan, G. B.,** Ultrastructure of the pars intermedia of the adult sheep hypophysis, *Cell Tiss. Res.,* 217, 211, 1981.
35. **Saland, L. C.,** Effects of reserpine administration on the fine structure of the rat pars intermedia, *Cell Tiss. Res.,* 194, 115, 1978.
36. **Perry, R. A., Robinson, P. M., and Ryan, G. B.,** Ultrastructure of the pars intermedia of the developing sheep hypophysis, *Cell Tiss. Res.,* 224, 369, 1982.
37. **Forbes, M. S.,** Observations on the fine structure of the pars intermedia in the lizard, *Anolis carolinensis, Gen. Comp. Endocrinol.,* 18, 146, 1972.
38. **Larsson, L., Rodríguez, E. M., and Meurling, P.,** Control of the pars intermedia of the lizard, *Anolis carolinensis.* II. Ultrastructure of the intact intermediate lobe, *Cell Tiss. Res.,* 198, 411, 1979.

39. **van Eys, G. J. J. M., and Peters, P. T. W.**, Evidence for a direct role of α-MSH in morphological background adaptation of the skin in *Sarotherodon mossambicus*, *Cell Tiss. Res.*, 217, 361, 1981.
40. **Ito, T., Yeh, W.-H., and Nishiyama, K.**, Secretory activity of the intraocular homotransplanted pars intermedia of the frog, *Rana nigromaculata*, *Neuroendocrinology*, 43, 689, 1986.
41. **Hungtington, T. and Hadley, M. E.**, Evidence against mass action direct feedback control of melanophore-stimulating hormone (MSH) release, *Endocrinology*, 96, 472, 1974.
42. **Baker, B. I.**, The cellular source of melanocyte-stimulating hormone in *Anguilla* pituitary, *Gen. Comp. Endocrinol.*, 19, 515, 1972.
43. **van Eys, G. J. J. M.**, Structural changes in the pars intermedia of the cichlid teleost *Sarotherondon mossambicus* as a result of background adaptation and illumination. I. The MSH-producing cells, *Cell Tiss. Res.*, 208, 99, 1980.
44. **van Eys, G. J. J. M.**, Structural changes in the pars intermedia of the cichlid teleost *Sarotherondon mossambicus* as a result of background adaptation and illumination. II. The PAS-positive cells, *Cell. Tiss. Res.*, 210, 171, 1980.
45. **Ball, J. N. and Batten, T. F. C.**, Pituitary and melanophore responses to background in *Poecilla latipinna* (Telostei): role of the pars intermedia PAS cell, *Gen. Comp. Endocrinol.*, 44, 233, 1981.
46. **Olivereau, M., Olivereau, J. M., and Aimar, C.**, Control of MSH- and PAS- positive cells in the pars intermedia of the goldfish, in *Current Trends in Comparative Endocrinology*, Lofts, B. and Holmes, W. N., Eds., Hong Kong University Press, Hong Kong, 1985, 145.
47. **van Vossel, A., van Vossel-Daeninck, J., Dierickx, K., and Vandesande, F.**, Electron microscopic immunocytochemical demonstration of separate mesotocinergic and vasotocinergic nerve fibres in the pars intermedia of the amphibian hypophysis, *Cell Tiss. Res.*, 178, 175, 1977.
48. **Schimchowitsch, S., Stoeckel, M. E., Klein, M. J., Garaud, J. C., Schmitt, G., and Porte, A.**, Oxytocin-immunoreactive nerve fibers in the pars intermedia of the pituitary in the rabbit and hare, *Cell Tiss. Res.*, 228, 255, 1983.
49. **Schimchowitsch, S., Palacios, J. M., Stoeckel, M. E., Schmitt, G., and Porte, A.**, Absence of inhibitory dopaminergic control of the rabbit pituitary gland intermediate lobe, *Neuroendocrinology*, 42, 71, 1986.
50. **Hopkins, C. R.**, Localization of adrenergic fibers in the amphibian pars intermedia by electron microscope autoradiography and their selective removal by 6-hyroxydomaine, *Gen. Comp. Endocrinol.*, 16, 112, 1971.
51. **de Volcanes, B. and Weatherhead, B.**, Stereological analysis of the effects of 6-hydroxydopamine on the ultrastructure of the melanocyte-stimulating hormone cell of the pituitary of *Xenopus laevis*, *Gen. Comp. Endocrinol.*, 28, 205, 1976.
52. **Stoll, G., Martin, R., and Voigt, K.-H.**, Control of peptide release from cells of the intermediate lobe of the rat pituitary, *Cell Tiss. Res.*, 236, 561, 1984.
53. **Ham, J., McFarthing, K. G., Toogood, C. I. A., and Smyth, D. G.**, Influence of dopaminergic agents on β-endorphin processing in the rat pars intermedia, *Biochem. Soc. Trans.*, 12, 927, 1984.
54. **Powers, C. A.**, Dopaminergic regulation of glandular kallikrein in the intermediate lobe of the rat pituitary, *Neuroendocrinology*, 43, 368, 1986.
55. **Vincent, S. R. Hokfelt, T., and Wu, J-.Y.**, GABA neuron systems in hypothalamus and pituitary gland. Immunohistochemical demonstration using antibodies against glutamate decarboxylase, *Neuroendocrinology*, 34, 117, 1982.
56. **Verburg-Van Kemenade, B. M. L., Tappaz, M., Paut, L., and Jenks, B. G.**, GABAergic regulation of melanocyte-stimulating hormone secretion from the pars intermedia of *Xenopus laevis*: immunocytochemical and physiological evidence, *Endocrinology*, 118, 260, 1986.
57. **Léránth, C., Palkovits, M., and Krieger, D. T.**, Serotonin immunoreactive fibers and terminals in the intermediate lobe of the rat pituitary — light and electron microscopic studies, *Neuroscience*, 9, 289, 1983.
58. **Levitin, H. P.**, Monoaminergic control of MSH release in lizard, *Anolis carolinensis*, *Gen Comp. Endocrinol.*, 41, 279, 1980.
59. **Larsson, L.**, Control of the pars intermedia of the lizard, *Anolis carolinensis*. V. Extracellular transfer and cellular uptake of horseradish peroxidase in the neuro-intermediate lobe, *Cell Tiss. Res.*, 214, 1, 1981.
60. **Rinehart, J. E. and Farquhar, M. G.**, Electron microscopic studies of the anterior pituitary gland, *J. Histochem. Cytochem.*, 93, 93, 1953.
61. **Rinehart, J. E. and Farquhar, M. G.**, The fine vascular organization of the anterior pituitary gland. An electron microscopic study with histochemical correlation, *Anat. Rec.*, 122, 207, 1955.
62. **Perryman, E. K.**, Permeability of the amphibian pars intermedia to peroxidase injected intravascularly, *Cell Tiss. Res.*, 193, 297, 1976.
63. **Weatherhead, B.**, Cytology of the neuro-intermediate lobe of the tuatara, *Sphenodon punctatus* gray, *Z. Zellforsch.*, 119, 21, 1971.
64. **Forbes, M. S.**, Fine structure of the stellate cells in the pars distalis of the lizard, *Anolis carolinensis*, *J. Morph.*, 136, 227, 1972.
65. **Pearson, A. K. and Licht, P.**, Embryology and cytodifferentiation of the pituitary gland in the lizard, *Anolis carolinensis*, *J. Morph.* 144, 85, 1974.

66. **Vila-Porcile, E.,** Le réseau des cellules folliculo-stellaires et les follicules de l'adénohypophyse du rat (Pars distalis). *Z. Zellforsch.,* 129, 328, 1972.
67. **Vila-Porcile, E. and Oliver, L.,** The problem of the folliculo-stellate cells in the pituitary gland, in *Ultrastructure of Endocrine Cells and Tissues,* Motta, P. M., Ed., Martinus Nijhoff Publishers, Boston, 1984, 64.
68. **Perryman, E. K. and Bagnara, J. T.,** Extravascular transfer within the anuran pars intermedia, *Cell Tiss. Res.,* 193, 297, 1978.
69. **deBold, A. J., deBold, M. L., and Kraicer, J.,** Structural relationships between parenchymal and stromal elements in the pars intermedia of the rat adenohypophysis as demonstrated by extracellular space markers, *Cell Tiss. Res.,* 207, 347, 1980.
70. **Saland, L. C.,** Extravascular spaces of rat pars intermedia as outlined by lanthanum tracer, *Anat. Rec.,* 196, 355, 1980.
71. **Krisch, B. and Buchheim, W.,** Access and distribution of exogenous substances in the intercellular clefts of the rat adenohypophysis, *Cell Tiss. Res.,* 236, 439, 1984.
72. **Soji, T.,** Cytochemical studies on their alkaline phosphatase (ALPase) and adenosine triphosphatase (ATPase) activities in folliculo-stellate cells of rat anterior pituitary, *Acta Histochem. Cytochem.,* 17, 69, 1984.
73. **Gracia-Navarro, F., Gonzalez-Reyes, J. A., Guerrero-Callejas, F., and Garcia-Herdugo, G.,** An electron microscopic study of stellate cells and cavities in the pars distalis of frog pituitary, *Tissue Cell,* 15, 729, 1983.
74. **Barberini, F. and Correr, S.,** Ultrastructure of Rathke's pituitary cleft, in *Ultrastructure of Endocrine Cells and Tissues,* Motta, P. M., Ed., Martinus Nijhoff Publishers, Amsterdam, 1984, 57.
75. **Leatherland, J. F. and Renfree, M. B.,** Ultrastructure of the nongranulated cells and morphology of the extracellular spaces in the pars distalis of adult and pouch-young tammar wallabies *(Macropus eugenii), Cell Tiss. Res.,* 227, 439, 1982.
76. **Ciocca, D. R. and Gonzalez, C. B.,** The pituitary cleft of the rat: an electron microscopic study. *Tissue Cell,* 10, 725, 1978.
77. **Cardell, R. R., Jr.,** The ultrastructure of stellate cells in the pars distalis of the salamander pituitary gland, *Am. J. Anat.,* 126, 429, 1969.
78. **Castel, M.,** Ultrastructure of the anuran pars intermedia following severance of hypothalamic connection, *Z. Zellforsch.,* 131, 545, 1972.
79. **Båge, G. and Fernholm B.,** Ultrastructure of the pro-adenohypophysis of the river lamprey, *Lampetra fluviatilis,* during gonad maturation, *Acta Zool.* (Stockholm), 56, 95, 1975.
80. **Perryman, E. K.,** Stellate cells as phagocytes of the anuran pars distalis, *Cell Tiss. Res.,* 231, 143, 1983.
81. **Perryman, E. K., deVellis, J., and Bagnara, J. T.,** Phagocytic activity of the stellate cells in the anuran pars intermedia, *Cell Tiss. Res.,* 208, 85, 1980.
82. **Stokreef, J. C., Reifel, C. W., and Shin, S. H.,** A possible role for folliculo-stellate cells of anterior pituitary following estrogen withdrawal from primed male rats, *Cell Tiss. Res.,* 243, 255, 1986.
83. **Semoff, S. and Hadley, M. E.,** Localization of ATPase activity to the glial-like cells of the pars intermedia, *Gen. Comp. Endocrinol.,* 35, 329, 1978.
84. **Bambauer, H. J., Ueno, S., Umar, H., and Ueck, M.,** Ultracytochemical localization of Ca^{++}-ATPase in pituicytes of the neurohypophysis of the guinea pig, *Cell Tiss. Res.,* 237, 491, 1984.
85. **Bambauer, H. J., Ueno, S., Umar, H., and Ueck, M.,** Histochemical and cytochemical demonstration of Ca^{++}-ATPase activity in the stellate cells of the adenohypophysis of the guinea pig, *Histochemistry,* 83, 195, 1985.
86. **Ferrar, N., Goldsmith, P., Fujii, D., and Weiner, R.,** Culture and characterization of follicular cells of the bovine anterior pituitary and pars tuberalis, in *Methods in Enzymology Vol. 124, Hormone Action, Part J. Neuroendocrine Peptides,* Conn. P. M., Ed., Academic Press, Orlando, Fla., 1986, 245.
87. **Tomiko, S. A., Taraskevich, P. S., and Douglas, W. W.,** Potassium-induced secretion of melanocyte-stimulating hormone from isolated pars intermedia cell signals participation of voltage-dependent calcium channels in stimulus-secretion coupling, *Neuroscience,* 6, 2259, 1981.
88. **Taleb, O., Trousland, J. Demeneix, B. A., and Feltz, P.,** Characterization of calcium and sodium currents in porcine pars intermedia cells, *Neurosci. Lett.,* 66, 55, 1986.
89. **Cota, G.,** Calcium channel currents in pars intermedia cells of the rat pituitary gland, *J. Gen. Physiol.,* 88, 83, 1986.
90. **Cocchai, D. and Miani, N.,** Immunocytochemical localization of the brain specific S-100 protein in the pituitary gland of the adult rat, *J. Neurocytol.,* 9, 771, 1980.
91. **Nakajima, T., Yamaguchi, H., and Takahashi, K.,** S-100 protein in folliculostellate cells of the rat pituitary anterior lobe, *Brain Res.,* 191, 523, 1980.
92. **Shirasawa, N., Yamaguchi, S., and Yoshimura, F.,** Granulated folliculo-stellate cells and growth hormone cells immunostained with anti-S100 protein serum in the pituitary glands of the goat, *Cell Tiss. Res.,* 237, 7, 1984.

93. **Velasco, M. E., Roessmann, U., and Gambetti, P.,** The presence of glial fibrillary acidic protein in the human pituitary gland, *J. Neuropathol. Exp. Neurol.*, 41, 150, 1982.
94. **Baes, M., Allaerts, W., and Denef, C.,** Evidence for functional communication between folliculo-stellate cells and hormone-secreting cells in perifused anterior pituitary cell aggregates, *Endocrinology*, 120, 685, 1987.
95. **Correr, S. and Motta, P. M.,** The rat pituitary cleft: a correlated study by scanning and transmission electron microscopy, *Cell Tiss. Res.*, 215, 515, 1981.
96. **Correr, S. and Motta, P. M.,** A scanning electron microscopic study of "supramarginal cells" in the pituitary cleft of the rat, *Cell Tiss. Res.*, 241, 275, 1985.
97. **Rapp, J. P. and Bergon, L.,** Characteristics of pituitary colloid proteins and their correlation with blood pressure in the rat, *Endocrinology*, 101, 93, 1977.
98. **Davis, M. D., Haas, H. L., and Lichtensteiger, W.,** The hypothalamohypophyseal system in vitro: electrophysiology of the pars intermedia and evidence for both excitatory and inhibitory inputs, *Brain Res.*, 334, 97, 1985.
99. **Cooper, V., Pintar, J. E., and Levitt, P.,** Localization of monoamine oxidase-B immunoreactivity in the neonate and adult rat pituitary gland, *Soc. Neurosci. Abstr.*, 9, 704, 1983.
100. **Nunez, E. A. and Gershon, M. D.,** Specific paracrystalline structures of rough endoplasmic reticulum in the follicular (stellate) cells of the dog adenohypophysis, *Cell Tiss. Res.*, 215, 215, 1981.
101. **Perryman, E. K. and Weatherhead, B.,** unpublished observations.

Chapter 3

PITUITARY MELANOTROPIN BIOSYNTHESIS

Robert M. Dores

TABLE OF CONTENTS

I.	Introduction	26
II.	Organization of the Pro-Opiomelanocortin (POMC) Gene	26
III.	Proteolytic Processing of POMC	28
	A. Mammalian Pituitary Systems	28
	B. Nonmammalian Pituitary Systems	30
IV.	C-Terminal Amidation of Alpha-Melanotropin	31
V.	N-Terminal Acetylation of Alpha-Melanotropin	32
	A. N-Acetyltransferase Activity in Mammals	32
	B. N-Acetyltransferase Activity in Nonmammalian Vertebrates	32
VI.	Conclusions	33
References		34
Acknowledgments		34

I. INTRODUCTION

Tissue-specific differential post-translational processing mechanisms represent a set of strategies for generating discrete subsets of polypeptide end-products from a common precursor. The pituitary pro-opiomelanocortin (POMC) biosynthetic pathway is one of the best-understood examples of this strategy. This review will focus on the posttranslational processing of POMC in corticotropic cells of the pars distalis and melanotropic cells of the pars intermedia.

By 1970, it was clear that several polypeptides that exhibit some degree of melanotropic activity were present in the pituitary gland. These polypeptides included: alpha-melanotropin (alpha-MSH), adrenocorticotropin [ACTH(1-39)], beta-lipotropin (beta-LPH), and beta-melanotropin (beta-MSH) (for review see Reference 1). Analyses of the primary sequences of these polypeptides indicated that alpha-MSH had the same sequence as the N-terminal thirteen amino acids of ACTH(1-39), and the sequence of beta-MSH is identical to an internal sequence in beta-LPH. Furthermore, the melanotropin core sequence, Met-Glu-His-Phe-Arg-Trp-Gly is present in both the corticotropin family (ACTH and alpha-MSH) and the lipotropin family (beta-LPH and beta-MSH). However, the significance of these sequence homologies within and between the two apparent families of polypeptides was confounded by the anatomical localization of these polypeptides. In mammals, ACTH(1-39) and beta-LPH are major end-products of the pars distalis, whereas, alpha-MSH and in some species, beta-MSH, are major end-products of the pars intermedia.

Since these observations, several studies have shown that these two families of polypeptides are contained within the same gene, pro-opiomelanocortin (POMC). Furthermore, as a result of differential posttranslational processing mechanisms this gene gives rise to distinct sets of end-products in the pars distalis and the pars intermedia. The objectives of this review are to give an overview of the structure of the POMC gene and to discuss recent developments in our understanding of the mechanisms involved in the differential posttranslational processing of this polypeptide hormone precursor in the pituitary gland.

II. ORGANIZATION OF THE POMC GENE

The isolation and characterization of POMC mRNA from the bovine pituitary, unequivocally established that corticotropin-related peptides and lipotropin-related peptides are synthesized from a common precursor.[2] The organization of the POMC gene, as determined from studies on several species of mammals,[3-5] is shown in Figure 1. The POMC gene contains 3 exons and 2 introns. Exons 1 and 2 code for, respectively, the 5′ untranslated region of POMC mRNA, and the signal peptide plus a portion of the N-terminal of the precursor. Exon 3 codes for all of the biologically interesting sequences in the precursor. Based on this organizational scheme it is clear that rearrangement of exons as a result of differential post-transcriptional processing is not a viable mechanism for generating different sets of end-products in this system.[6] However, an examination of the polyprotein that is derived from this gene, (Figure 2) reveals the strategic positioning of sets of basic amino acids that serve as potential recognition sites for proteolytic cleavage events.[7] The full list of potential end-products which can be derived from this precursor is shown in Figure 2. It should be appreciated that only subsets of these products are produced in any given tissue as a result of differential post-translational processing mechanisms. For the pituitary systems, this point will be expanded in the next section. A remarkable feature of this precursor is the repetition of the melanotropin core sequence Met-Glu-His-Phe-Arg-Trp-Gly. Hence, there are three "melanotropin" sequences found in POMC; gamma-MSH, alpha-MSH, and beta-MSH. The following sections will consider the post-translational processing events associated with each of these melanotropins.

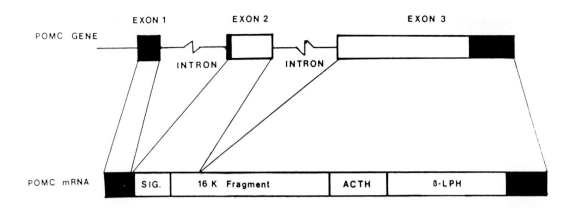

FIGURE 1. Schematic representation of the POMC gene in mammals. The shaded regions on the POMC gene correspond to untranslated regions on the POMC mRNA. The schematic is based on Chang et al.,[3] Druin and Goodman,[4] and Uhler et al.[5] Abbreviations: SIG, Signal Peptide Sequence; ACTH: adrenocorticotropin; β-LPH: beta-lipotropin.

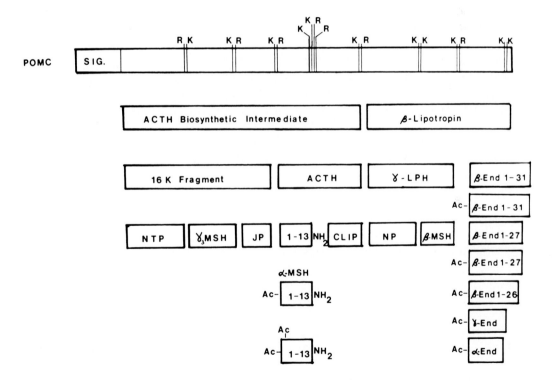

FIGURE 2. Diagram of POMC and the potential end-products of POMC. The figure is based on the sequence of bovine POMC mRNA.[2] Abbreviations: R, arginine; K, lysine; ACTH: adrenocorticotropin; 1-13NH$_2$, ACTH(1-13)amide; Ac-1-13NH$_2$, monoacetylated alpha-melanotropin; NTP, N-terminal peptide of 16 K fragment; γ$_3$MSH, gamma$_3$-melanotropin; JP, Joining peptide; CLIP, corticotropin-like intermediate lobe peptide; NP, N-terminal peptide of β-LPH; β-MSH, beta-melanotropin; β-End 1-31, beta-endorphin(1-31); Ac-β-End 1-31; N-acetyl-beta-endorphin(1-31); β-End 1-27, beta-endorphin(1-27); Ac-β-End 1-27, N-acetyl-beta-endorphin(1-27); Ac-β-End 1-26, N-acetyl-beta endorphin(1-26); Ac-γ-End, N-acetyl-gamma-endorphin: Ac-α-End, N-acetyl-alpha-endorphin.

In addition to studies on mammals, POMC mRNA has also been isolated and characterized from two species of nonmammalian vertebrates: the amphibian *Xenopus laevis*[8] and the teleost, *Onchorynchus keta*.[9] For both species, there is evidence of primary sequence substitutions in several of the potential end-products. However, in terms of general organization, the POMC mRNA of these nonmammalian vertebrates is remarkably similar to mammalian POMC mRNA. Both nonmammalian mRNA sequences have a "16 K fragment" region, an ACTH region, and a beta-LPH region. In the case of *Onchorynchus keta*, it would appear that the "16 K fragment" region has been significantly modified, and the gamma-MSH sequence is absent.[9,10] Since gamma-MSH has been isolated and characterized in a species of cartilaginous fish,[11] it seems likely that the gamma-MSH sequence has been lost secondarily in *Onchorynchus keta*.

III. PROTEOLYTIC PROCESSING OF POMC

A. Mammalian Pituitary Systems

Prior to the isolation and characterization of POMC mRNA from bovine pituitary,[2] several biochemical studies had presented evidence that corticotropin-related peptides and lipotropin-related peptides were synthesized from a common precursor (for reviews see References 12 to 14). Studies on the POMC systems in the pituitary of mammals have established that processing of POMC occurs as a series of discrete, ordered proteolytic cleavage events at sites of paired basic amino acids (i.e., Lys-Arg; Arg-Lys; Lys-Lys; Lys-Lys-Arg-Arg). In the corticotropic cells of the pars distalis, POMC is initially cleaved to yield ACTH biosynthetic intermediate and beta-LPH. ACTH biosynthetic intermediate is rapidly proteolytically cleaved to yield 16 K fragment and ACTH(1-39).[15-18] ACTH(1-39) is a major end-product in the pars distalis. There is now increasing evidence that 16 K fragment undergoes further proteolytic processing in the pars distalis. These processing events will be discussed in conjunction with the processing of 16 K fragment in the pars intermedia. In mammals, beta-LPH is a major end-product in the pars distalis.[13] However, approximately one third of the beta-LPH in these cells is converted to gamma-LPH and nonacetylated beta-endorphin(1-31).[19] The proteolytic processing events observed in the pars distalis result in the generation of ACTH (1-39), a hormone that regulates glucocorticoid production in the adrenal cortex,[20] and nonacetylated beta-endorphin(1-31), a polypeptide capable of binding to opiate receptors.[21]

In the pars intermedia, POMC is initially proteolytically cleaved to yield ACTH biosynthetic intermediate and beta-LPH; ACTH biosynthetic intermediate is sequentially cleaved to 16 K fragment and ACTH(1-39).[19,22,23] Up to this point, the processing events in the pars intermedia are identical to the processing events in the pars distalis. However, in this tissue ACTH(1-39), 16 K fragment, and beta-LPH are not end-products of the pathway, but serve as biosynthetic intermediates.

ACTH(1-39) is proteolytically cleaved at a set of basic amino acids located at positions 15, 16, and 17 to yield ACTH(1-13)amide and CLIP [corticotropin-like intermediate lobe peptide; ACTH(18-39)]. In mammals, ACTH(1-13)amide will undergo N-terminal actylation to yield alpha-MSH, [N-acetyl-ACTH(1-13)amide], and N,O-diacetyl-ACTH(1-13)amide as major end-products.[24-27] These processing events will be dicussed in greater detail in a later section. The functional significance of these processing events is striking. The proteolytic conversion of ACTH(1-39) to ACTH(1-13)amide greatly diminishes the glucocorticoid-stimulating activity associated with the former polypeptide.[20] However, ACTH(1-13)amide is a potent stimulator of physiological color, whereas, ACTH(1-39) is a very weak melanotropin.[29]

The biosynthetic intermediate, 16 K fragment contains the sequence of gamma-melanotropin (gamma-MSH). The existence of this third melanotropin sequence was only revealed following the characterization of bovine POMC mRNA.[2] Studies on the rat pituitary POMC systems[30] have shown that in the pars distalis, and to a greater extent in the pars intermedia, 16 K fragment undergoes proteolytic processing. These reactions yield the N-terminal peptide,[31] gamma$_3$-MSH,[32] and joining peptide[33] (Figure 2) as end-products. In the rat, gamma$_3$-MSH is a glycosylated twenty-five amino acid polypeptide which has the following sequence: Lys-Tyr-Val-Met-Gly-His-Phe-Arg-Trp-Asp-Arg-Phe-Gly-Arg-Arg-Asn-Gly-Ser-Ser-Ser-Ser-Gly-Val-Gly-Gly-Ala-Ala-Gln. N-linked glycosylation occurs at the asparagine residue at position 16. In rat 16 K fragment, the gamma$_3$-MSH sequence is preceded by an Arg-Lys sequence and is followed by a Lys-Arg sequence.[4] Proteolytic cleavage at the later site results in the complete removal of the Lys-Arg residues. However, proteolytic processing at Arg-Lys cleavage site occurs C-terminal to the Arg, thus, Lys remains as the N-terminal residue of rat gamma$_3$-MSH.[30,32] This is an unusual cleavage event for the POMC biosynthetic pathway. Normally, proteolytic cleavages in this pathway involve the complete removal of the proteolytic cleavage recognition site.[12] It is interesting that proteolytic cleavage does not occur at the Arg-Arg residues at position 14 and 15. Currently, there is no satisfactory explanation for why this cleavage does not occur.

Similar observations have also been made in mouse pituitary systems. In particular, studies on mouse AtT-20 tumor cells reveal that the processing of 16 K fragment to Joining Peptide, and presumably gamma$_3$-MSH, is quite extensive in these tumor cells.[34] Furthermore, studies by Bennett[35] on the mouse pars intermedia indicate the presence of an O-linked carbohydrate moeity on the threonine at position 45. It is speculated that this carbohydrate group may influence processing at the Arg-Lys pair at position 49 and 50. This may account for the presence of Lys$_{50}$ as the N-terminal residue of Lys-gamma$_3$-MSH.

The function of this ''melanotropin'' is unclear. Synthetic bovine gamma-MSH analogs have been tested in a variety of melanotropin bioassay systems.[36,37] In general, these analogs were weakly active in the melanotropin bioassay system. However, none of these analogs were glycosylated, and none of these analogs were synthesized with Lys as the N-terminal residue. It was only in later studies that it was appreciated that gamma-MSH in mammals has Lys as the N-terminal residue.[32] Studies that have utilized mild tryptic digests of 16 K fragment[38] or synthetic Lys-gamma$_3$-MSH[39] have shown that gamma$_3$-MSH potentiates the steroidogenic action of ACTH on the adrenal cortex. Further studies need to be done to correlate the processing of 16 K fragment in the pars distalis and the pars intermedia with respect to chronic stress paradigms.

In the pars intermedia, beta-LPH is initially proteolytically cleaved to yield gamma-LPH and beta-endorphin(1-31). Beta-endorphin(1-31) will undergo a series of post-translational modifications to yield a set of N-terminally acetylated, C-terminally truncated peptides (for review see References 40 to 42). The sequence of the melanotropin, beta-MSH, is located at the C-terminal of gamma-LPH. In some species of mammals, this sequence is preceded by a pair of basic amino acids. In these species, gamma-LPH will be cleaved to yield beta-MSH.[2,3] In the mouse and the rat, single point mutations have altered the proteolytic cleavage recognition site that precedes the beta-MSH sequence in these species, and hence, beta-MSH is not released as an end-product.[4,5] Beta-MSH has been sequenced in a number of vertebrates and there is evidence for considerable sequence variability among species.[1] In contrast to gamma-MSH, beta-MSH is a relatively potent stimulator of physiological color change.[29] Although both alpha-MSH and beta-MSH may be present in equimolar amounts in some species, there is little information on the role these melanotropins play in the coordinate regulation of physiological color change.

The numerous studies on the POMC biosynthetic pathway in the pituitary of mammals reinforce the conclusion that unique sets of proteolytic processing enzymes must be present in the pars distalis and pars intermedia. The task of characterizing the various proteolytic processing enzymes has proved to be a difficult undertaking. The status of proteolytic processing enzymes in the POMC biosynthetic pathway has recently been reviewed by Eipper et al.[43]

The subcellular localization of the various proteolytic processing steps in the POMC biosynthetic pathway has also been investigated. There is general agreement that the proteolytic cleavage of ACTH(1-39) and beta-LPH occurs primarily in secretory granules. Ultracentrifugation studies by Glembotski,[44] indicate that although the proteolytic processing of POMC may begin in the Golgi, alpha-MSH and beta-endorphin are primarily generated from their respective biosynthetic intermediates in secretory granules. Immunohistochemical analysis of thin sections of the pituitary support these observations. Stoeckel and colleagues[45,46] have used end-product-specific antisera to gamma-MSH and alpha-MSH to investigate this issue. These studies indicate that N-terminally acetylated alpha-MSH is only detected in secretory granules. However, gamma-MSH is initially detected in Golgi as well as secretory granules. These observations would suggest that the cleavage of ACTH biosynthetic intermediate to 16 K fragment and ACTH(1-39) is initiated in the Golgi, but this reaction goes to completion in the secretory granules.

The generalizations that have been made with respect to the POMC biosynthetic pathway are based on studies of adult mammalian pituitary systems. Studies on the ontogeny of the pre- and postnatal pars distalis of the rat, indicate deviations from the adult processing scheme. Anatomical studies by Khachaturian et al.,[47] revealed the co-localization of alpha-MSH-related immunoreactivity and beta-endorphin-related immunoreactivity in cells of the pars distalis at embryonic day 18. This co-localization persists during early postnatal development and disappears by postnatal day 21. Biochemical analyses of the developing pars distalis confirm these observations.[48] Approximately 10% of the ACTH-related material in the postnatal day 1 pars distalis is ACTH(1-13)amide. By postnatal day 21 this percentage had dropped dramatically. Thus, although the pars distalis during early postnatal development performs proteolytic processing events characteristic of the pars intermedia, there was no evidence of N-acetylation of ACTH(1-13) amide occuring to any appreciable extent in the pars distalis.[48] Furthermore, there is no evidence that the pars intermedia generates novel end-products during early postnatal development.[48,49] The physiological role that ACTH(1-13)amide may be playing in the developing neonate is unclear at this time.

B. Nonmammalian Pituitary Systems

Studies on the POMC system in the pars distalis of nonmammalian vertebrates have focused on the anatomical distribution of ACTH-related peptides in corticotropes and the efficacy of pars distalis extracts on the adrenal cortex glucocorticoid system (for review see Reference 1). By contrast, there is considerably more information on the processing of POMC in the pars intermedia of nonmammalian vertebrates. Pulse/chase analyses have been performed on pars intermedia cells of reptiles,[50] amphibians,[51-57] and fish.[58] These studies are in general agreement that the initial cleavage of POMC results in the generation of beta-LPH and ACTH biosynthetic intermediate. In these species, beta-LPH is a biosynthetic intermediate that gives rise to beta-endorphin-sized peptides. ACTH biosynthetic intermediate is sequentially processed to ACTH, which in turn, is processed to various alpha-MSH-related forms. Thus, the overall proteolytic processing scheme in the pars intermedia of nonmammalian vertebrates is very similar to the mammalian pathway.

Alpha-MSH has been characterized in several species of nonmammalian vertebrates.[59,60] Comparison of primary sequences underscore the high degree of conservation of the alpha-MSH sequence during evolution. Subtle sequence changes have been observed in dogfish alpha-MSH,[61] salmon alpha-MSH,[62] and amphibian alpha-MSH.[63] By contrast, the sequence of beta-MSH varies considerably across species (see the chapter by Kawauchi in this volume). Gamma$_3$-MSH-related immunoreactivity has been detected immunohistochemically in representatives of all major classes of vertebrates.[64] However, with the exception of mammals,

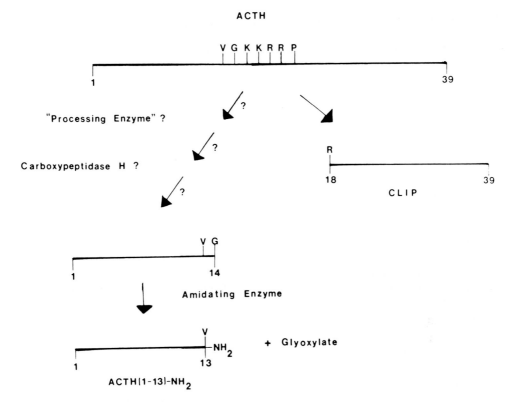

FIGURE 3. Hypothetical scheme for the conversion of ACTH(1-39) to ACTH(1-13)amide. Abbreviations: V, valine; G, glycine; K, lysine; R, arginine; P, proline; ACTH, adrenocorticotropin; CLIP, corticotropin-like intermediate lobe peptide.

gamma-MSH has only been characterized in the pituitary of the dogfish.[11] The gamma-MSH sequence appears to be absent from the salmon pituitary POMC systems.[10]

IV. C-TERMINAL AMIDATION OF ALPHA-MSH

The proteolytic cleavage of ACTH(1-39) in the pars intermedia involves a complex series of steps that terminate with the amidation of ACTH(1-14).[65] In the sequence of ACTH(1-39) the proteolytic recognition site is a block of basic amino acids at position 15-18 (Figure 3). Apparently as a result of steric hinderance, the proline residue at position 19 protects the C-terminal of Arg_{18} from proteolytic attack. Thus, CLIP, the C-terminal portion of ACTH(1-39), has Arg_{18} as the N-terminal residue. Because of this interaction, the proteolytic processing of ACTH(1-39) may occur at either residues 15, 16, or 17. Processing at these sites could result in the generation of several potential intermediates including: ACTH(1-17), ACTH(1-16), or ACTH(1-15). Figure 3 presents a potential mechanism to explain these cleavage events. In this model an as yet unidentified "proteolytic processing enzyme" would cleave at the C-terminal of residue 17 to generate ACTH(1-17). Carboxypeptidase H, an enzyme specific for substrates that terminate with a basic amino acid,[66] would sequentially remove Arg_{17}, Lys_{16}, and Lys_{15} to generate the glycine-extended intermediate ACTH(1-14). Carboxypeptidase H is present in pars intermedia secretory granules.[72] This model represents just one possible mechanism for generating ACTH(1-14). Other mechanisms which involve ACTH(1-16) or ACTH(1-15) as the initial proteolytic cleavage product are equally possible.

Although there is no concensus on the mechanism for removing the internal basic amino acid sequences in ACTH(1-39), it is clear that glycine-extended ACTH(1-14) is the substrate for the amidating enzyme. This enzyme, which is referred to as peptidyl-glycine alpha-amidation monooxygenase,[92] is a granule-associated, copper-dependent, mixed-function oxidase,[67,68] that requires ascorbate as a co-factor.[69-70] Smyth and his colleagues[71] had previously shown that a C-terminal glycine residue on a polypeptide substrate serves as an amide donor for the penultimate amino acid in the sequence of the polypeptide. As the amidating enzyme "splits" glycine, glyoxylate is formed as a by-product of the reaction (Figure 3). The granule-associated amidating enzyme is present in both corticotropic cells and melanotropic cells.[34,72] The substrate for the amidating enzyme in corticotropic cells is joining peptide, whereas, both joining peptide and ACTH(1-14) would be substrates for this enzyme in melanotropic cells.

Phylogenetic studies indicate that amidated forms of alpha-MSH are present in amphibians,[63,73] and teleosts.[60,74] In addition, amidated forms of alpha-MSH have been detected immunologically in reptiles,[75] holoestian fish,[76] and cartilaginous fish.[77] It would appear that the amidation of alpha-MSH is a reaction that has been conserved throughout evolution. A nonamidated form of alpha-MSH has been characterized in the pituitary of the salmon, *Onchorynchus keta*.[62] This form of alpha-MSH, which has been designated alpha-MSH 2, has the sequence, N-acetyl-Ser-Tyr-Ser-Met-Glu-His-Phe-Arg-Trp-Gly-Lys-Pro-Ile-Gly-His-OH. The initial twelve amino acids of this novel form of alpha-MSH are identical to salmon alpha-MSH 1,[60] however, the presence of a His residue at position 15 apparently blocks the amidation of this peptide.

V. N-TERMINAL ACETYLATION OF ALPHA-MSH

A. N-Acetyltransferase Activity in Mammals

The N-acetylation of alpha-MSH and beta-endorphin in the pars intermedia has a significant effect on the biological activity of these polypeptides. This post-translational modification selectively enhances the biological activity of alpha-MSH,[78,79] but decreases the opiate receptor binding activity of beta-endorphin.[80,81]

In mammals, the N-acetylation of alpha-MSH involves the sequential addition of acetyl groups to the amino group and the hydroxyl group of Ser_1 of ACTH(1-13)amide.[24] Pulse/chase studies have shown that in rat pars intermedia cells, ACTH(1-13)amide is first acetylated N-terminally to generate alpha-MSH, [N-acetyl-ACTH(1-13)amide]. Alpha-MSH, in turn, is then O-acetylated to yield N,O-diacetyl-ACTH(1-13)amide as the final end-product in the pathway.[82] These results indicate that ACTH(1-39) does not undergo N-acetylation prior to the proteolytic processing events at positions 15 through 17. Furthermore, the amidation of ACTH(1-14) must occur before the N-acetylation of ACTH(1-13)amide can begin. Under steady-state conditions, N-acetyl-ACTH(1-14) has not been detected in vivo in the pars intermedia of mammals.

The acetylation of ACTH(1-13)amide is catalyzed by a granule-associted acetyltransferase.[82-85] The acetyltransferase activity has a pH optimum between pH 6.5 and pH 7.[82,84] Unlike the granule-associated amidating enzyme, the granule-associated N-acetyltransferase activity is found in melanotropic cells, but not in corticotropic cells.[82] The acetyl donor in this reaction is acetyl-CoA.[83] Based on kinetic arguments, it appears that in the bovine and the rat pars intermedia, the same enzyme may acetylate both ACTH(1-13)amide and beta-endorphin(1-31).[82,84,86] In addition, there is evidence that the levels of granule-associated N-acetyltransferase activity can be correlated with the expression of the POMC gene in the pars intermedia.[87]

B. N-Acetyltransferase Activity in Nonmammalian Vertebrates

In the pituitaries of mammals, N-acetylation of alpha-MSH is restricted to the pars intermedia. This is a secretory granule-associated post-translational processing event that couples the N-acetylation of ACTH(1-13)amide with the N-acetylation of beta-endorphin. N-acetylated forms of alpha-MSH and beta-endorphin have also been detected in several species of nonmammalian vertebrates. However, a survey of the recent literature indicates a variety of strategies with respect to this post-translational processing event.

In the pituitary of the cartilaginous fish, *Squalus acanthias,* only nonacetylated ACTH(1-13)amide[61] and several forms of nonacetylated beta-endorphin[88] have been isolated. Furthermore, in the pars intermedia of the reptile, *Anolis carolinensis,* multiple forms of nonacetylated beta-endorphin predominate.[89,90] At present, the degree of acetylation of alpha-MSH in this species is not known.

With respect to bony fish, studies have been done on two species of teleosts and one species of holostean fish. Studies currently in progress on the pars intermedia of the holostean fish, *Amia calva,* indicate that N-acetylated forms of alpha-MSH and beta-endorphin are major end-products in this system.[76] In the teleost, *Carassius auratus,* both mono- and di-acetylated forms of alpha-MSH have been isolated from the pars intermedia.[74] Currently, there is no information on the forms of beta-endorphin in this species. In the pituitary of the salmon, *Onchorynchus keta,* the situation is more complex. Only N-acetylated forms of beta-endorphin have been isolated from the pituitary gland of this species.[91] As previously described, two forms of alpha-MSH, that differ in primary structure, have been characterized in the pituitary of the salmon: alpha-MSH 1[60] and alpha-MSH 2.[62] The first 12 amino acids in both forms of salmon MSH are identical. Salmon alpha MSH 1 has the same sequence as synthetic ACTH(1-13)amide. In salmon alpha-MSH 2, the residues Ile-Gly-His are present at positions 13, 14, and 15. These two forms of alpha-MSH also differ in terms of N-acetylation. Salmon alpha-MSH 1 is nonacetylated, whereas, salmon alpha-MSH 2 is N-acetylated. The discrepancy in the degree of N-acetylation between the two forms of alpha-MSH in the salmon pituitary raises the possibility that distinct substrate-specific N-acetyltransferases may be present in this species.

The N-acetylation of alpha-MSH in amphibians has received a great deal of attention in recent years. Martens and co-worker[63] have reported that the N-acetylation of ACTH(1-13)amide in the pars intermedia of the amphibian, *Xenopus laevis,* occurs as a co-secretory event. Following short-term pulse labeling these authors detected radiolabeled ACTH(1-13)amide in tissue extracts and could not detect radiolabeled N-acetyl-ACTH(1-13)amide in these extracts. Co-secretory N-acetylation of ACTH(1-13)amide has also been observed in the pars intermedia of the amphibian, *Rana ridibunda.*[73] In addition to the co-secretory pathway, Goldman and Loh have shown that ACTH(1-13)amide is also N-acetylated post-translationally in the pars intermedia of *Xenopus laevis.* Current steady-state analyses of extracts of *Xenopus* pars intermedia have detected N-acetylated forms of beta-endorphin in this tissue.[93] The coordination of the post-translational and co-secretory N-acetylation pathways in anuran amphibians will be an interesting topic for future research.

These examples illustrate that nonmammalian vertebrate pituitary systems provide a variety of models for studying different strategies with respect to the N-acetylation of POMC-related end-products. To date, these studies have focused on the isolation and characterization of the forms of alpha-MSH and beta-endorphin in these species. Future studies will need to focus on the isolation and characterization of the N-acetyltransferase activities in these species.

VI. CONCLUSIONS

Analyses of the sequences of POMC mRNA in several species of vertebrates underscores the high degree of conservation of the POMC gene across phylogeny. In mammals, the POMC gene has a rather simple organization (Figure 1) that precludes post-transcriptional exon reshuffling as a mechanism for generating tissue-specific sets of end-products. However, as revealed by the pituitary POMC systems, differential post-translational processing can provide mechanisms for the production of unique sets of end-products with distinct biological activities in the pars distalis and the pars intermedia. Studies on mammalian systems have provided a considerable amount of new information on two post-translational end-product modifications: alpha-amidation and N-acetylation. This has resulted in the purification of the first pituitary post-translational processing enzyme, peptidyl-glycine alpha-amidation monooxygenase.[94]

It is now time to focus on the isolation and characterization of the tissue-specific proteolytic processing enzymes in the pituitary. Since the tissue-specific proteolytic processing enzymes dictate which POMC end-products will be present in a given region of the pituitary, it becomes clear that an understanding of the regulation of these processing enzymes is as important as the mechanisms that regulate POMC gene expression.

In this review, numerous references have been made to the diversity of the POMC pituitary systems in nonmammalian vertebrates. A renewed effort to understand the POMC biosynthetic pathway in these species will provide future models that will expand our understanding of post-translational processing mechanisms.

ACKNOWLEDGMENTS

I wish to thank Catherine A. Sei and Tami C. Steveson for their assistance in the writing of this review. This work was supported by NIH grant DK36587.

REFERENCES

1. **Baker, B.,** The evolution of ACTH, MSH and LPH — Structure, Function and Development, *Hormone Evolution,* Vol. 2, Barrington, London, 1980, 643.
2. **Nakanishi, S., Inoue, A., Kita, T., Nakamura, M., Chang, A. C. Y., Cohen, S. N., and Numa, S.,** Nucleotide sequence of cloned cDNA for bovine corticotropin/beta-lipotropin precursor, *Nature (London),* 278, 423, 1979.
3. **Chang, A. C. Y., Cochet, M., and Cohen, S. N.,** Structural organization of human genomic DNA encoding the pro-opiomelanocortin peptide, *Proc. Natl. Acad. Sci. U.S.A.,* 77, 4890, 1980.
4. **Drouin, J. and Goodman, H. M.,** Most of the coding region of rat ACTH/beta-LPH precursor gene lacks intervening sequences, *Nature (London),* 288, 610, 1980.
5. **Uhler, M., Herbert, E., D'Eustachio, P., and Ruddle, F. D.,** The mouse genome contains two nonallelic pro-opiomelanocortin genes, *J. Biol. Chem.,* 258, 9444, 1983.
6. **Douglass, J., Civelli, O., and Herbert, E.,** Polyprotein gene expression: Generation of diversity of neuroendocrine peptides, *Annu. Rev. Biochem.,* 53, 665, 1984.
7. **Docherty, K. and Steiner, D. F.,** Post-translational proteolysis in polypeptide hormone biosynthesis, *Annu. Rev. Physiol.,* 44, 625, 1982.
8. **Martens, G. J. M., Civelli, O., and Herbert, E.,** Nucleotide sequence of cloned cDNA for pro-opiomelanocortin in the amphibian *Xenopus laevis, J. Biol. Sci.,* 260, 13685, 1985.
9. **Soma, G-I., Kitahara, N., Nishizawa, T., Nanami, H., Kotake, C., Okazaki, H., and Andoh, T.,** Nucleotide sequence of a cloned cDNA for proopiomelanocortin precursor of chum salmon, *Onchorynchus keta, Nucleic Acids Res.,* 12, 8029, 1984.
10. **Kawauchi, H., Akiyoshi, T., and Ken-Ihi, A.,** Gamma-melanotropin is not present in an N-terminal peptide of salmon pro-opiocortin, *Int. J. Pept. Protein Res.,* 18, 223, 1982.
11. **McLean, C. and Lowry, P. J.,** Natural occurrence but lack of melanotrophic activity of gamma-MSH in fish, *Nature (London),* 290, 341, 1981.
12. **Eipper, B. A. and Mains, R. E.,** Structure and function of proadrenocorticotropin/endorphin and related peptides, *Endocr. Rev.,* 1, 247, 1980.
13. **Krieger, D. T., Liotta, A. S., Brownstein, M. J., and Zimmerman, E. A.,** ACTH, beta-lipotropin and related peptides in brain, pituitary and blood, *Rec. Prog. Horm. Res.,* 36, 277, 1980.
14. **Chrétien, M. and Seidah, N. G.,** Chemistry and biosynthesis of proopiomelanocortin: ACTH, MSH's, endorphins and related peptides, *Mol. Cell. Endocrinol.,* 21, 101, 1981.

15. **Mains, R. E., Eipper, B. A., and Ling, N.,** Common precursor to corticotropins and endorphins, *Proc. Natl. Acad. Sci. U.S.A.,* 74, 3014, 1977.
16. **Roberts, J. L. and Herbert, E.,** Characterization of a common precursor to corticotropin and beta-lipotropin: cell free synthesis of the precursor and identification of corticotropin peptides in the molecules, *Proc. Natl. Acad. Sci. U.S.A.,* 74, 4826, 1977.
17. **Eipper, B. A. and Mains, R. E.,** Analysis of the common precursor to corticotropin and endorphin, *J. Biol. Chem.,* 253, 5732, 1978.
18. **Mains, R. E. and Eipper, B. A.,** Coordinate synthesis and secretion of corticotropins and endorphins by mouse pituitary tumor cells, *J. Biol. Chem.,* 253, 651, 1978.
19. **Eipper, B. A. and Mains, R. E.,** Existence of a common precursor to ACTH and endorphin in the anterior and intermediate lobes of the rat pituitary, *J. Supramol. Struct.,* 8, 247, 1978.
20. **Rhamachandran, J.,** The structure and function of adrenocorticotropin, in *Hormonal Proteins and Polypeptides,* Vol. 2, Li, C. H., Ed., Academic Press, New York, 1973, 1.
21. **Li, C. H.,** Beta-endorphin: synthetic analogs and structure-activity relationships, in *Hormonal Proteins and Polypeptides,* Vol. 10, Li, C. H., Ed., Academic Press, New York, 1981, 2.
22. **Gianoulakis, C. Seidah, N. G., Routheir, R., and Chrétien, M.,** Biosynthesis and characterization of adrenocorticotropoic hormone, alpha-melanocyte-hormone, and an N-terminal fragment of the adrenocorticotropic/beta-lipotropin precursor from rat pars intermedia, *J. Biol. Chem.,* 245, 11903, 1979.
23. **Mains, R. E. and Eipper, B. A.,** Synthesis and secretion of corticotropin, melanotropins, and endorphins from rat intermediate pituitary cells, *J. Biol. Chem.,* 254, 7885, 1979.
24. **Rudman, D., Chawla, R. K., and Hollins, B. M.,** N,O-Diacetylserine alpha-melanocyte stimulating hormone, a naturally occurring melanotropic peptide, *J. Biol. Chem.,* 254, 10102, 1979.
25. **Browne, C. A., Bennett, H. P. J., and Solomon, S.,** Isolation and characterization of corticotropin- and melanotropin-related peptides from the neurointermediary lobe of the rat pituitary by reverse phase liquid chromatography, *Biochemistry,* 20, 4538, 1981.
26. **Buckley, D. I., Houghton, R. A., and Ramachardran, J.,** Isolation of alpha-melanotropin and N,O-diacetyl serine$_1$ alpha-melanotropin from porcine pituitary extracts, *Intl. J. Pept. Protein Res.,* 17, 508, 1981.
27. **Evans, C. J., Lorenz, R., Weber, E., and Barchas, J. D.,** Variants of alpha-melanocyte stimulating hormone in rat brain and pituitary: evidence that acetylated alpha-MSH exists only in the intermediate lobe of the pituitary, *Biochem. Biophys. Res. Comm.,* 106, 910, 1982.
28. **O'Donohue, T. L., Handlemann, G. E., Chaconas, T., Miller, R. L., and Jacobowitz, D. M.,** Evidence that N-acetylation regulates the behavioral activity of alpha-MSH in the rat and central nervous system, *Peptides,* 2, 333, 1981.
29. **Hadley, M. E., Heward, C. B., Hruby, V. J., Sawyer, T. K., and Yang, Y. C. S.,** Biological actions of melanocyte-stimulating hormone, in *Peptides of the Pars Intermedia,* Evered, D. and Lawrenson, G., Ed., Pitman Press, Bath, England, 1981, 244.
30. **Pederson, R. C., Ling, N., and Brownie, A. C.,** Immunoreactive gamma-melanotropin in rat pituitary and plasma: a partial characterization, *Endocrinology,* 110, 825, 1982.
31. **Bennett, H. P. J.,** Isolation and characterization of the 1 to 49 amino-terminal sequence of pro-opiomelanocortin from bovine posterior pituitaries, *Biochem. Biophys. Res. Commun.,* 125, 229, 1984.
32. **Browne, C. A., Bennett, H. P. J., and Solomon, S.,** The isolation and characterization of gamma$_3$-melanotropin from the neurointermediary lobe of the rat pituitary, *Biochem. Biophsy. Res. Commun.,* 100, 336, 1981.
33. **Seidah, N. G., Rochemont, J., Hamelin, J., Benjannet, S., and Chrétien, M.,** The missing fragment of the pro-sequence of human pro-opiomelanocortin: sequence and evidence for C-terminal amidation, *Biochem. Biophys. Res. Commun.,* 102, 710, 1981.
34. **Eipper, B. A., Park, L., Keutmann, H. T., and Mains, R. E.,** Amidation of joining peptide, a major pro-ACTH/endorphin product peptide, *J. Biol. Chem.,* 261, 8686, 1986.
35. **Bennett, H. P. J.,** Biosynthetic fate of the amino-terminal fragment of pro-opiomelanocortin within the intermediate lobe of the mouse pituitary, *Peptides,* 7, 615, 1986.
36. **Ling, N., Ying, S., Minick, S., and Guillemin, R.,** Synthesis and biological activity of four gamma-melanotropin peptides derived from the crytic region of the adrenocorticotropin/beta-lipotropin precursor, *Life Sci.,* 25, 1773, 1979.
37. **O'Donohue, T. L., Handlemann, G. E., Loh, Y. P., Olton, D. S., Liebowitz, J., and Jacobowitz, D. M.,** Comparison of biological and behavioral activities of alpha- and gamma-melanocyte stimulating hormones, *Peptides,* 2, 101, 1981.
38. **Pederson, R. C. and Brownie, A. C.,** Adrenocortical response to corticotropin is potentiated by part of the amino-terminal region of pro-corticotropin/endorphin, *Proc. Natl. Acad. Sci. U.S.A.,* 77, 2239, 1980.

39. **Farese, R. V., Ling, N. C., Sabir, M. A., Larson, R. E., and Trudeau, W. L., III,** Comparison of the effects of adrenocorticotropin and Lys-gamma₃-melanocyte-stimulating hormone on steroidogenesis, cAMP production, and phospholipid metabolism in rat adrenal fasciculata-reticularis cells in vitro, *Endocrinology,* 112, 129, 1983.
40. **Zakarian, S. and Smyth, D. G.,** Review Article: Distribution of beta-endorphin related peptides in rat pituitary and brain, *Biochem. J.,* 202, 561, 1982.
41. **O'Donohue, T. L. and Dorsa, D. M.,** The opiomelanotropinergic neuronal and endocrine systems, *Peptides,* 3, 352, 1982.
42. **Mains, R. S. and Eipper, B. A.,** Differences in the processing of beta-endorphin in the rat anterior and intermediate pituitary, *J. Biol. Chem.,* 256, 5683, 1981.
43. **Eipper, B. A., May, V., Cullen, E. I., Sato, S. M., Murthy, S. N., and Mains, R. E.,** Co-translational and post-translational processing in the production of bioactive peptides, in *Psychopharmacology, the Third Generation of Progress,* Meltzer, H., Bunney, W., Coyle J., Davis, J., Copin, I., Schuster, R., Schader, R., and Simpson, J., Eds., Raven Press, New York, 1986, in press.
44. **Glembotski, C. C.,** Acetylation of alpha-melanotropin and beta-endorphin in the rat intermediate pituitary, *J. Biol. Chem.,* 257, 10493, 1982.
45. **Stoekel, M. E., Schimchowitsch, S., Garaund, J. C., Schmitt, G., Vaudry, H., and Porte, A.,** Immunocytochemical evidence of intragranular acetylation of alpha-MSH in the melanotrophic cells of the rabbit, *Cell Tiss. Res.,* 230, 511, 1983.
46. **Stoekel, M. E., Schimchowitsch, S., Garaud, J. C., Schmitt, G., Vaudry, H., Klein, M. J., and Porte, A.,** Immunocytochemical evidence for intragranular processing of pro-opiomelanocortin in the melanotropic cells of the rabbit, *Cell Tiss. Res.,* 242, 365, 1985.
47. **Khachaturian, H., Alessi, N. E., and Lewis, M. E.,** Ontogeny of opioid and related peptides in the rat brain and pituitary, *Life Sci., (Suppl. 1)* 33, 61, 1983.
48. **Sato, S. M. and Mains, R. E.,** Posttranslational processing of proadrenocorticotropin/endorphin-derived peptides during postnatal development in the rat pituitary, *Endocrinology,* 117, 773, 1985.
49. **Seizinger, B. R., Hollt, V., and Herz, A.,** Postnatal development of beta-endorphin-related peptides in rat anterior and intermediate pituitary lobes: evidence for contrasting development of proopiomelanocortin processing, *Endocrinology,* 115, 136, 1984.
50. **Dores, R. M.,** Evidence for a common precursor for alpha-MSH and beta-endorphin in the intermediate lobe of the pituitary of the reptile, *Anolis carolinensis, Peptides,* 3, 925, 1982.
51. **Loh, Y. P. and Gainer, H.,** Biosynthesis, processing and control of release of melanotropic peptides in the neurointermediate lobe of *Xenopus laevis, J. Gen. Physiol.,* 70, 37, 1977.
52. **Pezalla, P. D., Seidah, N. G., Benjannet, S., Crine, P., Lis, M., and Chrétien, M.,** Biosynthesis of beta-endorphin, beta-lipotropin, and the putative ACTH-LPH precursor in the frog pars intermedia, *Life Sci.,* 23, 2281, 1978.
53. **Loh, Y. P.,** Immunological evidence for two common precursors to corticotropins, endorphins, and melanotropin in the neurointermediate lobe of the toad pituitary, *Proc. Natl. Acad. Sci. U.S.A.,* 76, 796, 1979.
54. **Martens, G. J. M., Jenks, B. G., and van Overbeeke, A. P.,** Analysis of peptide biosynthesis in the neurointermediate lobe of *Xenopus laevis* using high performance liquid chromatography: occurrence of small bioactive products, *Comp. Biochem. Physiol.,* 67B, 493, 1980.
55. **Martens, G. J. M., Biermans, P-P. J., Jenks, B. G., and van Overbeeke, A. P.,** Biosynthesis of two structurally different pro-opiomelanocortins in the pars intermedia of the amphibian pituitary gland, *Eur. J. Biochem.,* 126, 17, 1982.
56. **Martens, G. J. M., Jenks, B. G., and van Overbeeke, A. P.,** Biosynthesis of pairs of peptides related to melanotropin, corticotropin and endorphin in the pars intermedia of the amphibian pituitary gland, *Eur. J. Biochem.,* 122, 1, 1982.
57. **Vaudry, H., Jenks, B. G., and van Overbeeke, A. P.,** Biosynthesis, processing and release of pro-opiomelanocortin related peptides in the intermediate lobe of the pituitary gland of the frog *(Rana ridibunda), Peptides,* 5, 905, 1984.
58. **Rodrigues, K. T., Jenks, B. G., and Sumpter, J. P.,** Biosynthesis of pro-opiomelanocortin-related peptides in the neurointermediate lobe of the pituitary of the rainbow trout, *(Salmo gairdneri), J. Endocr.,* 98, 271, 1983.
59. **Lowry, P. J. and Scott, A. P.,** The evolution of vertebrate corticotrophin and melanocyte-stimulating hormone, *Gen. Comp. Endocrinol.,* 26, 16, 1975.
60. **Kawauchi, H. and Muramoto, K.,** Isolation and primary structure of melanotropins from salmon pituitary glands, *Int. J. Pept. Protein Res.,* 14, 373, 1979.
61. **Lowry, P. J. and Chadwick, A.,** Purification and amino acid sequence of melanocyte-stimulating hormone from the dogfish, *Squalus acanthias, Biochem. J.,* 118, 713-718, 1970.
62. **Kawauchi, H., Adachi, Y., and Tsubokawa, N.,** Occurrence of a new melanocyte-stimulating hormone in the salmon pituitary gland, *Biochem. Biophys. Res. Comm.,* 96, 1508, 1980.

63. **Martens, G. J. M., Jenks, B. G., and van Overbeeke, A. P.,** N,alpha-acetylation is linked to alpha-MSH release from pars intermedia of the amphibian pituitary gland, *Nature (London)*, 94, 558, 1981.
64. **Estivariz, F. E., Iturriza, F., Hope, J., and Lowry, P. J.,** Immunohistochemical demonstration of pro-gamma-MSH-like substances in the pituitary gland of various vertebrate species, *Gen. Comp. Endocrinol.*, 46, 1, 1982.
65. **Glembotski, C. C., Eipper, B. A., and Mains, R. E.,** ACTH(1-14) hydroxy-related molecules in primary cultures of rat intermediate pituitary cells: identification and role in the biosynthesis of alpha-melanotropin, *J. Biol. Chem.*, 258, 7299, 1983.
66. **Fricker, L. D. and Snyder, S. H.,** Purification and characterization of enkephalin convertase, *J. Biol. Chem.*, 258, 10950, 1983.
67. **Eipper, B. A., Mains, R. E., and Glembotski, C. C.,** Identification in pituitary tissue of a peptide alpha-amidation activity that acts on glycine-extended peptides and requires molecular oxygen, copper, and ascorbic acid, *Proc. Nat. Acad. Sci. U.S.A.*, 80, 5144, 1983.
68. **Eipper, B. A., Glembotski, C. C., and Mains, R. E.,** Bovine intermediate pituitary alpha-amidation enzyme: preliminary characterization, *Peptides*, 4, 921, 1983.
69. **Glembotski, C. C.,** The characterization of the ascorbic acid-mediated alpha-amidation of alpha-melanotropin in cultured intermediate pituitary lobe cells, *Endocrinology*, 118, 1461, 1986.
70. **May, V. and Eipper, B. A.,** Regulation of peptide amidation in cultured pituitary cells, *J. Biol. Chem.*, 260, 16224, 1986.
71. **Bradbury, A. F., Finnie, M. D. A., and Smyth, D. G.,** Mechanism of C-terminal amide formation by pituitary enzymes, *Nature (London)*, 296, 686, 1983.
72. **Mains, R. E. and Eipper, B. A.,** Secretion and regulation of two biosynthetic enzyme activities, peptidyl-glycine alpha-amidating monooxygenase and a carboxypeptidase, by mouse pituitary corticotropic tumor cells, *Endocrinology*, 115, 1683, 1984.
73. **Jenks, B. G., Verburg van Kemenade, B. M. L., Tonon, M. C., and Vaudry, H.,** Regulation of biosynthesis and release of pars intermedia peptides in *Rana ridibunda*: dopamine affects both acetylation and release of alpha-MSH, *Peptides*, 6, 913, 1985.
74. **Follenius, E., Van Dorsselaer, A., and Meunier, A.,** Separation and partial characterization by high-performance liquid chromatography and radioimmunoassay of different forms of melanocyte-stimulating hormone from fish (Cyprinidae) neurointermediate lobes, *Gen. Comp. Endocrinol.*, 57, 198, 1985.
75. **Dores, R. M.,** Localization of multiple forms of ACTH- and beta-endorphin-related substances in the pituitary of the reptile, *Anolis carolinensis*. *Peptides*, 3, 913, 1982.
76. **Dores, R. M., Crim, J. W., and Kawauchi, H.,** Isolation of alpha-MSH and beta-endorphin from the pituitary of the holostean fish, *Amia calva*, *Am. Zool.*, 26, abstract, 1986.
77. **Denning-Kendall, P. A., Sumpter, J. P., and Lowry, P. J.,** Peptides derived from pro-opiocortin in the pituitary gland of the dogfish, *Squalus acanthias*. *J. Endocrinol.*, 9, 381, 1982.
78. **Schwyzer, R. and Eberle, A.,** On the molecular mechanisms of alpha-MSH receptor interactions, in *Frontiers in Hormone Research*, Vol. 4, van Wimersma Greidanus, Bj., Ed., S. Karger, Basel, 1971, 1380.
79. **O'Donohue, T. L., Handelmann, G. E., Miller, R. L., and Jacobowitz, D. M.,** N-acetylation regulates the behavioral activity of alpha-melanotropin in a multineurotransmitter neuron, *Science*, 215, 1125, 1982.
80. **Deakin, J. F., Dostrovsky, J. O., and Smyth, D. G.,** Influence of N-terminal acetylation and C-terminal proteolysis on the analgesic activity of beta-endorphin, *Nature (London)*, 279, 74, 1980.
81. **Akil, H., Young, E., Watson, S. J., and Coy, D. H.,** Opiate binding properties of naturally occurring N- and C-terminus modified beta-endorphins, *Peptides*, 2, 289, 1981.
82. **Glembotski, C. C.,** Characterization of the peptide acetyltransferase activity in bovine and rat intermediate pituitaries responsible for the acetylation of beta-endorphin and alpha-MSH, *J. Biol. Chem.*, 257, 10501, 1982.
83. **Woodford, T. A. and Dixon, J. E.,** The N-acetylation of corticotropin and fragments of corticotropin by a rat pituitary N-acetyltransferase, *J. Biol. Chem.*, 254, 4993, 1979.
84. **Chappell, M. C., Loh, Y. P., and O'Donohue, T. L.,** Evidence for an opiomelanotropin acetyltransferase in the rat pituitary neurointermediate lobe, *Peptides*, 3, 405, 1982.
85. **O'Donohue, T. L.,** Identification of endorphin acetyltransferase in rat brain and pituitary gland, *J. Biol. Chem.*, 258, 2163, 1983.
86. **Gibson, T. R. and Glembotski, C. C.,** Acetylation of alpha-MSH and beta-endorphin by rat neurointermediate pituitary secretory granule-associated acetyltransferase, *Peptides*, 6, 615, 1985.
87. **Millington, W. R., O'Donohue, T. L., Chappell, M. C., Roberts, J. L., and Mueller, G. P.,** Coordinate regulation of peptide acetyltransferase activity and proopiomelanocortin gene expression in the intermediate lobe of the rat pituitary, *Endocrinology*, 118, 2024, 1986.
88. **Lorenz, R. G., Tyler, A. N., Faull, K. F., Makk, G., Barchas, J. D., and Evans, C. J.,** Characterization of endorphins from the pituitary of the spiny dogfish, *Squalis acanthias*, *Peptides*, 7, 119, 1986.

89. **Dores, R. M. and Surprenant, A.,** Biosynthesis of multiple forms of beta-endorphin in the reptile intermediate pituitary, *Peptides,* 4, 889, 1983.
90. **Dores, R. M.,** Further characterization of the major forms of reptile beta-endorphin, *Peptides,* 4, 897, 1983.
91. **Kawauchi, H., Tsubokawa, A., Kanezawa, A., and Kitagawa, K.,** Occurrence of two different endorphins in the salmon pituitary, *Biochem. Biophsy. Res. Comm.,* 92, 1278, 1980.
92. **Goldman, M. E. and Loh, Y. P.,** Intracellular acetylation of desacetyl alpha-MSH in the *Xenopus laevis* neurointermediate lobe, *Peptides,* 5, 1129, 1984.
93. **Rothenberg, M. and Dores, R. M.,** The isolation of beta-endorphin in the intermediate pituitary of the amphibian, *Xenopus laevis, Am. Zool.,* 26, abstract, 1986.
94. **Murthy, A. S. N., Mains, R. E., and Eipper, B. A.,** Purification and characterization peptidyl-glycine alpha-amidation monooxygenase from bovine neurointermediate pituitary, *J. Biol. Chem.,* 261, 1815, 1986.

Chapter 4

THE MELANOTROPIC PEPTIDES: STRUCTURE AND CHEMISTRY

Hiroshi Kawauchi

I. Introduction 40

II. Melanophore-Stimulating Hormones 40
    A. Isolation and Structure 40
        1. α-MSH 40
        2. β-MSH 42
        3. γ-MSH 43
    B. Structure-Activity Relationship 43
        1. Bioassay 43
        2. Active Site 43

III. Melanin-Concentrating Hormone 45
    A. Isolation and Structure 45
        1. Salmon MCH 46
        2. Site of Origin 46
        3. Phylogenetic Distribution 47
    B. Structure-Activity Relationships 48
        1. Bioassay 48
        2. Melanophorotropic Activity 48
        3. Active Site 48

IV. Conclusions 49

Acknowledgment 50

References 50

I. INTRODUCTION

Many lower vertebrates change body color in response to their background coloration by dispersing and concentrating melanin granules within the integumentary melanophores. Smith[1] and Allen[2] independently reported pituitary control of skin coloration in the frog in 1916. Subsequently, Smith and Smith[3] showed a marked influence of extracts of the intermediate lobe of the pituitary on pigmentation of hypophysectomized tadpoles. The active principle was termed intermedin, melanocyte-stimulating hormone (MSH), or melanotropin. Lee and Lerner[4] demonstrated that two MSHs occur together in single individuals, and named the very basic peptide as α-MSH and the slightly basic peptide as β-MSH. α-MSH consists of 13 amino acids with blocked N- and C-termini. β-MSH consists of 18 amino acid residues. α- and β-MSHs share substantial homology with each other, and with corticotropin (ACTH) and β-lipotropin (LPH).

It is now firmly established that these peptides are derived from a common precursor molecule, termed pro-opiomelanocortin (POMC) or pro-opiocortin. In 1979, Nakanishi et al.[5] elucidated the entire nucleotide sequence of cDNA for the bovine prohormone which has repeating structural units of the common MSH core. The third MSH, γ-MSH, is found in the cryptic N-terminal portion of the precursor. The structural relationship between these molecules is shown in Figure 1. α-MSH is located at the N-terminal of ACTH, residues 1-13. β-MSH is located at the middle of β-LPH, residues 41-58. γ-MSH is in the N-terminal peptide, residues 51-64 of POMC. It is generally accepted that the prohormone is synthesized in both the corticotrophs of the pars distalis and the melanotrophs of the pars intermedia, but processed differently in the two lobes. In the corticotrophs, ACTH and β-LPH are the final products, whereas in the melanotrophs, ACTH is further processed to α-MSH and corticotropin-like intermediate lobe peptide (CLIP), and β-LPH is processed to the N-terminal fragment, β-MSH and β-endorphin (EP).

While it has long since been established that MSHs are responsible for melanin dispersion, a substance antagonistic to MSH has remained unknown until recently. In 1931, Hogben and Slome[6] postulated a dual hormonal control of color change by two antagonistic pituitary melanophorotropic hormones in amphibians. No supporting evidence for the dual hormonal concept has been found in amphibians, although melanin-concentrating hormone (MCH) has long since been recognized in teleost pituitary glands. Early attempts to separate MCH from MSHs proved unsuccessful. Thus, the hypothesis of an MCH was ignored for over two decades. Baker and Ball[7] re-evoked the dual hormonal control concept in 1975. In 1983, MCH from chum salmon pituitaries was characterized by Kawauchi et al.[8] In addition, salmon MCH was proved to be a movel heptadecapeptide synthesized in hypothalamus and transferred to the neurointermediate lobe of teleosts.[9]

II. MELANOPHORE-STIMULATING HORMONES

A. Isolation and Structure
1. α-MSH

During the 1950s, two melanocyte-stimulating hormones, α- and β-MSH, were identified in extracts from the posterior-intermediate lobes of mammalian pituitary glands. α-MSH was first isolated from the porcine pituitary by Lee and Lerner,[4] and its primary structure was determined by Harris and Lerner.[10] The hormone was later found and chemically identified in several mammalian species: bovine by Li,[11] ovine by Lee et al.,[12] equine by Dixon and Li,[13] macacus by Lee et al.,[14] and camel by Li et al.[15] In every mammal which has been studied, the amino acid sequence of α-MSH is identical; a tridecapeptide with an acetylated N-terminus and an amidated C-terminus (Figure 2).

Although MSHs of amphibians have never been sequenced. MSHs of *Xenopus laevis* have been extensively characterized by Martens and Jenks.[16-18] They performed in vitro pulse-

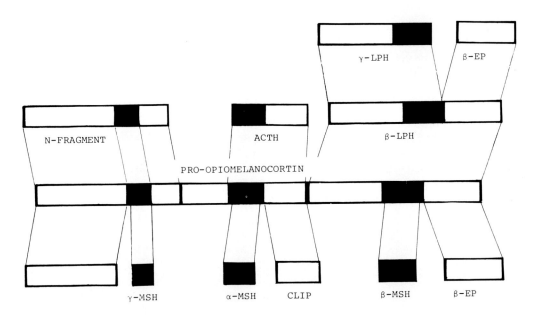

FIGURE 1. Pro-opiomelanocortin: precursor molecule of melanotropins. ACTH, corticotropin; CLIP, corticotropin-like intermediate lobe peptide; EP, endorphin; LPH, lipotropin; MSH, melanotropin.

```
α-MSH                                            5                  10
mammals, ostrich, & salmon I       Ac-Ser Tyr Ser Met Glu His Phe Arg Trp Gly Lys Pro Val-NH₂
salmon II                          Ac-Ser Tyr Ser Met Glu His Phe Arg Trp Gly Lys Pro Ile Gly His-OH
dogfish                             H-Ser Tyr Ser Met Glu His Phe Arg Trp Gly Lys Pro Met-OH
                                                                                        └NH₂

β-MSH                                            5                  10              15
salmon I                            H-Asp Gly  -  Ser Tyr Lys Met Asn His Phe Arg Trp Ser Gly Pro Pro Ala Ser-OH
salmon II                           H-Asp Gly  -  Ser Tyr Arg Met Gly His Phe Arg Trp Gly Ser Pro Thr Ala Ile-OH
dogfish: Scyliorhinus canicula      H-Asp Gly Ile Asp Tyr Lys Met Gly His Phe Arg Trp Gly Ala Pro Met Asp Lys-OH
dogfish: Squalus acanthias          H-Asp Gly Asp Asp Tyr Lys Phe Gly His Phe Arg Trp Ser Val Pro Leu-OH
bovine & ovine                      H-Asp Ser Gly Pro Tyr Lys Met Glu His Phe Arg Trp Gly Ser Pro Pro Lys Asp-OH
porcine                             H-Asp Glu Gly Pro Tyr Lys Met Glu His Phe Arg Trp Gly Ser Pro Pro Lys Asp-OH
equine                              H-Asp Glu Gly Pro Tyr Lys Met Glu His Phe Arg Trp Gly Ser Pro Arg Lys Asp-OH
bovine & ovine                      H-Asp Ser Gly Pro Tyr Lys Met Glu His Phe Arg Trp Gly Ser Pro Pro Lys Asp-OH
camel I                             H-Asp Gly Gly Pro Tyr Lys Met Glu His Phe Arg Trp Gly Ser Pro Pro Lys Asp-OH
camel II                            H-Asp Gly Gly Pro Tyr Lys Met Gln His Phe Arg Trp Gly Ser Pro Pro Lys Asp-OH
macacus                             H-Asp Glu Gly Pro Tyr Arg Met Glu His Phe Arg Trp Gly Ser Pro Pro Lys Asp-OH
```

FIGURE 2. Amino acid sequences of α- and β-melanotropins.

chase experiments with the neurointermediate lobe and subsequently analysed the secreted products or fragments from enzymatic digests of these products by reverse-phase high-performance liquid chromatography (HPLC). Their results clearly demonstrate that the amphibian secrete α-MSH that is identical to the mammalian hormone.

α-MSH of an elasmobranch, dogfish (*Squalus acanthias*), was isolated and chemically characterized by Lowry and Chadwick,[19] and Bennett et al.[20] The shark peptide differs from mammalian α-MSH in an unacetylated N-terminus and replacement of the C-terminal Val with a Met residue, half of which were not amidated.

Recently, isolation and structural studies of α-MSH and related peptides from the teleost, chum salmon (*Oncorhynchus keta*), were described by Kawauchi et al.[21-24] It is particularly interesting that the pituitary of chum salmon produces two different molecules of α-MSH. α-MSH I has a structure identical to that of mammalian α-MSH. However, the predominant form is not acetylated at the N-terminus. α-MSH II consists of 15 amino acid residues with an acetylated N-terminus and a free C-terminus. When compared with mammalian $ACTH_{1-15}$, 13 residues of α-MSH II are identical and the two variant residues, Ile^{13} and His^{15}, retain physicochemical characteristics of the replaced residues. Furthermore, the identification of several new intermediate peptides allows us to speculate about the maturational process of α-MSH. ACTH may be cleaved between the 17th and 18th residues as suggested by Scott et al.[25] Although the first intermediate, $ACTH_{1-17}$, has not been isolated, it is probably trimmed at the C-terminus by a carboxypeptidase, α-MSH II or N-Ac-ACTH II_{1-15} may be derived from the intermediate, $ACTH\ II_{1-15}$, by acetylation at the N-terminus without amidation at the C-terminus. The next probable intermediate, $ACTH\ I_{1-14}$, was also identified in salmon pituitary extracts. Whereas salmon $ACTH\ I_{1-13}$ has not been found, dogfish $ACTH_{1-13}$ was identified to be one of the predominant molecules among MSH-related peptides.[20] N-des-Ac-α-MSH or $ACTH_{1-13}$-NH_2 was the predominant form in both salmon and dogfish. On the contrary, in mammals, α-MSH is predominant. The des-acetylated form is also found in the camel and bovine pituitary.[15] These results suggest that amidation of the C-terminus of $ACTH_{1-13}$ precedes acetylation of the N-terminus. A possible role of modification at the termini may be for protection against exopeptidase attack. In fact, occurrence of α-MSH I_{3-13}-NH_2, α-MSH I_{7-13}-NH_2, and α-MSH II_{3-15} suggests that further processing of α-MSHs takes place at the N-terminal portion of the des-acetylated form by aminopeptidases.

2. β-MSH

β-MSH was isolated from porcine pituitary glands by Porath et al.,[26] and its amino-acid sequence was determined by Geschwind et al.,[27,28] and Harris and Roos.[29] Subsequently, β-MSHs were isolated and characterized from bovine by Geschwind et al.,[30] ovine by Lee et al.,[4] human by Harris,[31] macacus by Lee et al.,[14] and camel by Li et al.[15]

Mammalian β-MSHs consist of 18 amino acid residues, except for the human hormone. Human β-MSH has an additional tetrapeptide, Ala-Glu-Lys-Lys, attached at the N-terminus of monkey β-MSH. However, the adult human pituitary gland is unusual in not having an intermediate lobe and is believed to lack β-MSH. Therefore, recovery of human β-MSH from adult pituitary extracts is thought to be an artifact of the isolation procedure. Indeed, Barat et al.[32] showed that highly purified calf brain cathepsin D selectively splits Ala^{36}-Ala^{37} and Leu^{77}-Phe^{78} and suggested that the formation of human β-MSH from β-LPH is due to the action of the enzyme during isolation procedure.

β-MSHs from two elasmobranchs have been identified by Love and Pickering[33] and Bennett et al.[34] Dogfish (*Scyliorhinus caniculus*) β-MSH consists of 18 residues, whereas the peptide of the spiny dogfish (*Squalus acanthias*) is 2 residues shorter than that of the other species. Two molecular variants of β-MSH from the chum salmon (*Oncorhynchus keta*) were characterized by Kawauchi et al.[21,23,24,35] Both molecules of salmon β-MSH consist of 17 amino-acid residues, one residue less than the hormones of other species. In contrast to the variety of salmon α-MSH peptides, only β-MSH II_{1-15} has been identified as a derivative. The existence of diprolinyl at the 14th and 15th residues in β-MSH I may prevent cleavage by a carboxypeptidase, and the substitution of threonine for proline at residue 15, may explain the processing of β-MSH II, β-MSH of a dogfish, *Squalus acanthias*, seems to be a similar derivative.

The amino-acid sequences of MSHs are compared in Figure 2. The structure of α-MSH is highly conserved. The variation of β-MSHs between species is higher than that of α-MSH. However, it is striking that the tetrapeptide His-Phe-Arg-Trp in all species has been conserved during evolution from elasmobranchs to primates. The conserved core has been identified as the active site of melanotropin as discussed below.

3. γ-MSH

The elucidation of the entire base sequence of the cDNA coding for mammalian POMC revealed that they have repeating structural units of a common MSH core. The N-terminal peptides which contain a segment of γ-MSH have been characterized from several species; the human peptide consists of 76 amino-acid residues including a γ-MSH segment between residues 51 and 63;[36-39] the bovine peptide is a γ-MSH-like peptide, residues 51-61[40] in the bovine POMC; the rat peptide is a γ-MSH-like peptide corresponding to 25 residues between positions 50 and 74.[41] All of these mammalian peptides are glycosylated.

In lower vertebrates, evidence for a γ-MSH-like peptide in the pituitary of *Xenopus laevis* has been reported by Martens et al.[42] In addition, a dogfish γ-MSH has been isolated and the structure has been elucidated. The peptide consists of 12 residues with the common MSH core.[43] In contrast, the N-terminal peptide of salmon POMC consists of 76 amino acid residues as in the case of the human peptide.[44] Sequence comparison between the mammalian and the salmon peptides revealed that the salmon peptide is lacking in the counterpart of γ-MSH segment. The structural characteristics of salmon POMC have been confirmed by elucidation of the nucleotide sequence of cDNA for salmon POMC. No MSH-like sequence was found at the N-terminal region of salmon POMC.[45] However, there is significant sequence homology at the N-terminal 44 residues between mammalian and salmon N-terminal peptides (Figure 3). This observation suggests that γ-MSH may have a function in some, but not all vertebrates classes. The N-terminal-conserved region may be physiologically more important than the γ-MSH segment. Indeed, melanotropic activities of several synthetic peptides related to bovine γ-MSH and dogfish γ-MSH are not significant.[46-48] An N-terminal peptide of the bovine POMC residues 1-49 has also been identified.[49-50] Lowry et al.[51] and Estivaritz et al.[52] reported that peptides derived from the N-terminal, non-γ-MSH portion of the human peptide are potent stimulators of rat adrenal DNA synthesis and mitosis in vitro and in vivo.

B. Structure-Activity Relationship
1. Bioassay

Methods for biological assay of melanotropins can be classified into in vivo and in vitro methods. In vivo procedures use either intact or hypophysectomized frogs and lizards. After microscopic observation of the dispersion of melanin granules within the melanophores, melanotropic activity is evaluated according to the melanophore index of Hogben and Slome.[6] In vitro procedures are performed with isolated segments of skin which are immersed in a solution containing hormone, and the degree of darkening is evaluated photometrically. Shizume et al.[53] introduced a quantitative in vitro assay procedure by measuring light reflection from the skin mounted on metal rings by a photometric reflection meter. Tilders et al.,[54] reported an in vitro assay procedure for the quantitative determination of melanotropic activity using skin segments of the lizard, *Anolis carolinensis*. α-MSH is the most active in assays involving melanosome dispersion in amphibian skin. The activities of β-MSH, ACTH, and γ-MSH are approximately 1, 2, and 4, orders less potent, respectively, than that of α-MSH.

2. Active Site

Soon after elucidation of the structure of α-MSH, the peptide was synthesized. Since then, a large number of analogous MSH-like peptides have been synthesized by changing the length of the polypeptide chain and replacing one or more amino acids.

Hofmann et al.,[55] first demonstrated that the entire 13 amino acid sequence of α-MSH is not required for melanotropic activity by synthesis of a bioactive peptide, Met-Glu-His-Phe-Arg-Trp-Gly. It was also found that Ser-Tyr-Ser-Met-Glu-His-Phe-Arg has no bioactivity, whereas a biologically active peptide was created by the addition of Trp-Gly- to the C-

44 *The Melanotropic Peptides*

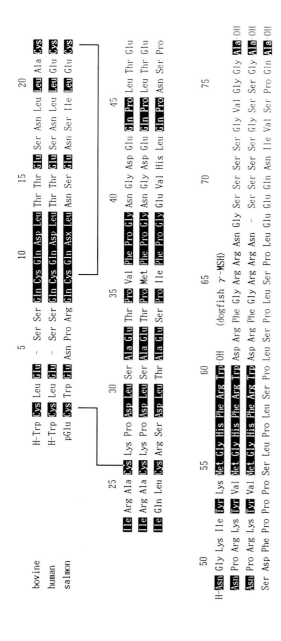

FIGURE 3. Amino acid sequences of amino-terminal peptides of pro-opiomelanocortins.

terminal end of the peptide. Furthermore, Yajima et al.[56] found that the C-terminal heptapeptide of α-MSH, Phe-Arg-Trp-Gly-Lys-Pro-Val-NH$_2$, has no melanotropic activity, while Hofmann and Yajima[57] showed addition of His to the N-terminus generates activity. These results strongly suggest that His^6 and Trp^{10} are essential for bioactivity. Since melanotropic activity of His-Phe-Arg-Trp-Gly synthesized by Schwyzer and Li[58] is of the same order of potency as that of His-Phe-Arg-Trp reported by Eberle et al.,[59] it appears that the tetrapeptide is the minimally required sequence for activity, i.e., an active site. It is striking that this particular tetrapeptide segment has been conserved through the evolution of vertebrates (Figure 2).

Contribution of other parts of α-MSH to melanotropic activity has also been investigated by chain elongation of peptides, including the active site sequence. For example, addition of H-Ser-Try-Ser to the N-terminus of Met-Glu-His-Phe-Arg-Trp-Gly does not increase the activity. However, addition of Lys-Pro-Val-NH$_2$ to the C-terminus of the heptapeptide resulted in a 10^3-fold increase in activity. Furthermore, Eberle et al.[59] and Eberle and Schwyzer,[60] reported that the C-terminal tripeptide exhibited low, but significant bioactivity. From these results, they postulated that Lys-Pro-Val-NH$_2$ could be another active site. Removal of the acetyl group from the N-terminus and deamidation of the C-terminus of α-MSH, also decreased bioactivity.[61]

Similarly, the relationship between chain length and biological activity of β-MSH has been investigated and resulted in similar conclusions. Thus, overall it can be concluded that the core sequence conserved throughout vertebrates, His-Phe-Arg-Trp, appears to be the essential structure for melanotropic activity. However, expression of full biological activity required additional components of the molecule, especially the C-terminal tripeptide.

It has been repeatedly shown that treatment of pituitary extracts, MSH, and ACTH with alkaline pH and heat enhances their melanosome-dispersing activity in two ways: by potentiation (increase in maximal activity) and prolongation (increase in duration) of the response. Enzymatic digestion of alkali-heat treated α-MSH revealed racemization of His^6, Arg^8, and Met^4. The racemization of Phe^7 is relatively more extensive in MSH. Thus, the activity of alkali-heat treated α-MSH can be explained as the consequence of increased resistance to proteolysis afforded by racemization. The quantitative studies on the extent of racemization of the amino acids in α-MSH has been studied by Engel et al.[62] Comparison of melanotropic activities of synthetic stereoisomers of His-Phe-Arg-Trp-Gly showed that analogues replaced with either D-Phe or D-Trp, or both, exhibit higher activity than that of the all-L-pentapeptides. The pentapeptides containing either D-His or D-Arg, or all, are inactive. Moreover, [Nle^4, D-Phe] α-MSH$_{4-10}$NH$_2$ exhibits prolongation of the melanosome-dispersing activity without alkali-heat treatment.[63] This compound is an agonist in the frog assay, but somewhat weaker in the lizard assay. In addition, Wilkes et al.,[64] found that Ac-[Nle^4,D-Phe^7]-α-MSH$_{4-10}$NH$_2$ and Ac-[Nle^4, D-Phe^7]-α-MSH$_{4-11}$NH$_2$ are agonists in the lizard assay, whereas only Ac[Nle^4,D-Phe^7]α-MSH$_{4-11}$NH$_2$ has a prolonged effect in the frog and lizard assays. These results suggest that residues 4, 7, and 11 are important in determining prolonged activity in the assay and that the enhancement of biological activity is not only due to an increased resistance to proteolysis, but also to a stabilized conformational feature of the melanotropins. Sawyer et al.,[65] introduced a disulfide bridge by replacing the Met^4 and Gly^{10} residues with cysteine residues. The resulting [Cys^4,Cys^{10}]α-MSH was originally reported as a superagonist, but activity of the peptide was revised recently.

III. MELANIN-CONCENTRATING HORMONE

A. Isolation and Structure

There have been a few attempts to separate two antagonstic melanophorotropic hormones from teleost pituitaries. Melanin-concentrating activity from catfish pituitaries was found in

```
                          5                    10                   15
H-Asp-Thr-Met-Arg-Cys-Met-Val-Gly-Arg-Val-Tyr-Arg-Pro-Cys-Trp-Glu-Val-OH
```

FIGURE 4. Amino acid sequence of melanin-concentrating hormone.

an alcohol-insoluble fraction by Enami[66] in 1955, while the activity from the killifish pituitary was found in the alcohol-soluble fraction by Pickford and Atz.[67] Adsorption chromatography of a catfish pituitary extraction an aluminum oxide column separated melanin-concentrating and -dispersing activities by stepwise elution with decreasing concentrations of acetone in water by Imai.[68] Melanin-concentrating activity was found in the adsorbed fraction and melanin-dispersing activity was found in the nonadsorbed fraction. In 1975, Baker and Ball[7] subjected extracts of trout neurointermediate lobes to polyacrylamide gel electrophoresis. Melanin-dispersing activity determined by the *Anolis* skin assay, was found in three fractions, and melanin-concentrating activity, measured on *Poecilia* melanophores, was found in a separate fraction. Westerfield et al.[69] reported that killifish MCH was a net positively charged peptide since it bound to CM cellulose, but not to DEAE cellulose. The molecular weight was estimated to be 3,500 to 13,000 by a dialysis technique. Baker and Rance[70] came to a similar conclusion concerning the physicochemical properties of rainbow trout MCH, except for the molecular weight. Trout MCH eluted slightly later than MSH on Bio-Gel P4, suggesting a molecular weight of less than 2,000 daltons. The isoelectric point was estimated to be greater than 9.5.

1. Salmon MCH

MCH was first isolated from the chum salmon (*Oncorhynchus keta*) pituitary by Kawauchi et al.[8] in 1983. Melanin-concentrating activity was monitored by a tilapia scale in an in vitro bioassay. Salmon MCH was found to be a heptadecapeptide with one disulfide bridge. The complete amino acid sequence of the peptide was elucidated and is shown in Figure 4. Several groups have confirmed the structure by organic synthesis of the peptide corresponding to the amino acid sequence.[71-73] Comparison of the amino acid sequence of salmon MCH with those of all known brain and pituitary proteins revealed that it is a novel peptide. However, there is a slight similarity to the carboxy-terminal disulfide loop of salmon prolactin, but not to that of mammalian prolactin.

2. Site of Origin

Melanin-concentrating activity was found in both the pituitaries and hypothalami of catfish (*Parasilurus*) by Enami.[66] The activity in the proximal pars distalis was approximately 4 or 5 times that of the posterior hypothalamus. Since hypophysectomy did not cause loss of MCH in hypothalamic extracts and transection of the pituitary stalk gradually decreased MCH content in the hypothalamus, MCH could be a neurosecretory substance produced in the nucleus lateralis tuberis (NLT) of the hypothalamus and transported to Übergangsteil for storage.[66] However, in most cases, MCH activity was found only in the neurointermediate lobe of *Phoxinus phoxinus*,[74] *Salmo gairdneri*,[7] and *Anguilla anguilla*.[75]

Although other workers have failed to detect MCH bioactivity in the brain,[69,76] Rance and Baker again found bioactivity in the hypothalamus of rainbow trout.[77] Baker and Rance mapped the distribution of MCH bioactivity in the trout brain, and found the highest values in the caudal third of the hypothalamus as suggested by Enami.[66] Naito et al.[78] have confirmed the results of Baker and Rance[77] by immunostaining with a specific antiserum against synthetic MCH. Numerous positive-staining processes of MCH-neurons from the hypothalamus projected to the pituitary and were distributed within neurohypophysial tissues in the pars intermedia and in the pars distalis, though few in number. In the neurohypophysial tissue, immunoreactivity for MCH appeared to be concentrated in the nerve terminals surrounding the blood vessels and the pituitary cells. No pituitary cells showed any cross-reactivity. These results proved that MCH is a neuropeptide synthesized in the neurons of the NLT/pars lateralis and is subsequently released in the neurohypophysis. These immunohistochemical results were further confirmed by Kawazoe et al.[79] The presence of MCH in the hypothalamus and pituitary was also confirmed by radioimmunoassay with a specific antiserum to MCH. Moreover, only a trace amount of MCH was isolated from the whole salmon brain.

Kawazoe et al.,[79] also suggested the occurrence of an MCH precursor in the salmon hypothalamus, since immunoreactivity and bioactivity for MCH in a hypothalamic extract increased significantly after lysyl endopeptidase digestion. A number of peptide hormones are synthesized as prohormones, which are processed at sites of two or three basic amino acids, lysine and arginine, by trypsin-like enzymes. In order to clip MCH from the precursor, lysyl endopeptidase was employed since the lysyl residue is absent in the hormone. Fractions of a hypothalamic extract after gel filtration on Sephadex G-100® were treated similarly. Most of the immunoreactivity eluted in the retarded MCH fraction. Immunoreactivity, but not biological activity, was found in the high molecular-weight fractions. After treatment of each fraction with lysyl endopeptidese, the biological activity emerged concomitantly with immunoreactivity from the high molecular weight fractions, while both activities from the retarded fractions were not affected. Therefore, it was concluded that the higher molecular-weight fractions contained the MCH prohormone.

3. Phylogenetic Distribution

Enami[65] was unable to detect MCH bioactivity in extracts of hypothalami and pituitaries of the dog. However, Baker and Rance[70] detected significant bioactivity in hypothalamic extracts, but not in pituitary extracts of the rat, *Rana temporaria*, *Xenopus laevis*, and *Lampetra fluviatilis*.

MCH-like immunoreactivity has been demonstrated in the central nervous system of the rat by immunohistochemical studies with salmon MCH antiserum.[80-82] Cell bodies with MCH-like immunoreactivity were confined to the dorsolateral regions of the rat hypothalamus, but not within the paraventricular nucleus, superaoptic nucleus, dorso- and ventro-medial nuclei, or arcuate nucleus. MCH-like immunoreactive fibers were found throughout the brain, but were most prevalent in the hypothalamus, mesencephalon, and pons-medulla regions. The anatomical distribution of MCH-stained neurons appears to be different from that of cell bodies containing other neuropeptides in the rat hypothalamus. It is noteworthy, that Naito et al.[82] found that the distribution of MCH-positive neurons appeared almost identical to that of cells which stained with α-MSH antisera in the dorsolateral hypothalamic region. However, no MCH-positive perikaria were observed in the arcuate nucleus where the same neurons had α-MSH-like immunoreactivities in the rat. Thus, MCH immunoreactive cells appear, for the most part, to be distinct from MSH-positive cells.

An MCH-like substance in the rat hypothalamus has been characterized by radioimmunoassay (RIA)[81] and by an in vitro fish-scale bioassay[83] after fractionation of the hypothalamus by reverse-phase HPLC and gel filtration. In the salmon MCH RIA, the rat hypothalamic extract showed a dose-dependent and parallel displacement of the salmon MCH tracer. When tested on the teleost melanophores, a weak, but significant melanin-concentrating activity was found in fractions from the chromatographic procedures that comigrate with MCH-like immunoreactivity. The rat MCH-like substance eluted at a position different from that of salmon MCH. It appeared to be somewhat smaller in size and more hydrophobic than salmon MCH. Gel chromatography of the rat hypothalamic extract revealed that the rat MCH-like substance may have a smaller molecular weight than salmon MCH.

The physiological role of the MCH-like substance in the rat brain is unknown. However, its persistence in the mammalian hypothalamo-neurohypophysial system suggests a role in posterior pituitary function, distinct from the mediation of color change seen in lower vertebrates. Baker and colleagues demonstrated that salmon MCH (10 nmol) inhibits CRF-41-induced secretion of ACTH by the rat pituitary[84] and α-MSH secretion from teleost pituitary glands.[85] Zamir et al.[80] suggested that MCH may be important in the regulation of food and water intake. Kawauchi et al.[86] demonstrated that intravenous administration of salmon MCH induces GH secretion in rats. Although these data are provocative, further biochemical and physiological studies are necessary to determine the molecular structure of MCH-like substances and their functions among vertebrates.

B. Structure-Activity Relationships
1. Bioassay

Melanin-concentrating activity has been evaluated by in vitro assays using fish scales or skin. The fish-scale bioassay was introduced by Rance and Baker.[77] Melanin granules in the scale gradually become concentrated within 2 to 3 hr due to the release of catecholamines from nerve endings around melanophores.[87] This melanophore response can be prevented by the addition of an α-adrenergic blocking agent to the incubation solution. In this solution, melanin attains a fully dispersed distribution. For the MCH bioassay, scales were incubated in a Tyrode-Ringer® containing 0.1 mM phentolamine for 30 min at 8 °C for rainbow trout scales, and at 20°C for tilapia scales[8] and Chinese grass carp scales.[88] The scales with fully dispersed melanin granules were then incubated in a test solution containing the sample dissolved in Tyrode-phentolamine Ringer®; after 20 min, they were examined under a microscope. Melanin-concentrating activity was expressed as the maximal dilution or minimal concentration which caused the melanin granules to become fully aggregated in 50% of the melanophores on a tilapia scale (EC_{50}). The EC_{50} of salmon MCH was 1 nM in tilapia scales. The melanophores of Chinese grass carp were very responsive to salmon MCH; the EC_{50} was only 63 pM.[88] The fish-skin assay is similar to the frog and lizard skin MSH bioassays (see Chapter IX Volume II). Catfish skin has been employed for the assay[66,68,72,89] in which salmon MCH was effective at a concentration of 1 nM.

2. Melanophorotropic Activity

When tested in a variety of teleost species, salmon MCH was active at concentrations ranging from 10^{-9} to 10^{-10} M in: *Salmo gairdneri, Cyprinus carpio, Sebastes schlegeli, Hexagrammos otakii*,[8] *Xiphophorus helleri, Carassius auratus, Lebistes reticulatus, Pimephales promelas*,[72] *Hyposyomus* sp., and *Ptergoplichys* sp.,[89] *Chrysiptera cyanea*,[90] and *Chrysiptra hollisi*, Fowler.[91] MCH has been shown to have melanin-concentrating activity both in vitro and in vivo.[92] A single intraperitoneal injection at a dose of 1 pmol/g of body weight turned the body color of a black-background-adapted rainbow trout into a golden color within 4 min after the injection. The effect lasted for 3 to 4 hr. Chronic administration of MCH via an osmotic minipump induced melanin concentration in the skin melanophore of rainbow trout, prevented increased melanogenesis of the pituitary melanophore, and depressed the stress-induced secretion of ACTH.[93] MCH also inhibited the CRF-41-induced secretion of ACTH at a concentration of 10 pM in the pars distalis removed from unstressed trout.[84]

An antagonistic interaction between MSH and MCH has also been determined using the in vitro scale bioassay.[79] Tilapia scales were incubated in test solutions containing 50 nM MCH and various concentrations of salmon α-MSH. At an equimolar ratio, all the melanophores showed full melanin concentration within 20 min, followed by gradual dispersion. Even at higher concentrations of α-MSH, the effect of MCH appeared first, although the melanin granules did not become fully concentrated. The action of MCH was not blocked by α- and β-adrenoreceptor antagonists. It was concluded that the effects of MCH were direct and were not mediated indirectly through the action of adrenergic neurotransmitters released from nerve terminals. When skins were incubated in dibenamine, an α-adrenoreceptor antagonist, the response to norepinephrine, but not to MCH, was inhibited. These results indicate that melanosome aggregation in response to the two agonists is mediated directly rather than through the release of a catecholamine. Lightening induced by MCH was readily reversed by α-MSH as well as norepinephrine.[89]

MCH stimulates aggregation of melanosomes within melanophores of teleost fishes, whereas, paradoxically, this cyclic neuropeptide disperses melanosomes within melanophores of frogs and lizards.[72,89,91] MCH was only approximately 1/600th as potent as α-MSH in the frog and lizard skin bioassays, and in cultured melanophores of the tadpole. The duration of the dispersing effect of MCH was relatively short compared to that of α-MSH.[91] It is interesting to note that a cyclic analogue of α-MSH, [Cys^4,Cys^{10}]α-MSH, is an agonist of α-MSH in the frog and lizard skin assays as described above.

3. Active Site

The structure and activity relationships of MCH have been investigated by chemical modification, enzymatic digestion and synthesis of a few peptide analogues. A C-terminal cyclic peptide obtained by chymotryptic digestion of salmon prolactin (H-Arg-Cys-Arg-Ala-Thr-Lys-Met-Arg-Pro-Glu-Thr-Cys-OH) exhibited an EC$_{50}$ of 5×10^{-6} M, whereas intact salmon prolactin was inactive. Hypothalamic cyclic peptides such as somatostatin, Arg-vasopressin, Arg-vasotocin, oxytocin, and isotocin, showed no activity.

Several peptides analogous to MCH have been synthesized and tested for their ability to concentrate melanin in the teleost scale. Replacement of two methionines at positions 3 and 6 with norvaline, lowered the potency by a factor of 2.7, while replacement with propargylglycine lowered potency by a factor of about 7. Reduced and carboxamidmethylated MCH was an agonist of MCH, but had a 345-fold lower potency than MCH. Iodinated MCH showed similar low bioactivity.[88]

Chemical and enzymatic modifications of natural MCH have been conducted to deduce the essential amino acid residues for biological activity.[92] Four N-terminal residues were removed by Edman degradation. The resulting fragment peptide, MCH$_{5-17}$, was found to have full bioactivity. Carboxypeptidase Y digestion of MCH and MCH$_{5-17}$ created MCH$_{1-14}$ and MCH$_{5-14}$, respectively, both of which showed full bioactivity. Moreover, modification of the Trp residue by *o*-nitrophenylsulfenyl chloride did not alter the potency. However, reduction and carboxamidomethylation of MCH caused complete loss of bioactivity. Modification of residues in the S-S loop, such as Tyr with tetranitromethane and Arg with 1,2 cyclohexadion, reduced the activity significantly, while oxidation with hydrogen peroxide caused only partial loss (10%) of bioactivity. These results suggest that the conformation of the S-S loop is essential for the activity, and residues Tyr and Arg are located in the active site.

IV. CONCLUSIONS

MSHs stimulate dispersion of melanin granules within the integumentary melanophore. Three molecules, α-, β-, and γ-MSH have been identified from several mammals, a teleost, and elasmobranches. All of them are derived from a common precursor, pro-opiomelanocortin. The structures of α-MSHs are highly conserved, whereas those of β-MSHs show relatively large species heterogeneity. γ-MSH, which is present in mammals and elasmobranchs, but lacking in teleosts, does not significantly contribute to melanotropic activity. A tetrapeptide sequence, His-Phe-Arg-Trp, has been conserved within the structures of α-, β-, and γ-MSH, suggesting that this segment is essential for bioactivity. By extensive studies on the relationship of structure and function with a large number of synthetic analogues, the active site of α-MSH could be assigned to this conserved tetrapeptide segment.

Moreover, quantitative analyses of racemization of each residue in alkali-heat treated α-MSH and comparison of bioactivity between stereoisomers of α-MSH, suggested that enhancement and prolongation of melanotropic activity may be due to stabilization of a conformational feature of the active site.

An antagonistic melanophorotropic hormone to MSHs has been isolated from pituitaries of one species of teleosts, chum salmon. The hormone, MCH, is a novel heptadecapeptide with one disulfide bridge. It is synthesized as a prohormone in neurons in the basal hypothalamus and translocated to the neurohypophysis. MCH induces aggregation of melanin granules within the melanophores of all teleost fishes examined. However, the same hormone stimulates dispersion of melanin granules in reptiles and amphibians.

Chemical and enzymatic modification of MCH revealed that the configuration of the disulfide loop of the molecules is essential for both bioactivity and immunoreactivity. Furthermore, residues Tyr[11] and Arg[12] may play an important role for bioactivity.

MCH-like immunoreactivity in the rat hypothalamus coincided with MCH bioactivity when tested in the fish-scale bioassay. These results imply that the chemical characteristics of MCH have been conserved during evolution, whereas the biological function may have diverged significantly.

ACKNOWLEDGMENT

I thank Dr. Penny Swanson and Dr. Ichiro Kawazoe for their helpful suggestions and comments in the preparation of the manuscript.

REFERENCES

1. **Smith, P. E.**, Experimental ablation of the hypophysis in the frog embryo, *Science*, 44, 280, 1916.
2. **Allen, B. M.**, The results of extirpation of the anterior lobe of the hypophysis and the thyroid of *Rana pipiens* larvae, *Science*, 44, 755, 1916.
3. **Smith, P. E. and Smith, I. P.**, The response of hypophysectomized tadpole to the intraperitoneal injection of the various lobes and colloid of the bovine hypophysis, *Anal. Rec.*, 25, 150, 1923.
4. **Lee, T. H. and Lerner, A. B.**, Isolation of melanocyte-stimulating hormone from hog pituitary gland, *J. Biol. Chem.*, 221, 943, 1956.
5. **Nakanishi, S., Inoue, A., Kita, T., Nakamura, M., Chang, A. C., Cohen, F. M., and Numa, S.**, Nucleotide sequence of cloned cDNA for bovine corticotropin-β-lipotropin precursor, *Nature (London)*, 278, 423, 1979.
6. **Hogben, L. T. and Slome, D.**, The pigmentary effector system. VI, the dual character of endocrine coordination in amphibian colour change. *Proc. Roy. Soc. London, Ser.*, B108, 10, 1931.
7. **Baker, B. I. and Ball, J. N.**, Evidence for a dual pituitary control of teleost melanophores, *Gen. Comp. Endocrinol.*, 25, 147, 1975.
8. **Kawauchi, H., Kawazoe, I., Tsubokawa, M., Kishida, M., and Baker, B. I.**, Characterization of melanin-concentrating hormone from chum salmon pituitaries, *Nature (London)*, 305, 321, 1983.
9. **Naito. N., Nakai, Y., Kawauchi, H., and Hayashi, Y.**, Immunocytochemical identification of melanin-concentrating hormone in the brain and pituitary gland of the teleost fishes *Oncorhynchus keta* and *Salmo gairdneri*, *Cell Tiss. Res.*, 242, 41, 1985.
10. **Harris, J. I. and Lerner, A. B.**, Amino-acid sequence of the α-melanocyte-stimulating hormone, *Nature (London)*, 179, 1346, 1957.
11. **Li, C.H.**, The relation of chemical structure to biologic activity of pituitary hormone, *Lab. Invest.*, 8, 574, 1959.
12. **Lee, T. H., Lerner, A. B., and Buettner-Janusch, V.**, Melanocyte-stimulating hormones from sheep pituitary glands, *Biochim. Biophys. Acta*, 71, 706, 1963.
13. **Dixon, J. S. and Li, C. H.**, The isolation and structure of α-melanocyte-stimulating hormone from horse pituitaries, *J. Am. Chem. Soc.*, 82, 4568, 1960.
14. **Lee, T. H., Lerner, A. B., and Buettner-Janusch, V.**, The isolation and structure of α- and β-melanocyte-stimulating hormones from monkey pituitary glands, *J. Biol. Chem.*, 236, 1390, 1961.

15. **Li, C. H., Danho, W. O., Chung, D., and Rao, A. J.,** Isolation, characterization and amino-acid sequence of melanotropins from camel pituitary gland, *Biochemistry,* 14, 947, 1975.
16. **Martens, G.J.M., Jenks, B. G., and van Overbeeke, A. P.,** Analysis of peptides biosynthesis in the neurointermediate lobe of *Xenopus laevis* using high-performance liquid chromatography: occurrence of small bioactive products, *Comp. Biochem. Physiol.,* 67B, 493, 1980.
17. **Martens, G. J. M., Jenks, B. G., and van Overbeeke, A. P.,** Biosynthesis of pairs of peptides related to melanotropin, corticotropin and endorphin in the pars intermedia of amphibian pituitary gland, *Eur. J. Biochem.,* 122, 1, 1982.
18. **Martens, G. J. M., Jenks, B. G., and van Overbeeke, A. P.,** N α-acetylation is linked to α-MSH release from pars intermedia of the amphibian pituitary gland, *Nature (London),* 294, 558, 1981.
19. **Lowry, P. J. and Chadwick, A.,** Purification and amino acid sequence of melanocyte-stimulating hormone from the dogfish, *Squalus acanthias, Biochem. J.,* 118, 713, 1970.
20. **Bennett, H. P. J., Lowry, P. J., McMartin, C., and Scott, A. P.,** Structural studies of α-melanocyte-stimulating hormone and a novel β-melanocyte-stimulating hormone from the neurointermediate lobe of the pituitary of the dogfish, *Squalus acanthias, Biochem. J.,* 141, 439, 1974.
21. **Kawauchi, H. and Muramoto, K.,** Isolation and primary structure of melanotropin from salmon pituitary glands, *Intl. J. Peptide Protein Res.,* 14, 373, 1979.
22. **Kawauchi, H., Adachi, Y., and Tsubokawa, M.,** Occurrence of a new melanocyte-stimulating hormone in the salmon pituitary gland, *Biochem. Biophys. Res. Commun.,* 96, 1508, 1980.
23. **Kawauchi, H.,** Chemistry of proopiocortin-related peptides in the salmon pituitary, *Arch. Biochem. Biophys.,* 227, 343, 1983.
24. **Kawauchi, H., Kawazoe, I., Adachi, Y., Buckley, B. I., and Ramachandran, J.,** Chemical and biological characterization of salmon melanocyte-stimulating hormone, *Gen. Comp. Endocrinol.,* 53, 37, 1984.
25. **Scott, A. P., Lowry, P. J., Ratcliffe, J. C., Rees, L. H., and Landon, J.,** Corticotropin-like peptide in the rat pituitary, *J. Endocrinol.,* 61, 355, 1974.
26. **Porath, J., Roos, P., Landgrebe, F. W., and Mitchell, G. M.,** Isolation of a melanophore-stimulating peptide from pig pituitary gland, *Biochim. Biophys. Acta,* 17, 596, 1955.
27. **Geschwind, I. I., Li, C. H., and Barnafi, L.,** Isolation and structure of melanocyte-stimulating hormone from porcine pituitary glands, *J. Am. Chem. Soc.,* 78, 4494, 1956.
28. **Geschwind, I. I., Li, C. H., and Barnafi, L.,** Isolation and structure of melanocyte-stimulating hormone, *J. Am. Chem. Soc.,* 79, 620, 1957.
29. **Harris, J. I. and Roos, P.,** Amino acid sequence of A-melanophore-stimulating peptide, *Nature,* 178, 90, 1956.
30. **Geschwind, I. I., Li, C. H., and Barnafi, L.,** The isolation and structure of melanocyte-stimulating hormone from bovine pituitary glands, *J. Am. Chem. Soc.,* 79, 1003, 1957.
31. **Harris, J. I.,** Structure of a melanocyte-stimulating hormone from the human pituitary gland, *Nature (London),* 184, 167, 1959.
32. **Barat, E., Patthy, A., and Graf, L.,** Action of cathepsin D on human β-lipotropin: a possible source of human "β-melanotropin", *Proc. Natl. Acad. Sci. U. S. A.,* 76, 6120, 1979.
33. **Love, R. M. and Pickering, B. T.,** A β-MSH in the pituitary gland of the spotted dogfish (*Scyliorhinus caniculus*): Isolation and structure, *Gen. Comp. Endocrinol.,* 24, 398, 1974.
34. **Bennett, H. P. J., Lowry, P. J., McMartin, C., and Scott, A. P.,** Structural studies of α-melanocyte-stimulating hormone and a novel β-melanocyte-stimulating hormone from the neurointermediate lobe of the pituitary of the dogfish *Squalus acanthias, Biochem. J.,* 141, 439, 1974.
35. **Kawauchi, H., Adachi, Y., and Ishizuka, B.,** Isolation and structure of another β-melanotropin from salmon pituitary glands, *Intl. J. Peptide Protein Res.,* 16, 79, 1980.
36. **Estovariz, F. E., Hope, J., McLean, C., and Lowry, P. J.,** Purification and characterization of a γ-melanotropin precursor from frozen human pituitary glands, *Biochem. J.,* 191, 125, 1981.
37. **Seidah, N. G. and Chrétien, M.,** Complete amino acid sequence of a human pituitary glycopeptide: An important maturation product of pro-opiomelanocortin, *Proc. Natl. Acad. Sci. U. S. A.,* 78, 4236, 1981.
38. **Seidah, N. G., Rochemont, J., Hamelin, J., Lis, M., and Chrétien, M.,** Primary structure of the major human pituitary pro-opiomelanocortin NH_2-terminal glycopeptide, *J. Biol. Chem.,* 256, 7977, 1981.
39. **Bennett, H. P. J., Seidah, N. G., Benjannet, S., Solomon, S., and Chrétien, M.,** Reinvestigation of the disulfide bridge arrangement in human pro-opiomelanocortin N-terminal segment (hNT1-76), *Intl. J. Peptide Protein Res.,* 27, 306, 1986.
40. **Bohlen, P., Esch, F., Shibasaki, T., Baird, A., Ling, N., and Guillemin, R.,** Isolation and characterization of a γ_1-melanotropin-like peptide from bovine neurointermediate pituitary, *FEBS Lett.,* 128, 67, 1981.
41. **Browns, C. A., Bennett, H. P. J., and Solomon, S.,** The isolation and characterization of γ_3-melanotropin from the neurointermediate lobe of the rat pituitary, *Biochem. Biophys. Res. Commun.,* 100, 336, 1981.

42. **Martens, G. J. M., Jenks, B. G., and van Overbeeke, A. P.,** Biosynthesis of a γ-₃-melanotropin-like peptide in the pars intermedia of the amphibian pituitary gland, *Eur. J. Biochem.*, 58, 2106, 1975.
43. **McLean, C. and Lowry, P. J.,** Natural occurrence but lack of melanotropic activity of γ-MSH in fish, *Nature (London)*, 290, 341, 1981.
44. **Kawauchi, H., Takahasi, A., and Abe, K. I.,** Gamma-melanotropin is not present in an amino terminal peptide of salmon proopiocortin, *Intl. J. Peptide Protein Res.*, 18, 223, 1981.
45. **Soma, J., Kitahara, N., Nishizuka, T., Nanami, H., Kotake, C., Okazaki, H., and Andoh, T.,** Nucleotide sequence of a cloned cDNA for proopiomelanocortin precursor of chum salmon, *Oncorhynchus keta*, *Nucleic Acid Res.*, 12, 8029, 1984.
46. **Ling, N., Ying, S., Minick, S., and Guillemin, R.,** Synthesis and biological activity of four γ-melanotropin peptides derived from the cryptic region of the adrenocorticotropin/β-lipotropin precursor, *Life Sci.*, 25, 1773, 1979.
47. **Okamoto, K., Yasumura, K., Shimamura, S., Nakanishi, S., Numa, S., Imura, H., Tanaka, A., Nakamura, A., and Yajima, H.,** Synthesis of the dodecapeptide designated as bovine γ-melanotropin (γ-MSH), *Chem. Pharm. Bull.*, 28, 2839, 1982.
48. **Shimamura, S., Yasumura, K., Okamoto, K., Miyata, K., Tanaka, A., Nakamura, A., and Yajima, H.,** Solution synthesis of a heptacosapeptide known as bovine γ-₃-melanotropin (γ-₃-MSH), *Chem. Pharm. Bull.*, 30, 2433, 1982.
49. **Bennett, H. P. J.,** Isolation and characterization of the 1 to 49 amino-terminal sequence of pro-opiomelanocortin from bovine posterior pituitaries, *Biochem. Biophys. Res. Commun.*, 125, 229, 1984.
50. **James, S. and Bennett, H. P. J.,** Use of reverse-phase and ion-exchange batch extraction in the purification of bovine pituitary peptides, *J. Chromatogr.*, 326, 329, 1985.
51. **Lowry, P. J., Silas, L., McLean, C., Linton, E. A., and Estivariz, F. E.,** Pro-γ-melanocyte-stimulating hormone cleavage in adrenal gland undergoing compensatory growth, *Nature (London)*, 306, 70, 1983.
52. **Estivaritz, F. E., Iturriza, F., McLean, C., Hope, J., and Lowry, P. J.,** Stimulation of adrenal mitogenesis by N-terminal proopiocortin peptides, *Nature (London)*, 297, 419, 1982.
53. **Shizume, K., Lerner, A. B., and Fitzpatrick, T. B.,** *In vitro* bioassay for the melanocyte-stimulating hormone, *Endocrinology*, 54, 553, 1954.
54. **Tilders, F. J. H., van Delft, A. M. C., and Smelik, P. G.,** Reinvestigation and evaluation of an accurate high-capacity bioassay for melanocyte-stimulating hormone using the skin of *Anolis carolinensis in vitro*, *J. Endocrinol.*, 66, 165, 1975.
55. **Hofmann, K., Thompson, T. A., and Schwartz, E. T.,** Polypeptides. XI. Preparation of an octapeptide possessing melanocyte-stimulating activity, *J. Am. Chem. Soc.*, 79, 6087, 1957.
56. **Yajima, H., Okada, Y., Oshima, T., and Laude, S.,** Studies on peptides. XIII. Synthesis of two heptapeptides isolated from pituitary glands, *Chem. Pharm. Bull.*, 14, 707, 1966.
57. **Hofmann, K. and Yajima, H.,** Synthesis pituitary hormones, *Recent. Prog. Horm. Res.*, 18, 41, 1962.
58. **Schwyzer, R. and Li, C. H.,** A new hormone and its melanocyte-stimulating activity, *Nature (London)*, 182, 1669, 1958.
59. **Eberle, A., Fauchere, J-L., Tesser, G. I., and Schwyzer, R.,** Hormone-receptor interactions. Syntheses of α-melanotropin and information-carrying sequences using alkali-labile protecting groups, *Helv. Chim. Acta*, 58, 2106, 1975.
60. **Eberle, A. and Schwyzer, R.,** Hormone-receptor interactions. Demonstration of two message sequences (active sites) in α-melanotropin, *Hevl. Chim. Acta*, 58, 1528, 1975.
61. **Li, V. H. and Hemmasi, B.,** Solid-phase synthesis of seryltyrosylmethionylglutaminylhistidylphenyl-alanyltryptophylglycine hydrazide and its N$^\alpha$-tetrabutyloxcarbonyl derivative and their melanotropic activity, *J. Med. Chem.*, 15, 697, 1972.
62. **Engel, M. H., Sawyer, T. K., Hadley, M. E., and Hruby, V. J.,** Quantitative determination of amino acid racemization in heat-alkali-treated melanotropins: implications for peptide hormone structure-function studies, *Anal. Biochem.*, 116, 303, 1981.
63. **Sawyer, T. K., Sanfilippo, P. J., Hruby, V. J., Engel, M. H., Heward, C. B., Burnett, J. B., and Hadley, M. E.,** 4-Norleucine, 7-D-phenylalanine-α-melanocyte-stimulating hormone: A highly potent α-melanotropin with ultralong biological activity, *Proc. Natl. Acad. Sci. U. S. A.*, 77, 5754, 1980.
64. **Wilkes, B. C., Sawyer, T. K., Hruby, V. J., and Hadley, M. E.,** Differentiation of the structural features of melanotropins important for biological potency and prolonged activity *in vitro*, *Int. J. Peptide Protein Res.*, 22, 313, 1983.
65. **Sawyer, T. K., Hruby, V. J., Darman, P. S., and Hadley, M. E.,** [half-Cys^4,half-Cys^{10}]α-melanocyte-stimulating hormone: A cyclic α-melanotropin exhibiting superagonist biological activity, *Proc. Natl. Acad. Sci. U. S. A.*, 79, 1751, 1982.
66. **Enami, M.,** Melanophore-concentrating hormone (MCH) of possible hypothalamic origin in the catfish, *Parasilurus.*, *Science*, 121, 36, 1955.
67. **Pickford, G. E. and Atz, J.,** *The Physiology of the Pituitary Gland of Fishes*, Zoological Society, New York, 1975.

68. **Imai, K.,** Extraction of melanophore concentrating hormone (MCH) from the pituitary of fishes, *Endocrinol. Japan*, 5, 34, 1958.
69. **Westerfield, D. B., Pang, P. K. T., and Burns, J. M.,** Some characteristics of melanophore-concentrating hormone (MCH) from teleost pituitary glands, *Gen. Comp. Endocrinol.*, 42, 494, 1985.
70. **Baker, B. I. and Rance, T. A.,** Further observation on the distribution and properties of teleost melanin-concentrating hormone, *Gen. Comp. Endocrinol.*, 50, 423, 1983.
71. **Okamoto, K., Yasumura, K., Fujitani, K., Kiso, Y., Kawauchi, H., Kawazoe, I., and Yajima, H.,** Synthesis of the heptadecapeptide corresponding to the entire amino acid sequence of salmon melanin-concentrating hormone (MCH), *Chem. Pharm. Bull.*, 32, 2963, 1984.
72. **Wilkes, B. C., Hruby, V. J., Sherbrooke, W. C., Castrucci, A. M. L., and Hadley, M. E.,** Synthesis and biological actions of melanin concentrating hormone, *Biochem. Biophys. Res. Commun.*, 122, 613, 1984.
73. **Eberle, A. N., Atherton, E., Dryland, A., and Sheppard, R. C.,** Peptide synthesis. Part 9. Solid-phase synthesis of melanin-concentrating hormone using a continuous-flow polyamide method, *J. Chem. Soc. Perkin Transcr.*, 1, 361, 1986.
74. **Kent, A. K.,** Distribution of melanophore-aggregating hormone in the pituitary of the minnow, *Nature (London)*, 183, 544, 1959.
75. **Kent, A. K.,** The influence in sodium hydroxide on the activity of the color change factors of the teleost pituitary, *Gen. Comp. Endocrinol.*, 1, 409, 1961.
76. **Rance, T. and Baker, B. I.,** The teleost melanin-concentrating hormone — A pituitary hormone of hypothalamic origin, *Gen. Comp. Endocrinol.*, 37, 64, 1979.
77. **Gilham, I. D. and Baker, B. I.,** Evidence for the participation of a melanin-concentrating hormone in physiological colour change in the eel, *J. Endocr.*, 102, 237, 1984.
78. **Naito, N., Nakai, Y., Kawauchi, H., and Hayashi, Y.,** Immunocytochemical identification of melanin-concentrating hormone in the brain and pituitary gland of the teleost fishes, *Oncorhynchus keta* and *Salmo gairdneri.*, *Cell Tiss. Res.*, 242, 41, 1985.
79. **Kawazoe, I., Kawauchi, H., Hirano, T., and Naito, N.,** Characterization of melanin-concentrating hormone in teleost hypothalamus, *Gen. Comp. Endocrinol.*, in press.
80. **Skofitsch, G., Jacobowitz, D. M., and Zamir, N.,** Immunohistochemical localization of a melanin-concentrating hormone-like peptide in the brain, *Brain Res. Bull.*, 15, 635, 1985.
81. **Zamir, N., Skofitsch, G., Bannon, M., and Jacobowitz, D. M.,** Melanin-concentrating hormone: Unique peptide neuronal system in the rat brain and pituitary gland, *Proc. Natl. Acad. Sci. U. S. A.*, 83, 1528, 1986.
82. **Naito, N., Kawazoe, I., Nakai, Y., Kawauchi, H., and Hirano, T.,** Coexistence of immunoreactivity for melanin-concentrating hormone and α-melanocyte-stimulating hormone in the dorsolateral hypothalamus of the rat, *Neuroscience Lett.*, 70, 81, 1986.
83. **Kawazoe, I. and Kawauchi, H.,** Structure-activity studies on melanin-concentrating hormone. Presented at XIIIth International Pigment Cell Conference, Tucson, Arizona, October 5 to 9, 1986.
84. **Baker, B. I., Bird, D. J., and Buckingham, J. C.,** Salmonid melanin-concentrating hormone inhibits corticotropin release, *J. Endocr.*, 106, R5, 1985.
85. **Barber, L. D., Baker, B. I., Penny, J. C., and Eberle, A. N.,** Melanin-concentrating hormone (MCH) inhibits the release of α-MSH from teleost pituitary glands, *Gen. Comp. Endocrinol.*, in press.
86. **Kawauchi, H., Kawazoe, I., Hayasi, Y., Minamitake, Y., Tanaka, S., Naito, N., Nakai, M., Ide, H., and Hirano, T.,** Melanin-concentrating hormone: Chemistry and biology, in *Natural Products and Biological Activities*, Imura, H., Goto, T., Murachi, T., and Nakajima, T., Eds., Tokyo University Press, Tokyo, 1986, 303.
87. **Fujii, R. and Novales, R. R.,** Cellular aspects of the control of physiological color changes in fishes, *Am. Zool. Soc. Transcr.*, 9, 453, 1969.
88. **Baker, B. I., Eberle, A. N., Baumann, J. B., Siegrist, W., and Girard, J.,** Effect of melanin concentrating hormone on pigment and adrenal cells *in vitro.*, *Peptides*, 6, 1125, 1986.
89. **Wilkes, B. C., Hruby, V. J., Sherbrooke, W. C., Castrucci, A. M. L., and Hadley, M. E.,** Synthesis of a cyclic melanotropic peptide exhibiting both melanin-concentrating and -dispersing activities, *Science*, 224, 1111, 1984.
90. **Oshima, N., Kasukawa, H., Fujii, R., Wilkes, B. C., Hruby, V. J., Sherbrooke, W. C., Castrucci, A. M. L., and Hadley, M. E.,** Melanin concentrating hormone (MCH) effects on teleost (*Chrysiptera cyanea*) melanophores, *J. Exptl. Zool.*, 235, 175, 1985.
91. **Ide, H., Kawazoe, I., and Kawauchi, H.,** Fish melanin-concentrating hormone disperses melanin in amphibian melanophores, *Gen. Comp. Endocrinol.*, 58, 486, 1985.
92. **Kawazoe, I., Kawauchi, H., Hirano, T., and Naito, N.,** Structure-Activity relationship of melanin concentrating hormone, *Intl. J. Peptide Protein Res.*, in press.
93. **Baker, B. I., Bird, D. J., and Buckingham, J. C.,** Effects of chronic administration of melanin concentrating hormone (MCH) on ACTH, MSH and pigmentation in the trout, *Gen. Comp. Endocrinol.*, 63, 62, 1986.

Chapter 5

MELANOCYTE-STIMULATING HORMONE IN THE CENTRAL NERVOUS SYSTEM

Bibie M. Chronwall and Thomas L. O'Donohue

TABLE OF CONTENTS

I.	Introduction	56
II.	Distribution of α-MSH and Related Peptides in the Brain	56
	A. Distribution of Radioimmunoassayed Peptides	56
	B. Immunohistochemical Distribution of Cell Bodies	56
III.	Co-localization of POMC-Related Peptides	57
	A. Co-localization of POMC-Related Peptides with Other Neuroactive Substances	59
IV.	*In Situ* Hybridization Histochemistry	59
V.	Distribution of POMC-Containing Axons	60
	A. Distribution of α-MSH Axons	61
VI.	Intrinsic and Efferent POMC Projections	61
VII.	Conclusions	62
References		63

I. INTRODUCTION

Alpha-melanocyte-stimulating hormone (α-MSH) was first described in neurons in the brain in 1978.[1-3] The α-MSH immunoreactive material in the brain was later found to be comprised of both acetylated and desacetyl peptides, as occurs in the pituitary, but the brain contained relatively more of the desacetyl peptide.[4,5] The distribution of α-MSH was remarkably similar to that previously shown for β-endorphin. This fact, combined with the discovery that α-MSH and β-endorphin were derived from a common prohormone, pro-opiomelanocortion (POMC), suggested the existence of a multiple neurotransmitter neuron for the first time.[6,7] These findings have led to extensive studies of the POMC neuronal system and multiple neurotransmitters in the central nervous system.

The anatomical distribution of POMC-related peptides has been reviewed recently by Khachaturian et al.[8] Other aspects of this system are included in reviews by O'Donohue and Dorsa[6] and Akil and Watson.[9] Since these reviews appeared, technical advances have made it possible to study new anatomical aspects of neuronal systems. With immunohistochemistry, especially using primary antibodies produced in different species, coexistence of neuroactive substances has been studied, and it has been found to be the rule rather than the exception.[7] Through combination of immunohistochemistry and axonal tract tracing,[10] the projections of neurons containing a specific neuroactive substance have been elucidated. The incorporation of *in situ* hybridization (For reviews containing technical information, see References 11 to 15) and its combination with axonal tract tracing (For review, see Reference 16), has made it possible to study the dynamics of mRNA content in defined neuronal pathways. This approach should allow greater understanding of the physiological regulation of peptidergic systems in the brain.

II. DISTRIBUTION OF α-MSH AND RELATED PEPTIDES IN THE BRAIN

A. Distribution of Radioimmunoassayed Peptides

Radioimmunoassay for α-MSH in microdissected areas of discrete regions of rat brain show the highest concentrations in the median eminence and the dorsomedial nucleus, followed by the arcuate and the periventricular nuclei of both the hypothalamus and thalamus. The ventral part of the interstitial nucleus, ansa lenticularis, the paraventricular, posterior hypothalamic, and rhomboid nuclei contain moderate levels, as do the central gray and dorsal raphe (Table 1). Several studies show consistent data in rat,[2,17] cat,[18] and human.[19,20] Discrete regions of the rat brain show a diurnal rhythm of immunoreactive α-MSH content.[21]

B. Immunohistochemical Distribution of Cell Bodies

The pituitary was for some time considered the only biosynthetic site for POMC-related peptides. The POMC-related material found in the brain was thought to have been transported there and then absorbed. However, Krieger et al.,[22] found ACTH immunoreactivity using RIA in the CNS of normal as well as hypophysectomized rats, and similar results were found for α-MSH by other groups.[23,24] Neurons of the arcuate nucleus, that could be the CNS POMC biogenic site, were then described using antibodies against different POMC peptides.[1,25-32]

Neurons containing POMC-related products have now been localized to three CNS areas; the arcuate and peri-arcuate areas, the nucleus of the solitary tract and the dorsolateral hypothalamus. The arcuate nucleus is the primary site, (Figure 1a), each mid-level coronal section through the nucleus contains 100 to 150 neurons with a ventrolateral distribution within the nucleus. POMC neurons are also scattered laterally to the arcuate proper. The arcuate-periarcuate neurons are 10 to 15 μm in diameter, polygonal with 2 to 4 conspicuous dendrites; in comparison to other arcuate neurons, these neurons are large (Figure 1a,b).

Table 1
CONCENTRATIONS OF α-MSH IN MICRODISSECTED DISCRETE REGIONS OF RAT FOREBRAIN

Region	α-MSH concentration (pg/μg protein ± S.E.M.)
Piriform cortex	0.20 ± 0.09
Caudate nucleus	0.13 ± 0.03
Globus pallidus	0.39 ± 0.15
Nucleus accumbens	0.33 ± 0.05
Interstitial nucleus of the stria terminalis, dorsal	1.43 ± 0.27
Medial preoptic nucleus	3.75 ± 0.32
Periventricular nucleus (hypothalamus)	7.23 ± 0.49
Paraventricular nucleus	5.75 ± 0.35
Arcuate nucleus	8.72 ± 1.33
Median eminence	11.02 ± 1.60
Dorsomedial nucleus	8.95 ± 0.65
Posterior hypothalamic nucleus	4.32 ± 0.56
Ansa lenticularis	3.83 ± 0.56
Rhomboid nucleus	4.56 ± 0.60
Periventricular nucleus (thalamus)	7.19 ± 0.73
Central gray, caudal	3.44 ± 0.67
Dorsal raphe	4.15 ± 0.87

More recently, a less extensive group of POMC neurons has been described in the nucleus of the solitary tract using ACTH,[33,34] β-endorphin[35], α-MSH,[36] and γ-MSH[37] antibodies. The neurons are distributed in the caudal medulla from the area postrema rostrally to the medullary spinal cord junction, caudally. At the caudal level of the area postrema, they number 40 to 80 per coronal section, they appear polygonal with 2 to 3 dendrites or bipolar with their long axes horizontal in the coronal plane.

After heavy colchicine treatment, a second α-MSH-immunoreactive neuronal group in the hypothalamus has been reported[38-42] (Figure 2). These α-MSH cells are distributed in the dorsal hypothalamus over the top of the third ventricle, out laterally into the zona incerta, between the fornix and the mammillothalamic tract. They are twice as numerous as the arcuate POMC cells,[8] are not recognized by any other antibodies against POMC-related peptides, and when using antibodies that stain both the arcuate and the dorsal hypothalamic α-MSH neurons, the dorsal group appears comparatively dim (Figure 2 and Chronwall, B.,[66] unpublished observations). The dorsolateral hypothalamic α-MSH population stand out as being exceptional in other respects (see below).

The identification of authentic α-MSH in this region is yet to be established and Quinn and Weber[43] have presented evidence suggesting that this peptide is, in fact, not α-MSH.

III. CO-LOCALIZATION OF POMC-RELATED PEPTIDES

The arcuate POMC neurons have been described using antibodies to different POMC-related peptides. In addition, actual co-localization of these peptides has been established.[25,26,39,44,46]

Mezey et al.,[47] and Millington et al.,[48] studied the distribution of POMC-derived peptides in different hypothalamic nuclei and found interesting differences in the ratios of peptide products in different regions. There are also data suggesting the possibility of a difference in the processing of the β-endorphin depending on the area.[49] Agnati et al.,[50] have used a statistical approach to quantitate the coexistence of ACTH and β-endorphin immunoreac-

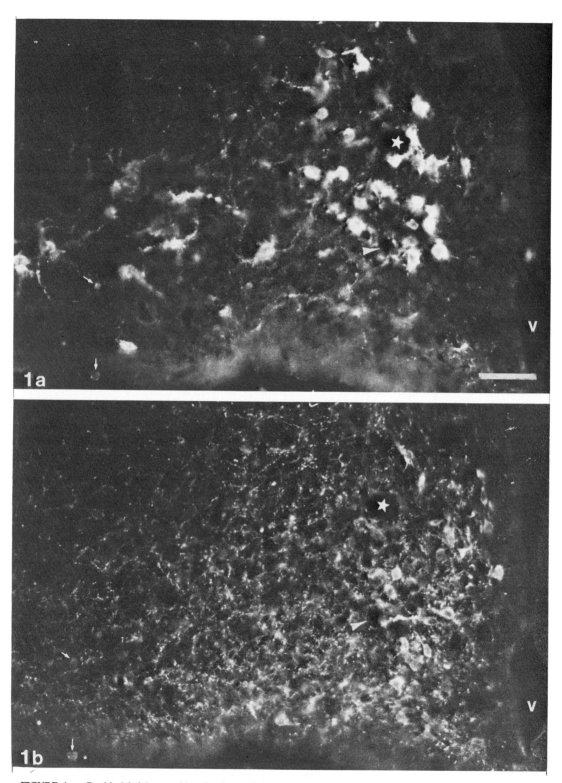

FIGURE 1. Double label immunohistochemistry of the arcuate nucleus. The same section is seen through a filter combination showing (a) rhodamine labeled α-MSH neurons and (b) fluorescein labeled neuropeptide Y neurons. There is no co-localization of the two peptides in one neuron. The following landmarks are useful when comparing neuronal positions; the shape of the third ventricle (v), star and arrowhead denoting blood vessels and small arrows pointing at artifacts visible in both filter combinations. Magnification bar = 10 μm.

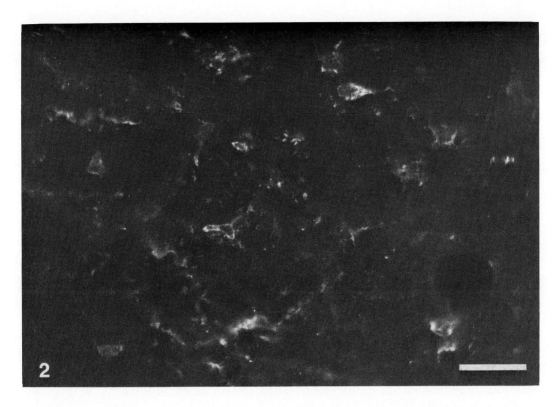

FIGURE 2. Neurons in the dorsolateral hypothalamus staining with an α-MSH antibody. Magnification bar = 50 μM.

tivity, and evidence has been obtained for a lack of balanced coexistence in mediobasal neurons.

No actual co-localization studies have been published for the POMC neurons in the caudal medulla, although these neurons have been reported to contain most of the known POMC peptides.

A. Co-localization of POMC-Related Peptides and other Neuroactive Substances

The arcuate nucleus contains a high number of transmitters and neuropeptides ($cf.$[51-53]). It is, however, now clear that the arcuate POMC neurons are anatomically distinct from those containing dynorphin or enkephalin as well as dopamine and, surprisingly enough, from those containing any other known neuroactive substance.[52-54]

No thorough study of coexistence has been published for the nucleus of the solitary tract, which also contains a high number of neuroactive substances. ($cf.$[51])

IV. *IN SITU* HYBRIDIZATION HISTOCHEMISTRY

Using *in situ* hybridization, the presence of POMC mRNA has been established in the intermediate lobe of the pituitary and in the corticotropes of the anterior lobe.[11,13,55] The POMC mRNA has also been quantitated in individual melanotropes as a function of up and down regulation of the dopamine receptor.[56] In the brain, POMC mRNA has been shown in the arcuate region (Figure 3)[13,55,57] At present, no other brain region shows POMC mRNA by *in situ* hybridization.[55] The absence of POMC mRNA in the dorsolateral hypothalamus supports the suggestion that these cells do not contain α-MSH.[43] The POMC message could also be present at undetectable levels in the dorsolateral hypothalamus and the nucleus of

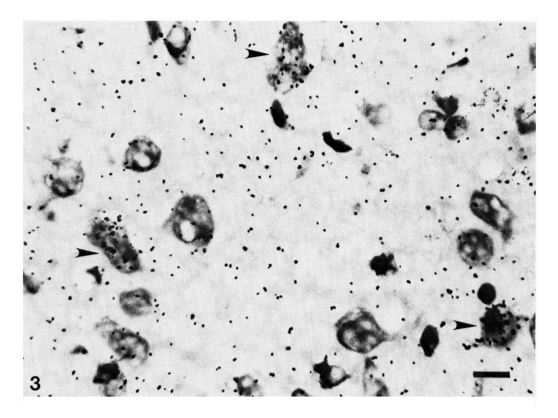

FIGURE 3. *In situ* hybridization of a POMC cDNA probe to POMC mRNA of arcuate neurons (arrowheads). Also, note several nonlabeled neurons, some of which are small. Silver grains over neuropil are due to nonspecific hybridization. Magnification bar = μm.

the solitary tract as the peptides seem to be present at low concentrations in these areas. The estrogen influence on POMC mRNA in the arcuate has been quantitated by dot-blot analysis,[58] which is a promising start to determine hormone-neuronal interactions at the molecular level. In the future, the resolution probably will be at the cellular level using *in situ* hybridization.

V. DISTRIBUTION OF POMC-CONTAINING AXONS

Immunohistochemistry has shown an extensive system of discrete varicose POMC-containing axons in the rat CNS (see References 2, 8, and 27 for detailed mapping). A comparatively dense fiber network has been observed in the medial nucleus and the periventricular, dorsomedial, and anterior hypothalamic nuclei. Moderate numbers were found in the lateral preoptic, paraventricular, and posterior hypothalamic nuclei and in the septal region. In the thalamus, a particularly dense distribution of fibers was found in the periventricular nucleus and moderate densities of fibers were located in the rhomboid nucleus. In the mid- and hindbrain, many fibers were observed in the mesencephalic central gray. In the pons, moderate numbers of fibers were found in the dorsal lateral tegmental nucleus, dorsal parabrachial nucleus, nucleus of the mesencephalic tract, trigeminal nerve and superior cerebellar peduncle. In the medulla, a moderate number of fibers was located in the nucleus of the solitary tract. These findings are in agreement with biochemical results for the distribution of α-MSH in microdissected brain regions.[2]

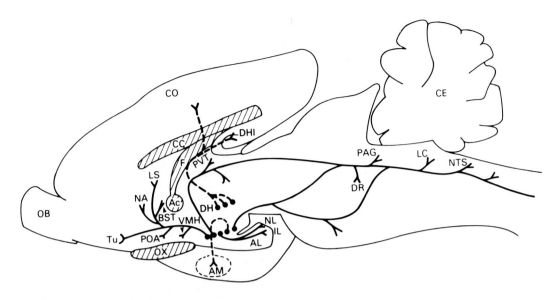

FIGURE 4. Schematic drawing of a sagittal section of rat brain showing the efferent pathways of the arcuate. Broken lines indicate projections lateral to the plane of section. AL, anterior lobe of the pituitary; AM, amygdala; BST, bed nucleus of the stria terminalis; CC, corpus callosum; CE, cerebellum; CO, cortex; DH, dorsolateral hypothalamus; DHI, dorsal hippocampus; DR, dorsal raphe; F, fornix; IL, intermediate lobe of the pituitary; LC, locus coeruleus; LS, lateral septum; NA, nucleus accumbens; NL, neural lobe; NTS, nucleus of the solitary tract; OB, olfactory bulb; OX, optic chiasma; PAG, periaqueductal gray; POA, preoptic area; PVT, periventricular nucleus of the thalamus; VMH, ventromedial hypothalamus.

A. Distribution of α-MSH Axons

Sensitive (nonspecific?) antisera against α-MSH stain fibers in the hippocampus and cortex.[27,40] Antisera against other POMC-related peptides do not stain fibers in these areas. These staining characteristics are similar to those of the α-MSH neurons in the dorsolateral hypothalamus.

VI. INTRINSIC AND EFFERENT POMC PROJECTIONS

Within the dense population of POMC neurons in the arcuate, there is electronmicroscopical evidence that ACTH-positive axons make synaptic contacts with other POMC-positive perikarya and dendrites[59,60] which indicates autoregulation within this population. Most commonly, however, POMC axons innervate other neuronal populations in the arcuate.[60]

From POMC fiber distribution maps, three general projection directions can be deduced (Figure 4); frontal to the preoptic and septal areas, caudal to the periaqueductal gray and the nucleus of the solitary tract, and lateral to the amygdala. When the arcuate nucleus is lesioned electrically or surgically, all immunoreactivity is depleted from areas where POMC is radioimmunoassayed in control animals.[2,17] Similarly, rats which have had a selective glutamate lesion of the arcuate nucleus during their neonatal period or have had their arcuate surgically deafferented, have markedly decreased or absent POMC fiber immunohistochemical staining.[60]

Through tract tracing, combined with immunohistochemistry, some of the efferent pathways of the arcuate POMC neurons have been confirmed. Parvo- as well as magnocellular paraventricular neurons, receive synaptic input from arcuate POMC neurons.[62] Preoptic and periaqueductal central gray cells are also innervated by arcuate POMC neurons.[52] The preoptic projection is ipsi- as well as contralateral, and comparatively, many neurons are doubly labeled by tract tracing and immunohistochemistry (Figure 5a,b). The more lateral

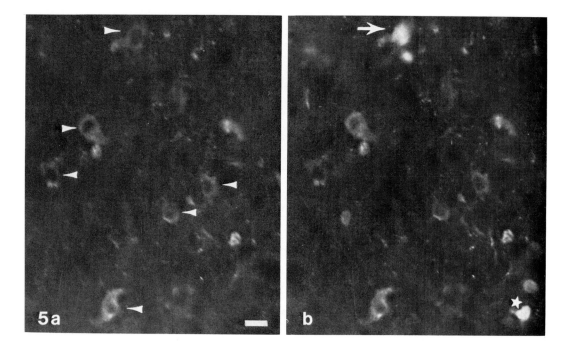

FIGURE 5. Combined immunohistochemistry and axonal tract tracing. The fluorescent dye Fast Blue was injected unilaterally into the preoptic area. 5a shows immunostained α-MSH neurons (arrowheads) in the arcuate. 5b is a double exposure of 5a and the same field seen through a filter combination visualizing Fast Blue. The field is contralateral of the third ventricle compared to the injection site. One neuron is doubly labeled for α-MSH and with Fast Blue (arrow). Other neurons are labeled for α-MSH only (cf. 5a) and one neuron with Fast Blue only (star), Magnification bar = 10 μm.

POMC population preferentially projects to the central gray, but a few have collaterals to the preoptic area as well (Chronwall B., et al.[67] in preparation). There is also a POMC projection to the nucleus of the solitary tract.[63] An interesting finding is that POMC fibers from the arcuate are surrounding the α-MSH-like perikarya in the dorsolateral hypothalamus. It remains to be established whether synaptic contacts are made or not. No similar studies have been performed on the POMC cells in the nucleus of the solitary tract.

A distinction between the pathways containing POMC-related peptides and those containing only α-MSH has been made using glutamate lesioning. The results show that fibers in the hippocampus and cortex are not affected by this lesion.[42] Retrograde axonal transport of HRP injected into the parietal cortex, label neurons in the zona incerta.[64] Likewise, fluorescent tracers injected into the hippocampal area labeled α-MSH neurons in the dorsolateral hypothalamus. Some of these α-MSH neurons have collateral projections to the hippocampus as well as the spinal cord.[65]

VII. CONCLUSIONS

Although it is now clear that α-MSH exists in the brain and a fairly good understanding of the anatomy has been achieved, very little is known of the physiological role of these neurons.

Through a combination of fast blue tract tracing and *in situ* hybridization, it has recently been possible to delineate a small population of arcuate neurons containing POMC mRNA that is defined by its projection to the preoptic nucleus.[57] This technique will make it possible to study regulation of gene expression in discrete subpopulations of neurons, which will be of great importance to further elucidate the function of the POMC system in the brain.

REFERENCES

1. **Jacobowitz, D. M. and O'Donohue, T. L.**, α-Melanocyte-stimulating hormone: Immunohistochemical identification and mapping in neurons of rat brain, *Proc. Natl. Acad. Sci. U.S.A.*, 12, 6300, 1978.
2. **O'Donohue, T. L., Miller, R. L., and Jacobowitz, D. M.**, Identification, characterization and stereotaxic mapping of intraneuronal α-melanocyte-stimulating hormone-like immunoreactive peptides in discrete regions of the rat brain, *Brain Res.*, 176, 101, 1979.
3. **Oliver, C. and Porter, J. C.**, Distribution and characterization of α-melanocyte-stimulating hormone in the rat brain, *Endocrinology*, 102, 697, 1978.
4. **O'Donohue, T. L., Handelmann, G. E., Chaconas, T., Miller, R. L., and Jacobowitz, D. M.**, Evidence that N-acetylation regulates the behavioral activity of α-MSH in the rat and human central nervous system, *Peptides*, 2, 333, 1981.
5. **O'Donohue, T. L., Handelmann, G. E., Miller, R. L., and Jacobowitz, D. M.**, N-acetylation regulates the behavioral activity of α-melanotropin in a multineurotransmitter neuron, *Science*, 215, 1125, 1982.
6. **O'Donohue, T. L. and Dorsa, D. M.**, The opiomelanotropinergic neuronal and endocrine systems, *Peptides*, 3, 353, 1982.
7. **O'Donohue, T. L., Millington, W. R., Handelmann, G. E., Contreras, P. C., and Chronwall, B. M.**, On the 50th anniversary of Dale's law: multiple neurotransmitter neurons, *Trends Pharmaco. Sci.*, 6, 305, 1985.
8. **Khachaturian, H., Lewis, M. E., Tsou, K., and Watson, S. J.**, β-Endorphin, α-MSH, ACTH, and related peptides, in *Handbook of Chemical Neuroanatomy*, Björklund, A. and Hökfelt, T., Eds., Elsevier, Amsterdam, 1985, 4.
9. **Stanley, J.**, Beta-endorphin and biosynthetically related peptides in the central nervous system, in *Handbook of Psychopharmacology*, Iversen, L., Iversen, S. D., and Synder, S. H., Eds., Plenum Press, New York, 1983, 16.
10. **Skirboll, L., Hökfelt, Norell, G., Phillipson, O., Kuypers, H. G. J., Bentivoglio, M., Catsman-Berrevoets, C. E., Visser, T. J., Steinbusch, H., Verhofstad, A., Cuello, A. C., Goldstein, M., and Brownstein, M.**, A method for specific transmitter identification of retrogradely labeled neurons: immunofluorescence combined with fluorescence tracing, *Brain Res. Rev.*, 8, 99, 1984.
11. **Lewis, M. E., Arentzen, R., and Baldino, F., Jr.**, Rapid, high-resolution in situ hybridization histochemistry with radioiodinated synthetic oligonucleotides, *J. Neurosci. Res.*, 16, 117, 1986.
12. **Lewis, M. E., Sherman, T. G., Burke, S., Akil, H., Davis, L. G., Arentzen, R., and Watson, S. J.**, Detection of proopiomelanocortin mRNA by *in situ* hybridization with an oligonucleotide probe, *Proc. Natl. Acad. Sci. U.S.A.*, 83, 1986, 5419.
13. **Bloch, B., Popovici, T., Le Guellec, D., Normand, E., Chouham, S., Guitteny, A. F., and Bohlen, P.**, *In situ* hybridization histochemistry for the analysis of gene expression in the endocrine and central nervous system tissues: a 3-year experience, *J. Neurosci. Res.*, 16, 183, 1986.
14. **Griffin, W. S. and Morrison, M. R.**, *In situ* hybridization/visualization and quantitation of genetic expression in mammalian brain, *Peptides*, 6, Suppl. 2, 89, 1985.
15. **Wilcox, J. N., Gee, C. E., and Roberts, J. L.**, *In situ* cDNA:mRNA hybridization: Development of a technique to measure mRNA levels in individual cells, *Methods Enzymol.*, 124, 510, 1986.
16. **Chronwall, B. M., Lewis, M. E., Schwaber, J. S., and O'Donohue, T. L.**, *In situ* hybridization combined with axonal tracing, in *Anatomical Tract Tracing Methods II*, Heimer, L. and Zaborszky, L., Eds., Plenum Press, New York, 1988.
17. **Eskay, R. L., Giraud, P., Oliver, C., and Brownstein, M. J.**, Distribution of α-melanocyte-stimulating hormone in the rat brain: evidence that α-MSH-containing cells in the arcuate region send projections to extra hypothalamic areas, *Brain Res.*, 178, 55, 1979.
18. **O'Donohue, T. L., Massari, V. J., Tizabi, Y., and Jacobowitz, D. M.**, Identification and distribution of α-melanotropin in discrete regions of the cat brain, *Brain Res. Bull.*, U.S.A., 4, 829, 1979.
19. **Emson, P. C., Corder, R., Ratter, S. J., Tomlin, S., Lowry, P. J., Ress, L. H., Arregui, A., and Rosser, M. N.**, Regional distribution of pro-opiomelanocortin-derived peptides in the human brain, *Neuroendocrinology*, 38, 45, 1984.
20. **O'Donohue, T. L. and Jacobowitz, D. M.**, Studies on α-melanotropin in the central nervous system, in *Polypeptide Hormones*, Beers, R. F., Jr. and Bassett, E. G., Eds., Raven Press, New York, 1980.
21. **O'Donohue, T. L., Miller, R. L., Pendleton, R. C., and Jacobowitz, D. M.**, A diurnal rhythm of immunoreactive α-melanocyte-stimulating hormone in discrete regions of the rat brain, *Neuroendocrinology*, 29, 281, 1979.
22. **Krieger, D. T., Liotta, A., and Brownstein, M. J.**, Presence of corticotropin in brain of normal and hypophysectomized rats, in *Proc. Natl. Acad. Sci. U.S.A.*, 74, 1977, 117.
23. **Vaudry, H., Tonon, M. C., Delarue, C., Vaillant, R., and Kraicer, J.**, Biological and radioimmunological evidence for melanocyte-stimulating hormones (MSH) of extrapituitary origin in the rat brain, *Neuroendocrinology*, 27, 9, 1978.

24. **O'Donohue, T. L., Holmquist, G. E., and Jacobowitz, D. M.**, Effect of hypophysectomy on α-melanotropin in discrete regions of the rat brain, *Neurosci. Lett.*, 14, 271, 1979.
25. **Watson, S. J., Barchas, J. D., and Li, C. H.**, β-Lipotropin localization of cells and axons in rat brain by immunocytochemistry, *Proc. Natl. Acad. Sci.*, 11, 5155, 1977.
26. **Bloch, B., Bugnon, C., Fellmann, D., Lenys, D., and Gouget, A.**, Neurons of the rat hypothalamus, *Cell Tiss. Res.*, 204, 1, 1979.
27. **Dube, D., Lissitzky, J. D., Leclerc, R., and Pelletier, G.**, Localization of α-melanocyte-stimulating hormone in rat brain and pituitary, *Endocrinology*, 102, 4, 1978.
28. **Zimmerman, L. A., Liotta, A., and Krieger, D. T.**, β-Lipotropin in brain: localization in hypothalamic neurons by immunoperoxidase technique, *Cell Tiss. Res.*, 186, 393, 1978.
29. **Bloom, F. E., Battenberg, E. L. F., Shibasaki, T., Benoit, R., Ling, N., and Guillemin, R.**, Localization of γ-melanocyte-stimulating hormone (γ-MSH) immunoreactivity in rat brain and pituitary, *Regulatory Peptides*, 1, 205, 1980.
30. **Pelletier, G.**, Ultrastructural immunohistochemical localization of adrenocorticotropin and beta-lipotropin in the rat brain, *J. Histochem. Cytochem.*, 27, 1046, 1979.
31. **Pelletier, G.**, Ultrastructural localization of a fragment (16K) of the common precursor for adrenocorticotropin (ACTH) and β-lipotropin (β-LPH) in the rat hypothalamus, *Neurosci. Lett.*, 16, 85, 1980.
32. **Osamura, R. Y., Komatsu, N., Watanabe, K., Nakai, Y., Tanaka, I., and Imura, H.**, Immunohistochemical and immunocytochemical localization of γ-melanocyte stimulating hormone (γ-MSH)-like immunoreactivity in human and rat hypothalamus, *Peptides*, 3, 771, 1982.
33. **Schwartzberg, D. G. and Nakane, P. K.**, ACTH-related peptide containing neurons within the medulla oblongata of the rat, *Brain Res.*, 276, 351, 1983.
34. **Joseph, S. A., Pilcher, W. H., and Bennett-Clarke, C.**, Immunocytochemical localization of ACTH perikarya in nucleus tractus solitarius: evidence for a second opiocortin neuronal system, *Neurosci. Lett.*, 38, 221, 1983.
35. **Khachaturian H., Alessi, N. E., Munfakh, N., and Watson, S. J.**, Ontogeny of opioid and related peptides in the rat CNS and pituitary: an immunocytochemical study, *Life Sci.*, 33, Suppl. I, 61, 1983.
36. **Yamazoe, M., Shiosaka, S., Yagura, A., Kawai, Y., Shibasaki, T., Ling, N., and Tohyama, M.**, The distribution of α-melanocyte stimulating hormone (α-MSH) in the central nervous system of the rat: An immunohistochemical study. II. Lower brain stem, *Peptides*, 5, 721, 1984.
37. **Kawai, Y., Inagaki, S., Shiosaka, S., Shibasaki, T., Ling, N., Tohyama, M., and Shiotani, Y.**, The distribution and projection of γ-melanocyte-stimulating hormone in the rat brain; an immunohistochemical analysis, *Brain Res.*, 297, 21, 1984.
38. **Watson, S. J., Akil, H., Richard, C. W., III, and Barchas, J. D.**, Evidence for two separate opiate peptide neuronal systems, *Nature*, 275, 226, 1978.
39. **Watson, S. J. and Akil, H.**, α-MSH in rat brain: occurrence within and outside of β-endorphin neurons, *Brain Res.*, 182, 217, 1980.
40. **Watson, S. J. and Akil, H.**, The presence of two α-MSH-positive cell groups in rat hypothalamus, *Eur. J. Pharmacol.*, 58, 101, 1979.
41. **Jegou, S., Tonon, M. C., Guy, J., Vaudry, H., and Pelletier, G.**, Biological and immunological characterization of α-melanocyte-stimulating hormone (α-MSH) in two neuronal systems of the rat brain, *Brain Res.*, 260, 91, 1983.
42. **Guy, J., Vaudry, H., and Pelletier, G.**, Differential projections of two immunoreactive α-melanocyte-stimulating hormone (α-MSH) neuronal systems in the rat brain, *Brain Res.*, 220, 199, 1981.
43. **Quinn, B. and Weber, E.**, Metorphamide and α-MSH antisera cross-react immunohistochemically with the same population of lateral hypothalamus neurons, *Soc. Neurosci. Abstr.*, 12, 407, 1986.
44. **Nilaver, G., Zimmerman, E. A., Defendini, R., Liotta, A. S., Krieger, D. T., and Brownstein, M. J.**, Adrenocorticotropin and β-lipotropin in the hypothalamus — localization in the same arcuate neurons by sequential immunocytochemical procedures, *J. Cell. Biol.*, 81, 50, 1979.
45. **Sofroniew, M. V.**, Immunoreactive α-endorphin and ACTH in the same neurons of the hypothalamic arcuate nucleus in the rat, *Am. J. Anat.*, 154, 283, 1979.
46. **Watson, S. J., Richard, C. W., III, and Barchas, J. D.**, Adrenocorticotropin in rat brain; immunocytochemical localization in cells and axons, *Science*, 200, 1180, 1978.
47. **Mezey, E., Kiss, J. Z., Mueller, G. P., Eskay, R., O'Donohue, T. L., and Palkovits, M.**, Distribution of the pro-opiomelanocortin derived peptides, adrenocorticotrope hormone, α-melanocyte-stimulating hormone and β-endorphin (ACTH, α-MSH, β-END) in the rat hypothalamus, *Brain Res.*, 328, 341, 1985.
48. **Millington, N. R., Mueller, G. P., and O'Donohue, T. L.**, Regional heterogeneity in the ratio of α-MSH:β-endorphin in rat brain, *Peptides*, 5:841-843, 1984.
49. **Zakarian, S. and Smyth, D. C.**, β-endorphin is processed differently in specific regions of rat pituitary and brain, *Nature*, 296, 250, 1982.

50. **Agnati, L. F., Fuxe, K., Locatelli, V., Benfenati, F., Zini, I., Panerai, A. E., El Etreby, M. F., and Hökfelt, T.**, Neuroanatomical methods for the quantitative evaluation of coexistence of transmitters in the nerve cells. Analysis of the ACTH- and β-endorphin immunoreactive nerve cell bodies of the mediobasal hypothalamus of the rat, *J. Neurosci. Methods*, 5, 203, 1982.
51. **Palkovits, M.**, Distribution of neuropeptides in the central nervous system: a review of biochemical mapping studies, *Prog. Neurobiol.*, 23, 151, 1984.
52. **Chronwall, B. M.**, Anatomy and physiology of the neuroendocrine arcuate nucleus, *Peptides*, 6, Suppl. 2, 1, 1985.
53. **Everitt, B. J., Meister, B., Hökfelt, T., Melander, T., Terenius, L., Rökaeus, A., Theodorsson-Norheim, E., Dockray, G., Edwardson, J., Cuello, C., Elde, R., Goldstein, M., Hemmings, H., Ouimet, C., Walaas, I., Greengard, P., Vale, W., Weber, E., Yen Wu, J., and Chang, K. J.**, The hypothalamic arcuate nucleus-median eminence complex: immunohistochemistry of transmitter, peptides and DARPP-32 with special reference to coexistence in dopamine neurons, *Brain Res. Rev.*, 11, 97, 1986.
54. **Bugnon, C., Block, B., Lenys, D., Gouget, A., and Fellmann, D.**, Comparative study of the neuronal populations containing β-endorphin, corticotropin and dopamine in the arcuate nucleus of the rat hypothalamus, *Neurosci. Lett.*, 14, 43, 1979.
55. **Gee, C. E., Chen, C.-L. C., and Roberts, J. L.**, Identification of proopiomelanocortin neurones in rat hypothalamus by *in situ* cDNA-mRNA hybridization, *Nature*, 306, 374, 1983.
56. **Chronwall, B. M., Millington, W. R., Griffin, W. S., Unnerstall, J. R., and O'Donohue, T. L.**, Histological evaluation of the dopaminergic regulation of pro-opiomelanocortin gene expression in the intermediate lobe of the rat pituitary involving *in situ* hybridization and ^{3}H-thymidine uptake measurement, *Endocrinology*, 120, 1201, 1987.
57. **Wilcox, J. N., Roberts, J. L., Chronwall, R. M., Bishop, J. F., and O'Donohue, T. L.**, Localization of proopiomelanocortin mRNA in functional subsets of neurons defined by their axonal projections, *J. Neurosci. Res.*, 16, 89, 1986.
58. **Wilcox, J. N. and Roberts, J. L.**, Estrogen decreases rat hypothalamic proopiomelanocortin messenger ribonucleic acid levels, *Endocrinology*, 117, 2392, 1985.
59. **Chen, Y. Y. and Pelletier, G.**, Demonstration of contacts between proopiomelanocortin neurons in the rat hypothalamus, *Neuro. Lett.*, 43, 271, 1983.
60. **Kiss, J. Z. and Williams, T. H.**, ACTH-immunoreactive boutons form synaptic contacts in the hypothalamic arcuate nucleus of rat: evidence for local opiocortin connections, *Brain Res.*, 263, 142, 1983.
61. **Pelletier, G., Leclerc, R., Saavedra, J. M., Brownstein, M. J., Vaudry, L. F., and Labrie, F.**, Distribution of β-lipotropin (β-LPH), adrenocorticotropin (ACTH) and α-melanocyte-stimulating hormone (α-MSH) in the rat brain. I. Origin of the extrahypothalamic fibers, *Brain Res.*, 433, 1980.
62. **Sawchenko, P. E., Swanson, L. W., and Joseph, S. A.**, The distribution and cells of origin of ACTH(1-39)-stained varicosities in the paraventricular and supraoptic nuclei, *Brain Res.*, 232, 365, 1982.
63. **Gray, T. S., O'Donohue, T. L., Watson, S. J., and Magnuson, D. J.**, Pro-opiomelanocortin and neuropeptide Y projections from arcuate and peri-arcuate hypothalamic areas to the nucleus tractus solitarious-dorsal vagal complex, *Soc. Neurosci. Abstr.*, 10, 432, 1984.
64. **Shiosaka, S., Shibasaki, T., and Tohyama, M.**, Bilateral α-melanocyte-stimulating hormonoergic fiber system from zona incerta to cerebral cortex: combined retrograde axonal transport and immunohistochemical study, *Brain Res.*, 309, 350, 1984.
65. **Köhler, C., Haglund, L., and Swanson, L. W.**, A diffuse α-MSH-Immunoreactive projection to the hippocampus and spinal cord from individual neurons in the lateral hypothalamic area and zona incerta, *J. Comp. Neurol.*, 223, 501, 1984.
66. **Chronwall, B. M.**, unpublished observations.
67. **Chronwall, B. M., et al.**, in preparation.

Chapter 6

THE PRO-OPIOMELANOCORTIN GENE IN *XENOPUS LAEVIS*: STRUCTURE, EXPRESSION, AND EVOLUTIONARY ASPECTS

Gerard J.M. Martens

TABLE OF CONTENTS

I.	Introduction	68
II.	Structural Organization of the POMC Gene in *Xenopus laevis*	70
	A. *Xenopus* POMC Gene Structure	70
	B. Evolutionary Aspects	70
III.	Primary Structures of *Xenopus* POMC mRNAs and Proteins	72
	A. *Xenopus* POMC mRNA and Protein Structures	72
	B. Evolutionary Aspects	74
IV.	Expression of the *Xenopus* POMC Gene	75
V.	Summary	76
	A. Concluding Remarks	76
	B. Future Prospects	78
VI.	Acknowledgments	79
	References	80

I. INTRODUCTION

Several peptides which can cause the dispersion of the black pigment, melanin, in dermal melanophores of the amphibian skin have been isolated from pituitary glands of a number of different species.[1-6] The smallest of these melanotropic peptides is α-melanophore-stimulating hormone (α-MSH), the amino acid sequence of which is essentially the same in all vertebrate species studied (see Reference 7). In many amphibians, α-MSH has an important physiological function because it mediates the process of background adaptation.[8,9] During this neuroendocrine reflex the pars intermedia of the pituitary gland functions as a neuroendocrine transducer cell, in that the neuronal input originating from the hypothalamus is integrated in the melanotroph cells of the pars intermedia, which ultimately leads to an output of α-MSH from this tissue. As a consequence of the process of background adaptation, the amphibian intermediate lobe melanotrophs have a high protein biosynthetic activity when animals are on a black background and a low synthetic activity in white-background-adapted animals. Details of this process are discussed by Jenks et al. (Chapter 8, this Volume). Because amphibians have the capacity to adapt to their background, the melanotroph cells of the amphibian pars intermedia constitute an interesting neuroendocrine model system to study the activation and inactivation of the genes involved in the cascade of events leading to this neuroendocrine reflex. For the study of the gene(s) coding for the melanotropic peptides, it was first of all necessary to establish the biosynthesis of MSH. Lowry and Scott[10] were the first to propose that α-MSH, the most potent MSH, is derived from adrenocorticotropic hormone (ACTH). At approximately the same time, structural evidence accumulated that ACTH itself is also derived from a prohormone ("big ACTH").[11,12] Since structural analyses of proteins give only circumstantial evidence for a possible precursor-product relationship, biosynthetic studies were performed to show the proteolytic conversion of the precursor to its end-products.[13] Subsequently, it was shown that in the rodent pituitary gland the biosynthetic precursor to ACTH, designated pro-ACTH, is also the prohormone for β-lipotropic hormone (β-LPH),[14-16] and the common prohormone was termed pro-ACTH/β-LPH. Within the sequence of β-LPH are the sequences of an opioid peptide, β-endorphin, and of a second MSH, namely that of β-MSH.[17] In the amphibian pars intermedia, a similar complexity of biosynthetic events became apparent when indications were obtained that amphibian MSH is derived from a stored precursor.[18] Biochemical studies demonstrated that in *Xenopus* pars intermedia, a precursor is synthesized which is processed to a number of smaller peptides including some with melanotropic activity.[19-21] Details concerning the biosynthesis and processing of the MSH precursor and aspects of the post-translational modifications and secretion of its cleavage products in the amphibian pars intermedia, are discussed elsewhere (Jenks et al., Chapter 8, this Volume).

With regard to the structural organization of the precursor to α-MSH, biochemical data showed that ACTH occupied a central position in the prohormone structure and β-LPH represented the carboxy-terminal portion of the protein (see Reference 22). The full characterization of the structure of the precursor protein, however, had to await new developments in molecular genetics. The introduction of recombinant DNA techniques in the study of neuroendocrine model systems was the most important of these recent developments. Application of molecular biological techniques, mainly involving cloning of cDNA or genomic DNA fragments and nucleotide sequence analysis of the cloned molecules, has led to the structural characterization of peptide hormone/neuropeptide precursors and of the genes coding for these proteins.[23] With these techniques, the complete structure of pro-ACTH/β-LPH was first determined in the cow by deducing the amino acid sequence from the nucleotide sequence of cloned intermediate lobe cDNA.[24] Subsequently, the sequences of the prohormones were established for several other mammalian species[25-29] and recently, a partial structure of salmon POMC has been reported.[30] Besides the sequences of α- and β-MSH in the primary structures of the mammalian proteins, an amino acid sequence is present which is strikingly similar to that of the two MSHs, and the peptide corresponding to this sequence was, therefore, named γ-MSH.[24] This single precursor protein can, thus, produce three MSH molecules which all contain the tetrapeptide His-Phe-Arg-Trp, a sequence partially responsible for melanotropic activity.[31] Since, in addition to these melanotropic peptides the prohormone can produce corticotropins and peptides with opiate activity, it was named pro-opiomelanocortin (POMC) in order to adequately represent the bioactivities of the products which can be generated from this precursor.[32]

As already mentioned, we are interested in using the amphibian intermediate lobe melanotroph as a neuroendocrine model system to study regulation of neuropeptide gene expression during a neuroendocrine reflex (i.e., background adaptation). An important prerequisite for such a study is knowledge of the structures of the amphibian melanotropic peptides and of their precursor. Although the production of melanotropic factors by the amphibian pars intermedia had been firmly established, the identity of the peptide(s) responsible for this bioactivity remained unknown for a long time. It was not until 1972 that immunological evidence indicated the presence of an α-MSH-like peptide[33] and several years later biosynthetic evidence became available for the presence of a POMC-like precursor in amphibians.[19-21,34,35] The recent application of recombinant DNA techniques to the amphibian POMC system summarized in this chapter, resulted in the isolation and characterization of POMC cDNA clones and of the POMC gene in the South African clawed toad, *Xenopus laevis*.[36-38] The availability of the structure of amphibian POMC and of its constituent bioactive peptides, is important in order to be able to understand evolutionary relationships of peptide hormones. Such an analysis is often hampered when peptide hormone/neuropeptide structures are only known for a restricted range of species. The complete amino acid sequence of POMC and the structural organization of the POMC gene are known for some mammalian species.[24-29,39-42] The mammalian POMC gene consists of three exons whereby the segments corresponding to the 5'-untranslated mRNA region and the amino-terminal protein-coding region are interrupted by introns. The amphibian, *Xenopus laevis*, is evolutionarily very different from mammals, in that amphibians diverged from the main line of vertebrate evolution some 350 million years ago.[43] A comparative analysis of the genes coding for POMC over a long evolutionary time-period is, for the following reasons, particularly interesting. The mammalian POMC protein contains three MSH sequences and the coding information for these sequences is located on one exon, the main exon 3 of the mammalian POMC gene. It has been speculated that the repetitive MSH-coding sequences may have evolved by a series of duplication and rearrangement events during POMC gene evolution[24] and possibly intervening sequences may have existed between these duplicated units. The MSH, ACTH, and endorphin domains within the POMC molecule constitute potential endproducts in precursor processing, and these peptides have very different biological functions.[23] Therefore, because functional domains within a protein may correlate with separate exons in the gene coding for this protein,[44] the possibility that early during evolution the main exon of the POMC gene might have contained intervening sequences, is worth considering. A further interesting characteristic of the polyprotein POMC is its differential regulation of gene expression in the two lobes of the pituitary gland. In the intermediate lobe of this gland the level of POMC gene expression is decreased by dopaminergic agonists, while this neurotransmitter does not affect POMC mRNA levels in the anterior pituitary gland.[45] Conversely, glucocorticoids alter POMC gene transcription in the anterior lobe, but these steroids do not influence the level of POMC mRNA in the intermediate pituitary gland.[46] In the regulation of gene transcription, 5'-flanking regions of a gene are often important.[47] Since transcriptional regulatory units are functional regions, the presence of evolutionary constraints on these regions will lead to conservation of their nucleotide sequences.[48,49] A comparative analysis of the nucleotide sequences upstream of the capping sites of POMC genes might, therefore, result in the identification of conserved and thus functionally significant segments.

This chapter deals with the application of the recombinant DNA approach in the study of *Xenopus* POMC gene expression and POMC gene evolution. The structural organization and evolutionary aspects of the *Xenopus* POMC gene are described in the following section. Next, the structures and evolutionary aspects of the *Xenopus* POMC mRNAs and proteins will be considered, and this section is followed by a description of the expression of the *Xenopus* POMC gene. In the last section, a summary will be given of the results concerning the *Xenopus* POMC gene and its expression, and this section includes a brief account of future research which will mainly deal with the molecular mechanisms underlying *Xenopus* POMC gene regulation.

II. STRUCTURAL ORGANIZATION OF THE POMC GENE IN *XENOPUS LAEVIS*

A. Xenopus POMC Gene Structure

As mentioned in the introduction, isolation and characterization of genes coding for peptide hormone/neuropeptide precursors had to await the advent of recombinant DNA technology, and application of these techniques mainly involves cDNA cloning procedures and cloning of genomic DNA fragments. The construction and screening of a cDNA library in order to isolate a cDNA clone corresponding to *Xenopus* POMC mRNA will be discussed in the next section of this chapter. With a POMC cDNA clone available, it is possible to isolate a cloned genomic DNA fragment containing the POMC gene from a *Xenopus laevis* λ-genomic library. Such a library consists of large genomic DNA fragments (16 to 20 kilo base pairs, kb) which are inserted into λ-phage vectors and the recombinant molecules are packaged into viable phage particles. After infection of bacterial cells, the genomic library can be screened by plaque hybridization.[50] Recently, a *Xenopus laevis* genomic fragment containing the POMC gene was isolated using a radioactively labeled *Xenopus* POMC cDNA clone as a hybridization probe in the library screening.[38] This hybridization-positive clone was characterized by restriction enzyme mapping of phage DNA in combination with blot hybridization using cDNA clones as hybridization probes. The segments of genomic DNA containing the exons and their surrounding regions were sequenced with the dideoxy chain termination method. It was found that the isolated genomic DNA fragment contained the complete *Xenopus* POMC gene which appeared to consist of three exons (exons 1 to 3) and two introns (introns A and B). The main exon 3, with a size of 1,001 base pairs (bp), encodes 350 nucleotides of the 3'-untranslated region of *Xenopus* POMC mRNA. The remainder of this exon codes for the major portion of the POMC protein (217 amino acids), including all bioactive domains. Exon 2 (143 bp) codes for 14 nucleotides of the 5'-untranslated mRNA region and contains the coding information for the signal peptide and for the first 18 amino acids of the amino-terminal region of the mature precursor protein. The remainder of the 5'-untranslated mRNA region is encoded by exon 1 (48 bp). Hence, the segment of the *Xenopus* POMC gene corresponding to the 5'-untranslated mRNA region, is interrupted by an intron (intron A) which appeared to have a size of approximately 2.6 kb. Intron B (size approximately 2.5 kb) separates the protein-encoding region corresponding to the amino-terminal portion of *Xenopus* POMC. All exon-intron boundaries in the *Xenopus* POMC gene conform to the well-defined consensus sequences for acceptor or donor sites in exon-intron splice junctions.[51] The entire protein-encoding region, thus, consists of 780 bp which encode 260 amino acids. The total length of *Xenopus* POMC mRNA is 1192 nucleotides, excluding the poly(A) tail. With the inclusion of the poly(A) tail this size agrees well with that determined by Northern blot analysis of *Xenopus* pituitary and brain RNA (1300 nucleotides).[36]

The 5'-flanking regions of most eukaryotic genes contain a transcriptional control element located about 20 to 30 bp upstream from the mRNA initiation site.[47] This element, the

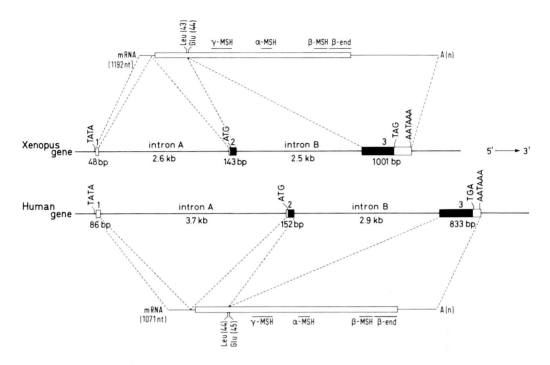

FIGURE 1. Schematic comparison of the *Xenopus* and human pro-opiomelanocortin genes and of the corresponding mRNAs. In the schematics of the gene structures, open and closed boxes represent exon sequences coding for the untranslated and translated mRNA regions, respectively. The positions within the genes of the TATA box sequences, the translational initiation (ATG), and termination (TAG or TGA) codons, and the polyadenylation signals (AATAAA) are indicated. In the schematics of the mRNA structures, open boxes and lines represent translated and untranslated mRNA regions, respectively. For reference, the locations of the mRNA sequences encoding the bioactive domains are indicated; β-end: β-endorphin. Numbering of amino acids (Leu and Glu residues) started with the initiative methionine.

TATA or Goldberg/Hogness box,[52] helps to determine the precise location of the transcription start site.[53] The *Xenopus* POMC gene also contains a TATA-like sequence (5'-TATATAA-3') 21 bp from the presumptive initiation point of POMC gene transcription.[38] A sequence reminiscent of a second less-well-defined control element, the CCAAT-box,[54,55] was also found in the toad gene. The presence of other possible regulatory elements in the 5'-flanking region of the POMC gene, based on a comparative analysis, will be discussed later in this section.

B. Evolutionary Aspects

Figure 1 shows a schematic comparison of the structures of the *Xenopus* and human POMC genes and their corresponding mRNAs. From this figure it is clear that the structural organizations of the two genes are remarkably similar. Previous studies have shown that the POMC gene structure was virtually unchanged during the whole of mammalian evolution.[25,26,29,39-42] It now appears that the POMC gene has been remarkably stable during 350 million years of vertebrate evolution. In all species thus far examined, the POMC gene consists of three exons with one main exon coding for all biologically active peptides. This means that if intervening sequences have ever existed between the regions coding for the bioactive domains or between the repetitive MSH units, these introns must have been spliced out of the gene early during the course of vertebrate evolution. A similar conclusion can be drawn for the proenkephalin gene because in mammals and *Xenopus*, the seven enkephalin-coding sequences present within this gene are also encoded by one main exon.[56-58] The sizes of the three exons of the *Xenopus*, human, bovine, mouse, and rat POMC genes are as follows. Exons 1 of the toad, human, bovine, mouse, and rat genes have a length of 48, 86, 108, 96, and 97 bp, respectively. Exons 2 are 143, 152, 152, 155, and 151 (181) bp, respectively; and exons 3 are 1,001, 833, 833, 677, and 688 bp, respectively, in these five species. Hence the sizes of exons 1 and 3 of the POMC genes vary considerably among vertebrates, mainly due to the variable regions corresponding to the 5'- and 3'-untranslated mRNA regions. The lengths of the two introns of the POMC genes also changed during vertebrate evolution. Intron A has a size of approximately 2.6, 3.7, 4, 3, and 3 kb, and intron B is approximately 2.5, 2.9, 2.2, 1.7, and 1.8 kb in *Xenopus*, human, bovine, mouse, and rat, respectively. The positioning of these introns within the genes is, however, essentially the same in the five species. In all cases, intron A separates the 5'-untranslated mRNA region and intron B interrupts the coding information for the amino-terminal region near the signal peptide of the POMC protein.

The 5'-flanking regions of many eukaryotic genes contain transcriptional regulatory units.[47] A comparative analysis of the regions upstream of the transcription start-sites of the *Xenopus*, human, bovine, and mouse POMC genes revealed several regions of more than 10 nucleotides which exhibit over 65% nucleotide sequence homology.[38] One of these conserved regions is centered around the TATA box sequence, reflecting the importance of this segment in gene regulation. Since the nucleotide sequence of part of a second conserved region is reminiscent of the viral core enhancer consensus sequence,[59] and this sequence is repeated twice in the 5'-flanking region of the *Xenopus* gene, this segment might well have an enhancer-like activity during POMC gene transcription. A third segment conserved among the *Xenopus* and mammalian 5'-flanking gene regions shows homology with sequences lying upstream of the capping sites of other glucocorticoid-controlled genes. It is, therefore, tempting to speculate that this region is involved in the glucocorticoid-regulated transcription of the POMC gene in the anterior pituitary gland. In the 3'-flanking region of a gene, nucleotide sequences important for correct 3'-end formation of pre-mRNA may be located.[60] For the POMC gene, such sequences are not evident since a comparative analysis of the 3'-flanking regions of the known POMC genes revealed that only a segment surrounding the polyadenylation signal was conserved between *Xenopus* and mammals.

From the next sections it will become evident that two POMC genes exist in the genome of *Xenopus laevis*. The gene described in this section corresponds to POMC gene transcript B, and we are currently isolating and characterizing the second *Xenopus* POMC gene.

III. PRIMARY STRUCTURES OF *XENOPUS* POMC mRNAs AND PROTEINS

A. Xenopus POMC mRNA and Protein Structures

In order to determine the primary structure of a neuropeptide precursor, the recombinant DNA approach has proven to be a powerful approach, especially the application of cDNA cloning procedures.[23] In this procedure, mRNA isolated from a tissue homogenate is transcribed into complementary DNA (cDNA) which, after enzymatic conversion into double-stranded DNA, is inserted into a bacterial vector. These chimeric molecules are introduced into bacterial cells and the resulting cDNA library is a reflection of the mRNA species present in that particular tissue. To prepare a cDNA library for the isolation of a cDNA clone corresponding to *Xenopus* POMC mRNA, the tissue of choice is the pituitary gland of a toad adapted to a black background because POMC biosynthetic activity in such a gland is very high (see Jenks et al., Chapter 8 this Volume). Hence, a cDNA library was constructed by homopolymeric (dC) tailing of cDNA produced from pituitary polyadenylated RNA and annealing of the dC-tailed cDNA to dG-tailed pBR322 vector DNA.[36] Screening of this library under hybridization conditions of low stringency using a mammalian POMC cDNA clone covering the complete protein-encoding region as a probe, did not result in hybridization-positive signals. Apparently, the homology between *Xenopus* and mammalian POMC mRNA is too low for the formation of sufficiently stable hybrids between the corresponding cDNA molecules. Therefore, a pool of synthetic oligonucleotides (tetradecamers) corresponding to the His-Phe-Arg-Trp-Gly sequence (third nucleotide of the Gly-codon was not included) of mammalian α-MSH was used in the cDNA library screening. With this probe, between 0.5 and 1% of the clones in the pituitary cDNA library appeared to be hybridization-positive.[36] This amount of positive signals is not surprising since one would expect for a rat pituitary cDNA library approximately 0.6% of the clones to be related to POMC mRNA (information for the calculation of this value was taken from Reference 61). Sequencing of several hybridization-positive cDNA clones revealed two groups of clones which corresponded to two *Xenopus* POMC gene transcripts, designated POMC mRNA-A and -B. Hence, in *Xenopus laevis*, two POMC genes exist and both of these genes are transcribed in the pituitary gland. The sequence homology between the two mRNAs and between the two proteins is 91 and 92%, respectively. Each of the two gene transcripts contains two polyadenylation signals (AAUAAA), but in each case only one of these is used, as concluded from Northern blot analysis of pituitary RNA and nucleotide sequencing of cDNA clones. This finding indicates that the presence of the AAUAAA sequence is not sufficient for correct 3' end formation of the mRNA. The POMC proteins A and B consist of 259 and 260 amino acids, respectively, which give calculated molecular weights of 29,844 and 29,883, respectively, for the preprohormones. Assuming a signal peptide sequence consisting of 25 amino acids, based on a comparative analysis with their mammalian counterparts, the mature proteins A and B have a molecular weight of 27,014 and 27,224, respectively. Each one of the two proteins contains one potential site for asparagine-linked N-glycosylation (Asn-X-Ser/Thr[62]) located within their γ-MSH regions. Biosynthetic studies with [^{3}H]glucosamine and *Xenopus* neurointermediate lobes revealed that this site is indeed used as a glycosylation site in both prohormones.[63] It is, therefore, surprising that glycosylation in the ACTH region of the *Xenopus* POMC protein, leading to the production of a 13K ACTH-like peptide, has been reported.[64] All bioactive domains in the *Xenopus* prohormone are flanked by pairs of basic amino acids which are thought to be recognition sites for proteolytic cleavage enzymes in the processing of precursor proteins.[23,65] Biosynthetic studies have shown that all of these sites are indeed cleavage sites during POMC processing in *Xenopus* neurointermediate lobes, except for the Arg-Arg and Lys-Lys pairs within the γ-MSH and β-endorphin regions (see Jenks et al. Chapter 8, this Volume). That Arg-Arg and Lys-Lys residues are not necessarily unused cleavage signals in *Xenopus* POMC, is illustrated by the biosynthesis of CLIP, carboxy-terminally flanked by an Arg-Arg pair, and of β-MSH, amino-terminally flanked by Lys-Lys. This circumstance raises intriguing questions concerning the reason(s) for differential recognition of the substrate by the proteolytic enzyme in order to achieve correct processing. In the processing of mammalian polyproteins, Arg-Arg and Lys-Lys are not common cleavage sites.[23] The amino-terminal residue of *Xenopus* α-MSH differs from that of all known α-MSH structures (Ser/Ala substitution). This finding is of great interest since this residue is acetylated in α-MSH and this acetylation greatly enhances the biological activity of the hormone.[66] In addition, it has been suggested that, at least in bovine pituitary, the presence of the amino-terminal Ser-Tyr sequence is required for binding of the hormone to the catalytic site of the acetylating enzyme, acetyltransferase.[67] Also, the amino-terminal Ser-Tyr-Ser-Met sequence of mammalian α-MSH is a potentiator sequence in that it increases the activity of the central "classical" MSH message in the melanophore assay.[68] The substitution of a serine for an alanine in the *Xenopus* peptide might, therefore, have considerable physiological implications, such as a different bioactivity of this peptide compared to that of the mammalian hormone. A synthetic *Xenopus* α-MSH peptide has to be made and its bioactivity examined to investigate possible changes in activity. The fact that the Ser/Ala substitution precludes the formation of N,O-diacetyl-α-MSH found in most species

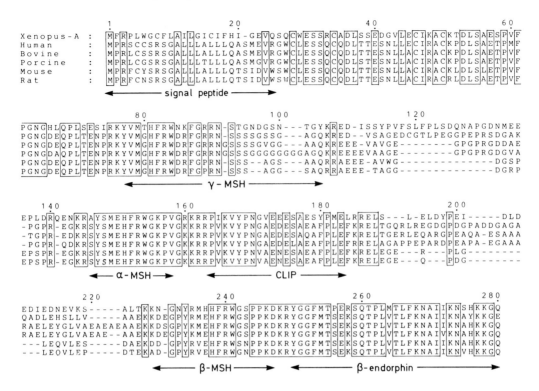

FIGURE 2. Alignment of the amino acid sequences of *Xenopus*, human, bovine, porcine, mouse and rat pro-opiomelanocortin. The one-letter amino acid notation is used. Amino acid residues are numbered beginning with the initiative methionine. Residues identical among the 6 species are boxed. Gaps(-) have been introduced to achieve maximum homology. The locations of the sequences of the signal peptide, γ-MSH, α-MSH, corticotropin-like intermediate lobe peptide (CLIP), β-MSH and β-endorphin are indicated by arrows below the sequences. Xenopus-A refers to Xenopus-pro-opiomelanocortin-A.

examined,[69-72] including in the amphibian, *Rana ridibunda*,[73] might also be of physiological importance. The inability to generate a diacetylated form of *Xenopus* α-MSH explains why we never found evidence for the presence of such a form among the melanotropic peptides in *Xenopus* neurointermediate lobes. *Xenopus* α-MSH is presumably α-amidated at its carboxy-terminus because within the precursor structure the hormone is carboxy-terminally flanked by Gly-Arg-Lys, a combined signal for proteolysis and amidation generally found in the conversion of precursor proteins to α-amidated bioactive peptides.[74] The message part of α-MSH (His-Phe-Arg-Trp) is conserved in the *Xenopus* α-, β-, and γ-MSH structures which reinforces the concept that this portion of the molecule is crucial for melanotropic activity.[31] Like in all known β-endorphin structures, the amino-terminal part of *Xenopus* β-endorphin consists of the sequence Tyr-Gly-Gly-Phe-Met, the met-enkephalin portion essential for binding of the endogenous opioid peptide to an opioid receptor.[75] It is, thus, very possible that *Xenopus* β-endorphin has opiate-like activity.

B. Evolutionary Aspects

Our cloning and characterization of several *Xenopus* POMC cDNAs revealed the entire protein-coding mRNA sequence and allows for a comparative analysis of the complete amino acid sequences of toad and mammalian POMC molecules. Figure 2 presents a comparison between *Xenopus* POMC-A and the human, bovine, porcine, mouse, and rat POMC structures. The distribution of the bioactive domains within the precursor proteins is remarkably similar among the six species. These domains are all flanked on both sides by pairs of basic amino acids (i.e., potential recognition sites of proteolytic cleavage enzymes[65]), except for the β-MSH sequences in rat and mouse which are flanked amino-terminally by only one basic residue. The bioactive peptides share a high degree of amino acid sequence homology. The α-MSH and β-endorphin moieties represent the most conserved segments within the precursor molecule which might reflect the physiological importance of these peptides in a wide range of species. The fact that the carboxy-terminal region of γ-MSH has diverged considerably during vertebrate evolution, indicates that this region does not have an important function. The spacer regions between the bioactive domains, one of which has been termed "joining peptide",[76] share an extremely low degree of amino acid sequence homology which suggests a structural rather than a functional role for these segments. The conservation of the acidic nature of the spacer regions is noteworthy and this hydrophilic characteristic of the spacers might be important for efficient proteolytic processing of the prohormone. In this respect, it is interesting to note that the spacer regions between the enkephalin units in the opioid peptide precursor proteins proenkephalin-A[56-58] and prodynorphin (proenkephalin-B),[77] are also acidic. In contrast to the spacers between the bioactive domains, the amino acid sequence of the spacer region between the signal peptide and the γ-MSH region is well conserved (Figure 2). In this segment, are nearly all of the cysteine residues of the POMC molecule. Such a cluster of cysteines is also found in the amino-terminal region of proenkephalin-A and prodynorphin. These clusters might be important for correct folding (formation of disulfide bridges) of the precursor proteins during prohormone processing.[78] It is not yet clear whether the relatively high degree of conservation of this spacer region is due to the presence of the cysteine residues or if its conservation reflects a biological function for the peptide comprising this region. With respect to the latter possibility, the suggestion that this peptide has adrenal mitogenic activity[79] should be mentioned. Finally, the extremely low degree of conservation of the signal peptide sequence of the POMC molecule, reflects the high variability in amino acid sequence generally observed for signal sequences.[80] A hydrophobic nature of signal peptides of secretory proteins is a general feature and appears to be necessary for transportation of the protein across the membranes of the endoplasmic reticulum.[81]

IV. EXPRESSION OF THE *XENOPUS* POMC GENE

A cDNA clone or a clone containing a genomic DNA fragment can be used as a hybridization probe for the detection and quantification of mRNA to study tissue-specific gene expression at the level of transcription. In mammals, the POMC gene is expressed in the anterior and intermediate lobes of the pituitary gland, in various brain regions, and other parts of the animal.[61,82,83] Northern blot analysis of pituitary, brain, and skin RNA revealed that in *Xenopus laevis* the POMC gene is expressed at high levels in the pituitary gland, at low levels in the brain, but not in the skin.[36] This latter observation shows that α-MSH cannot be produced by the skin, the tissue in which the target cells of the hormone are located.

The protein biosynthetic activity in the melanotrophs of the intermediate pituitary gland is high when *Xenopus* is on a black background and low in animals on a white background (see Jenks et al., Chapter 8, this Volume). Northern blot analysis showed that in fully black-background-adapted toads the level of POMC mRNA in the pars intermedia is at least 15 times higher than that in fully white-background-adapted animals.[84] Hence, the observed difference at the level of translation (POMC protein production) appears to be accompanied by a difference in the level of POMC mRNA. When a fully black-background-adapted animal was transferred to a white background, the levels of POMC mRNA in the *Xenopus* pars intermedia remained high for several days. In contrast, POMC mRNA levels in the pars intermedia increased relatively fast when a white-background-adapted animal was placed on a black background.[84] These findings are in line with previous results concerning changes in protein biosynthetic activity of the *Xenopus* pars intermedia during background adaptation.[85] It remains to be established whether the high level of POMC mRNA in a black animal adapting to a white background is due to a slow inactivation of the POMC gene (sustained POMC gene transcription) or whether it is the consequence of POMC mRNA stability. During background adaptation, the levels of POMC mRNA did not significantly change in the pars distalis of the pituitary gland.[84] These observations are in line with the concept that the pars intermedia and not the pars distalis of the pituitary gland, is the tissue that mediates the process of background adaptation.

From the analysis of *Xenopus* POMC cDNA clones it appears that two POMC genes are transcriptionally active in the pituitary gland.[37] In view of the number of cDNA clones corresponding to either POMC mRNA-A or -B, the level of transcription of the two genes appears to be virtually the same in this tissue. Also, SDS gel electrophoretic analysis of POMC biosynthesis in *Xenopus* neurointermediate lobes revealed that translation of the two gene transcripts results in the production of approximately equal amounts of the two sequentially different POMC proteins.[63] The fact that two POMC genes occur in *Xenopus laevis* is not surprising because its genome appears to have arisen by chromosome duplication and tetraploidization in an ancestor of *Xenopus laevis*. This genome duplication in the genus *Xenopus*, occurred some 30 million years ago and it has been postulated on the basis of the DNA content per cell and other chromosomal studies.[86,87] The absence of a genome duplication in *Xenopus tropicalis*,[86] a contemporary equivalent of the *Xenopus* ancestor, would indicate that in this species only a single POMC gene is present. The finding that in *Xenopus laevis* both POMC genes are more or less equally active is remarkable, because, in general, duplication of a gene may result in an increase of mutational frequency in one of the genes which might lead to a nonfunctional or a different gene, as discussed below. The existence of two POMC genes in the rat and mouse has been reported previously,[29,88] but for the rat this was an erroneous observation,[41] while the second mouse gene appeared to be a pseudogene.[42,88] For the salmon, the presence of two sequentially different POMC proteins has been suggested on the basis of amino acid sequence analysis of POMC-derived peptides.[89] Until now, however, the existence and expression of two POMC genes has only been demonstrated in *Xenopus laevis*.

The transcriptional, translational, and post-translational events occurring during POMC gene expression in the pars intermedia of *Xenopus laevis* are summarized in Figure 3. The two POMC genes, A and B, produce two similarly sized gene transcripts, POMC mRNA-A and -B. These transcripts are translated to give POMC-A and -B which have essentially the same molecular weight. The two precursor proteins are similarly processed which results in the biosynthesis of two forms of γ-MSH, two CLIPs, and two β-endorphins, but only one form of α-MSH and one β-MSH. All of these bioactive peptides appear to be concomitantly released.

V. SUMMARY

A. Concluding Remarks

From recently published molecular biological studies of neuroendocrine model systems it is clear that recombinant DNA techniques are very powerful and allow, in a relatively short period of time, impressive progress in the understanding of peptide hormone production.[23] Also, they provide us with tools to measure mRNA levels and to study mechanisms of gene regulation, as discussed below. Application of the recombinant DNA approach has allowed for the determination of the complete structure of POMC in a nonmammalian species, the amphibian *Xenopus laevis*, and for the examination of the *Xenopus* POMC gene structure. The remarkable conservation of the structures of the *Xenopus* POMC proteins, mRNAs, and

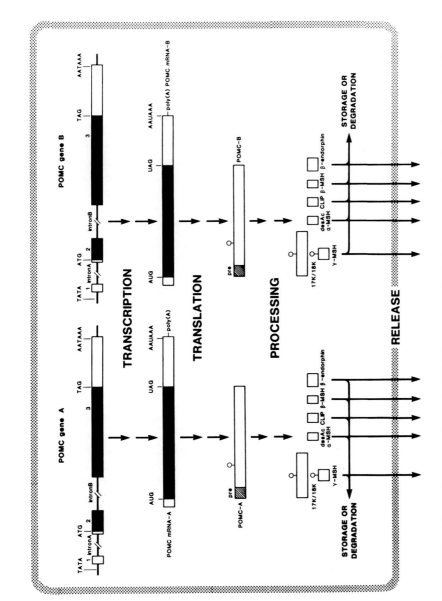

FIGURE 3. Model of pro-opiomelanocortin (POMC) gene expression in the pars intermedia of the pituitary gland of *Xenopus laevis*. In the schematics of the gene structures, open and closed boxes represent exon sequences coding for the untranslated and translated mRNA regions, respectively. The positions of the TATA box sequences, the translational initiation (ATG, AUG) and termination (TAG, UAG) codons, and the polyadenylation signals (AATAAA, AAUAAA) are indicated. Lollipops indicate positions of Asn-linked glycosylations. Pre indicates signal peptide sequence.

genes, when compared with their mammalian counterparts, reflects the importance of this multifunctional prohormone in both higher and lower vertebrates. The fact that in *Xenopus*, two nonallelic POMC genes exist and that both genes are expressed in the pituitary gland, are interesting findings. One might have expected that one of the POMC genes would be relatively free to mutate and thus give rise to either a new functional gene (coding for new hormones) or a nonfunctional gene (pseudogene). It is clear, however, that in *Xenopus*, the duplicate gene is still active. One wonders how long both POMC genes will remain transcriptionally active in *Xenopus*. It is unlikely that a mutation in one of the two genes, leading to a new gene or a pseudogene, will harm *Xenopus* since the other gene still produces α-MSH and other POMC-derived peptides which serve their physiological functions. Therefore, one of the *Xenopus* POMC genes might be temporarily freed from selective pressure and the *Xenopus* genes will probably be more amenable to mutations than the single POMC gene in mammals. In general, a region of DNA that is free to mutate seems to do so rapidly.[90] It is, thus, remarkable that both *Xenopus* POMC genes are still active after 30 million years of gene evolution, the time that elapsed after the genome duplication in the genus *Xenopus*. The duplication of the POMC gene may lead, in the future, to more diversification in the generation of bioactive peptides derived from POMC. Gene duplication is usually considered as one of the major mechanisms of evolution and functional diversification of peptide hormones,[90] and gene families encoding distinct, but related proteins are a common feature of eukaryotic genomes. Examples are the growth hormone gene family[91] and the opioid peptide precursor genes.[23] As a result of the duplication of the entire *Xenopus* genome, not only the POMC gene, but also genes coding for e.g., receptors and processing enzymes, have been duplicated. This provides a great source of diversification which may allow *Xenopus* to adapt more efficiently to a changing environment than many other amphibians. Finally, it is interesting to note that the length difference observed between the analogous introns A of *Xenopus* POMC genes A and B, at least partially reflects the acquisition or loss of repetitive nucleotide sequences.[92] Such a process of insertion/deletion causes a drift in the duplicated genes which may have a beneficial effect because it reduces the homology between the two genes and, thus, decreases the possibility for recombination events which might lead to a loss of the POMC gene.

B. Future Prospects

One of the future challenges will be to define the mechanisms underlying *Xenopus* POMC gene regulation. Gene regulation can be studied at the level of transcription by quantification of POMC mRNA changes during the process of background adaptation and in response to potential stimulatory or inhibitory factors. Hence, the factors that affect release of POMC-derived peptides from the pars intermedia of *Xenopus* (see Jenks et al. Chapter 8, this Volume) should be tested in vitro and in vivo to study their effect on POMC gene transcription. Such experiments could contribute to a better understanding of the relationship between the regulation of biosynthesis and release of peptide hormones. Levels of mRNA can be quantified by Northern blot or dot-blot analysis, solution hybridization, S1-mapping, and primer extension.[23] In addition to these assays, the technique of *in situ* hybridization histochemistry[93] can be used to study whether all melanotroph cells of the pars intermedia are regulated coordinately or if POMC gene transcription is only affected in a subpopulation of cells. Since some of the quantitative changes in the level of POMC mRNA might be due to mRNA stability rather than changes in gene transcription, one would also like to measure turnover of POMC mRNA. The above-mentioned techniques to quantify levels of mRNA do not provide information concerning this aspect. Nuclear transcription run-on experiments[94] should reveal whether an observed alteration in POMC mRNA levels is at least, in part, the result of a direct effect on POMC gene transcription.

A major goal of our research will be to delineate the molecular mechanisms involved in the activation and inactivation of the *Xenopus* POMC gene. To examine such mechanisms, the recently developed technique of gene transfer may be important. As the molecular mechanisms regulating gene expression are markedly different between prokaryotes and eukaryotes, a cloned eukaryotic gene should not be introduced into bacterial cells, but it must be preferably examined in a eukaryotic environment. The *Xenopus* oocyte transcription assay system is one of the first in vivo eukaryotic systems in which purified DNA was assembled into an apparently normal chromatin structure which resulted in correct DNA transcription.[95] Injection of the *Xenopus* POMC gene into the nucleus of *Xenopus* oocytes seems especially attractive as this concerns a homologous gene transcription assay system. It will be interesting to see whether the oocyte is able to translate the POMC gene transcript and convert the prohormone to its correct end-products. In vitro mutagenesis can be used to generate deletion and other mutants of the POMC gene. With presently available techniques it is possible to make in vitro virtually any type of insertion, deletion, or substitution in order to construct a mutant DNA template.[96] Mutations will be made in the segments of the 5'-flanking region of the POMC gene which are conserved between *Xenopus* and mammals (see above). Quantification of POMC gene transcription in the oocyte assay system will reveal the effect of the mutation on gene activity. The size of the *Xenopus* oocyte (10^5 to 10^6 times larger than a normal somatic cell) allows additional injection experiments in which the cloned POMC gene and other macromolecules (e.g., nuclear RNAs or proteins) are injected simultaneously. These co-injection studies, in which mutated genes can also be injected in conjunction with the potentially regulatory proteins or RNAs, may give valuable information concerning gene regulatory mechanisms. For instance, interaction of specific proteins with DNA regulatory sequences probably represents an important step in the control of gene activity.[97] One of the most important questions in cell biology is how a specialized cell selectively expresses only a subset of the genes which are at its disposal. The model system we are using provides a specialized cell (pars intermedia melanotroph), its function is clearly established (production of POMC/α-MSH for the purpose of background adaptation), a number of parameters concerning regulation of the secretory activity of this cell are known (see Jenks, et al.; Chapter 8, and Vaudry, et al. Chapter 9 this Volume), and one of the genes involved in proper cell functioning (the POMC gene) has been cloned and can be introduced into the germ line of *Xenopus* by means of a homologous gene-transfer system (DNA injections into fertilized *Xenopus* eggs). Such experiments with mutated POMC genes may reveal which gene regions are involved in tissue-specific gene expression. Construction of transgenic frogs can also help in the understanding of pars intermedia cell differentiation. It remains to be seen whether in these transformed animals an ''over'' expressed POMC gene ultimately leads to ''over'' produced and ''over'' released POMC-derived peptides, and whether increased blood levels of these peptides will affect neuroendocrine or other processes. Hence, the approaches described above may well result in a better understanding of regulatory events in secretory cells.

VI. ACKNOWLEDGMENTS

The author thanks Dr. B.G. Jenks for critical reading of the manuscript. This work was supported by the Netherlands Organization for the Advancement of Pure Research (Z.W.O.).

REFERENCES

1. **Harris, J. T. and Lerner, A. B.**, Amino acid sequence of the α-melanocyte-stimulating hormone, *Nature (London)*, 179, 1346, 1957.
2. **Dixon, J. S. and Li, C. H.**, The isolation and structure of α-melanocyte-stimulating hormone from horse pituitaries, *J. Am. Chem. Soc.*, 82, 4568, 1960.
3. **Li, C. H., Danoh, W. O., Chung, D., and Rao, A. J.**, Isolation, characterization, and amino acid sequence of melanotropins from camel pituitary glands, *Biochemistry*, 14, 947, 1975.
4. **Bennett, H. P. J., Lowry, P. J., McMartin, C., and Scott, A. P.**, Structural studies of alpha-melanocyte-stimulating hormone and a novel beta-melanocyte-stimulating hormone from the neurointermediate lobe of the pituitary of the dogfish, *Squalus acanthias, Biochem. J.*, 141, 439, 1974.
5. **Kawauchi, H. and Muramoto, K.**, Isolation and primary structure of melanotropin from salmon pituitary glands, *Intl. J. Peptide Protein Res.*, 14, 373, 1979.
6. **Burgers, A. C. J., Imai, K., and van Oordt, G. J.**, The amount of melanophore-stimulating hormone in single pituitary glands of *Xenopus laevis* kept under various conditions, *Gen. Comp. Endocrinol.*, 3, 53, 1963.
7. **O'Donohue, T. L. and Dorsa, D. M.**, The opiomelanotropinergic neuronal and endocrine systems, *Peptides*, 3, 353, 1982.
8. **Waring, H.**, *Color Change Mechanisms in Cold-Blooded Vertebrates*, Academic Press, New York, 1963.
9. **Bagnara, J. T. and Hadley, M. E.**, *Chromatophores and Color Change: The Comparative Physiology of Animal Pigmentation*, Prentice-Hall, Englewood Cliffs, New Jersey, 1973.
10. **Lowry, P. J. and Scott, A. P.**, The evolution of vertebrate corticotrophin and melanocyte stimulating hormone, *Gen. Comp. Endocrinol.*, 26, 16, 1975.
11. **Yalow, R. S. and Berson, S. A.**, Characteristics of "Big ACTH" in human plasma and pituitary extracts, *J. Clin. Endocrinol. Metab.*, 36, 415, 1973.
12. **Eipper, B. A. and Mains, R. E.**, High molecular weight forms of adrenocorticotropic hormone are glycoproteins, *J. Biol. Chem.*, 251, 4121, 1976.
13. **Mains, R. E. and Eipper, B. A.**, Biosynthesis of ACTH by mouse pituitary tumor cells, *J. Biol. Chem.*, 251, 4115, 1976.
14. **Mains, R. E., Eipper, B. A., and Ling, N.**, Common precursor to corticotropins and endorphins, *Proc. Natl. Acad. Sci. U.S.A.*, 74, 3014, 1977.
15. **Roberts, J. L. and Herbert, E.**, Characterization of a common precursor to corticotropin and β-lipotropin: identification of β-lipotropin peptides and their arrangement relative to corticotropin in the precursor synthesized in a cell-free system, *Proc. Natl. Acad. Sci. U. S. A.*, 74, 5300, 1977.
16. **Mains, R. E. and Eipper, B. A.**, Coordinate synthesis of corticotropins and endorphins by mouse pituitary tumor cells, *J. Biol. Chem.*, 253, 651, 1978.
17. **Li, C. H., Barnafi, L., Chrétien, H., and Chung, D.**, Isolation and amino acid sequence of β-LPH from sheep pituitary glands, *Nature (London)*, 208, 1093, 1965.
18. **Thornton, V. F.**, The effect of change of background color on the melanocyte-stimulating hormone content of the pituitary gland of *Xenopus laevis, Gen. Comp. Endocrinol.*, 17, 554, 1971.
19. **Loh, Y. P. and Gainer, H.**, Biosynthesis, processing and control of release of melanotropic peptides in the neurointermediate lobe of *Xenopus laevis, J. Gen. Physiol.*, 70, 37, 1977.
20. **Jenks, B. G. and van Overbeeke, A. P.**, Biosynthesis and release of neurointermediate lobe peptides in the aquatic toad, *Xenopus laevis*, adapted to black background, *Comp. Biochem. Physiol.*, 66C, 71, 1980.
21. **Martens, G. J. M., Jenks, B. G., and van Overbeeke, A. P.**, Analysis of peptide biosynthesis in the neurointermediate lobe of *Xenopus laevis* using high-performance liquid chromatography: occurrence of small bioactive products, *Comp. Biochem. Physiol.*, 67B, 493, 1980.
22. **Eipper, B. A. and Mains, R. E.**, Structure and biosynthesis of pro-adrenocorticotropin/endorphin and related peptides, *Endocr. Rev.*, 1, 1, 1980.
23. **Douglass, J., Civelli, O., and Herbert, E.**, Polyprotein gene expression: generation of diversity of neuroendocrine peptides, *Annu. Rev. Biochem.*, 53, 665, 1984.
24. **Nakanishi, S., Inoue, A., Kita, T., Nakamura, M., Chang, A. C. Y., Cohen, S. N., and Numa, S.**, Nucleotide sequence of cloned cDNA for bovine corticotropin-β-lipotropin precursor, *Nature (London)*, 278, 423, 1979.
25. **Change, A. C. Y., Cochet, M., and Cohen, S. N.**, Structural organization of human genomic DNA encoding the pro-opiomelanocortin peptide, *Proc. Natl. Acad. Sci. U.S.A.*, 77, 4890, 1980.
26. **Whitfeld, P. L., Seeburg, P. H., and Shine, J.**, The human pro-opiomelanocortin gene: organization, sequence, and interspersion with repetitive DNA, *DNA*, 1, 133, 1982.
27. **Boileau, G., Barbeau, C., Jeanotte, L., Chrétien, M., and Drouin, J.**, Complete structure of the porcine pro-opiomelanocortin mRNA derived from the nucleotide sequence of cloned cDNA, *Nucl. Acids Res.*, 11, 8063, 1983.

28. **Uhler, M. and Herbert, E.,** Complete amino acid sequence of mouse pro-opiomelanocortin derived from the nucleotide sequence of pro-opiomelanocortin cDNA, *J. Biol. Chem.*, 258, 257, 1983.
29. **Drouin, J. and Goodman, H.,** Most of the coding region of rat ACTH-β-LPH precursor gene lacks intervening sequences, *Nature (London)*, 288, 610, 1980.
30. **Soma, G.-I., Kitahara, N., Nishizawa, T., Nanami, H., Kotake, C., Okazaki, H., and Andoh, T.,** Nucleotide sequence of a cloned cDNA for proopiomelanocortin precursor of chum salmon, *Onchorynchus keta*, *Nucl. Acids Res.*, 12, 8029, 1984.
31. **Schwyzer, R. and Eberle, A.,** On the molecular mechanism of α-MSH receptor interactions, in *Frontiers of Hormone Research*, van Wimersma Greidanus, Tj. B., Ed., S. Karger, Basel, 1977, 18.
32. **Chrétien, M., Benjannet, S., Gossard, F., Gianoulakis, C., Crine, P., Lis, M., and Seidah, N. G.,** From β-lipotropin to β-endorphin and pro-opio-melanocortin, *Can. J. Biochem.*, 57, 1111, 1979.
33. **Shapiro, M., Nicholson, W. E., Orth, D. N., Mitchell, W. M., Island, D. P., and Liddle, G. W.,** Preliminary characterization of the pituitary melanocyte-stimulating hormones of several vertebrate species, *Endocrinology*, 90, 249, 1972.
34. **Martens, G. J. M., Jenks, B. G., and van Overbeeke, A. P.,** Biosynthesis of pairs of peptides related to melanotropin, corticotropin and endorphin in the pars intermedia of the amphibian pituitary gland, *Eur. J. Biochem.*, 122, 1, 1982.
35. **Vaudry, H., Jenks, B. G., Verburg-van Kemenade, B. M. L., and Tonon, M. C.,** Effects of tunicamycin on biosynthesis, processing and release of proopiomelanocortin-derived peptides in the intermediate lobe of the frog, *Rana ridibunda*, *Peptides*, 7, 163, 1986.
36. **Martens, G. J. M., Civelli, O., and Herbert, E.,** Nucleotide sequence of cloned cDNA for pro-opiomelanocortin in the amphibian, *Xenopus laevis*, *J. Biol. Chem.*, 260, 13685, 1985.
37. **Martens, G. J. M.,** Expression of two proopiomelanocortin genes in the pituitary gland of *Xenopus laevis*: complete structures of the two preprohormones, *Eur. J. Biochem.*, 165, 467, 1987.
38. **Martens, G. J. M.,** Structural organization of the proopiomelanocortin gene in *Xenopus laevis*: 5' end homologies within the toad and mammalian genes, *Nucleic Acids Res.*, (submitted).
39. **Nakanishi, S., Teranishi, Y., Watanabe, Y., Notake, M., Kikidani, H., Jingami, H., and Numa, S.,** Isolation and characterization of the bovine corticotropin/β-lipotropin precursor gene, *Eur. J. Biochem.*, 115, 429, 1981.
40. **Takahashi, H., Teranishi, Y., Nakanishi, S., and Numa, S.,** Isolation and structural organization of the human corticotropin-β-lipotropin precursor gene, *FEBS Lett.*, 135, 97, 1981.
41. **Drouin, J., Chamberland, M., Charron, J., Jeannotte, L., and Nemer, M.,** Structure of the rat pro-opiomelanocortin (POMC) gene, *FEBS Lett.*, 193, 54, 1985.
42. **Notake, M., Tobimatsu, T., Watanabe, Y., Takahashi, H., Mishina, M., and Numa, S.,** Isolation and characterization of the mouse corticotropin-β-lipotropin precursor gene and a related psuedogene, *FEBS Lett.*, 156, 67, 1983.
43. **Goodman, M., Moore, G. W., and Matsuda, G.,** Darwinian evolution in the geneology of hemoglobin, *Nature (London)*, 253, 603, 1975.
44. **Blake, C. C. F.,** Exons encode protein functional units, *Nature (London)*, 277, 598, 1979.
45. **Chen, C. L. C., Dionne, F. T., and Roberts, J. L.,** Regulation of the proopiomelanocortin mRNA levels in the rat pituitary by dopaminergic compounds, *Proc. Natl. Acad. Sci. U.S.A.*, 80, 2211, 1983.
46. **Birnberg, N., Lissitzky, J. C., Hinman, M., and Herbert, E.,** Glucocorticoids regulate pro-opiomelanocortin gene expression in vivo at the levels of transcription and secretion, *Proc. Natl. Acad. Sci. U.S.A.*, 80, 6982, 1983.
47. **Breathnach, R. and Chambon, P.,** Organization and expression of eukaryotic split genes coding for proteins, *Annu. Rev. Biochem.*, 50 349, 1981.
48. **Bell, G. I., Pictet, R. L., Rutter, W. J., Cordell, B., Tischer, E., and Goodman, H. M.,** Sequence of the human insulin gene, *Nature (London)*, 284, 26, 1980.
49. **Efstratiadis, A., Posakony, J. W., Maniatis, T., Lawn, R. M., O'Connell, C., Spritz, R. A., DeRiel, J. K., Forget, B. G., Weissman, S. M., Slightom, J. L., Blechl, A. E., Smithies, O., Baralle, F. E., Shoulders, C. C., and Proudfoot, N. J.,** The structure and evolution of the human β-globin gene family, *Cell*, 21, 653, 1980.
50. **Benton, W. D. and Davis, R. W.,** Screening λ-gt recombinant libraries by hybridization to single plaques in situ, *Science*, 196, 180, 1977.
51. **Mount, S.,** A catalogue of splice junction sequences, *Nucleic Acids Res.*, 10, 459, 1982.
52. **Goldberg, M. L.,** Ph.D. thesis, Stanford University, Palo Alto, California, 1979.
53. **Grosschedl, R. and Birnstiel, M. L.,** Identification of regulatory sequences in the prelude sequences of an H2A histone gene by the study of specific deletion mutants in vivo, *Proc. Natl. Acad. Sci. U.S.A.*, 77, 1432, 1980.
54. **Grosveld, G. C., Rosenthal, A., and Flavell, R. A.,** Sequence requirements for the transcription of the rabbit β-globin gene in vivo: the −80 region, *Nucleic Acids Res.*, 10, 4951, 1982.

55. **Dierks, P., van Ooyen, A., Cochran, M. D., Dobkin, C., Reiser, J., and Weismann, C.,** Three regions upstream from the cap site are required for the efficient and accurate transcription of the rabbit β-globin gene in mouse 3T6 cells, *Cell,* 32, 695, 1983.
56. **Noda, M., Teranishi, T., Takahashi, H., Toyosato, M., Notake, M., Nakanishi, S., and Numa, S.,** Isolation and structural organization of the human preproenkephalin gene, *Nature (London),* 297, 431, 1982.
57. **Rosen, H., Douglass, J., and Herbert, E.,** Isolation and characterization of the rat proenkephalin gene, *J. Biol. Chem.,* 259, 14309, 1984.
58. **Martens, G. J. M. and Herbert, E.,** Polymorphism and absence of leu-enkephalin sequences in proenkephalin genes in *Xenopus laevis, Nature (London),* 310, 251, 1984.
59. **Weiher, H., Konig, M., and Gruss, P.,** Multiple point mutations affecting the simian virus 40 enhancer, *Science,* 219, 626, 1983.
60. **Birnstiel, M. L., Busslinger, M., and Strub, K.,** Transcription termination and 3' processing: the end is in site!, *Cell,* 41, 349, 1985.
61. **Civelli, O., Birnberg, N., and Herbert, E.,** Detection and quantitation of pro-opiomelanocortin mRNA in pituitary and brain tissues from different species, *J. Biol. Chem.,* 257, 6783, 1982.
62. **Pless, D. D. and Lennarz, W. J.,** Enzymatic conversion of proteins to glycoproteins, *Proc. Natl. Acad. Sci. U.S.A.,* 74, 134, 1977.
63. **Martens, G. J. M., Biermans, P.-P., J., Jenks, B. G., and van Overbeeke, A. P.,** Biosynthesis of two structurally different pro-opiomelanocortins in the pars intermedia of the amphibian pituitary gland, *Eur. J. Biochem.,* 126, 17, 1982.
64. **Loh, Y. P.,** Immunological evidence for two common precursors to corticotropins, endorphins, and melanotropins in the neurointermediate lobe of the toad pituitary, *Proc. Natl. Acad. Sci. U.S.A.,* 76, 796, 1979.
65. **Steiner, D. F., Patzelt, C., Chan, S. J., Quinn, P. S., Tager, H. S., Nielsen, D., Lernmark, A., Noyes, B. E., Agarwal, K. E., Gabbay, K. H., and Rubenstein, A. H.,** Formation of biologically active peptides, *Proc. Roy. Soc. Lond.,* 210, 45, 1980.
66. **Guttmann, S. T. and Boissonnas, R. A.,** Influence of the N-terminal extremity of α-MSH on the melanophore stimulating activity of this hormone, *Experientia,* 17, 265, 1961.
67. **Glembotski, C. C.,** Characterization of the peptide acetyltransferase activity in bovine and rat intermediate pituitaries responsible for the acetylation of β-endorphin and α-melanotropin, *J. Biol. Chem.,* 257, 10501, 1982.
68. **Eberle, A. N.,** Structure and chemistry of peptide hormones of the intermediate lobe, in *Peptides of the Pars Intermedia,* Evered, D. and Lawrensen, G., Eds., Pitman Medical, Summit, New Jersey, 1981, 13.
69. **Rudman, D., Chawla, R. K., and Hollins, B. M.,** N,O-diacetyl-serine-α-melanocyte-stimulating hormone, a naturally occurring melanotropic peptide, *J. Biol. Chem.,* 254, 10102, 1979.
70. **Browne, C. A., Bennett, H. P. J., and Solomon, S.,** Isolation and characterization of corticotropin- and melanotropin-related peptides from the neurointermediate lobe of the rat pituitary by reversed-phase liquid chromatography, *Biochemistry,* 20, 4538, 1981.
71. **Follenius, E., van Dorsselaer, A., and Meunier, A.,** Separation and partial characterization by high-performance liquid chromatography and radioimmunoassay of different forms of melanocyte-stimulating hormone from fish (cyprinidae) neurointermediate lobes, *Gen. Comp. Endocrinol.,* 57, 198, 1985.
72. **Leenders, H. J., Janssens, J. J. W., Theunissen, H. J. M., Jenks, B. G., and van Overbeeke, A. P.,** Acetylation of melanocyte-stimulating hormone and β-endorphin in the pars intermedia of the perinatal pituitary gland in the mouse, *Neuroendocrinology,* 43, 166, 1986.
73. **Jenks, B. G., Verburg-van Kemenade, B. M. L., Tonon, M. C., and Vaudry, H.,** Regulation of biosynthesis and release of pars intermedia peptides in *Rana ridibunda:* dopamine affects both acetylation and release of α-MSH, *Peptides,* 6, 913, 1985.
74. **Eipper, B. A., Mains, R. E., and Glembotski, C. C.,** Identification in pituitary tissue of a peptide α-amidation activity that acts on glycine-extended peptides requires molecular oxygen, copper and ascorbic acid, *Proc. Natl. Acad. Sci. U.S.A.,* 80, 5144, 1983.
75. **Bradbury, A. F., Smyth, D. G., Snell, C. R., Birdsall, N. J., and Hulme, E. C.,** The C-fragment of lipotropin: an endogenous peptide with high affinity for brain opiate receptors, *Nature (London),* 260, 793, 1976.
76. **Seidah, N. G., Rochemont, J., Hamelin, J., Benjannet, S., and Chrétien, M.,** The missing fragment of the pro-sequence of human pro-opiomelanocortin: sequence and evidence for C-terminal amidation, *Biochem. Biophys. Res. Commun.,* 102, 710, 1981.
77. **Horikawa, S., Takai, T., Toyosato, M., Takahashi, H., Noda, M., Kakidani, H., Kubo, T., Hirose, T., Inayama, S., Hayashida, H., Miyata, T., and Numa, S.,** Isolation and structural organization of the human preproenkephalin-B gene, *Nature (London),* 306, 611, 1983.
78. **Comb, M., Seeburg, P. H., Adelman, J., Eiden, L., and Herbert, E.,** Primary structure of the human met- and leu-enkephalin precursor and its mRNA, *Nature (London),* 295, 663, 1982.

79. **Lowry, P. J., Estivariz, F. E., Silas, L., Linton, E. A., McLean, C., and Crombe, K.**, The case for pro-γ-MSH as the adrenal growth factor, *Endocr. Res.*, 10, 243, 1984.
80. **Von Heijne, G.**, Signal sequences, the limits of variation, *J. Mol. Biol.*, 184, 99, 1985.
81. **Blobel, G., Walter, P., Chang, C. N., Goldman, B. M., Erickson, A. H., and Lingappa, R.**, Translocation of proteins across membranes: the signal hypothesis and beyond, *Symp. Soc. Exp. Biol.*, 33, 37, 1979.
82. **Liotta, A.S., Gildersleeve, D., Brownstein, M. J., and Krieger, D. T.**, Biosynthesis in vitro of immunoreactive 31,000 dalton corticotropin/β-endorphin-like material by bovine hypothalamus, *Proc. Natl. Acad. Sci. U.S.A.*, 76, 1448, 1979.
83. **Odagiri, E., Sherrel, B. J., Mount, C. D., Nicholson, W. E., and Orth, D. N.**, Human placental immunoreactive corticotropin, lipotropin and β-endorphin: evidence for a common precursor, *Proc. Natl. Acad. Sci. U.S.A.*, 76, 2027, 1979.
84. **Martens, G. J. M., Weterings, K. A. P., van Zoest, I. D., and Jenks, B. G.**, Physiologically-induced changes in proopiomelanocortin mRNA levels in the pituitary gland of the amphibian, *Xenopus laevis*, *Biochem. Biophys. Res. Commun.*, submitted.
85. **Jenks, B. G., van Overbeeke, A. P., and McStay, B. F.**, Synthesis, storage, and release of MSH in the pars intermedia of the pituitary gland of *Xenopus laevis* during background adaptation, *Can. J. Zool.*, 55, 922, 1977.
86. **Bisbee, C. A., Baker, M. A., Wilson, A. C., Hadji-Azimi, I., and Fischberg, M.**, Albumin phylogeny for clawed frogs *(Xenopus)*, *Science*, 195, 785, 1977.
87. **Thiebaud, C. H. and Fischberg, M.**, DNA content in the genus *Xenopus*, *Chromosoma*, 59, 253, 1977.
88. **Uhler, M., Herbert, E., D'Eustachio, P., and Ruddle, F. D.**, The mouse genome contains two nonallelic pro-opiomelanocortin genes, *J. Biol. Chem.*, 258, 9444, 1983.
89. **Kawauchi, H., Tsubokawa, M., Kanerawa, A., and Kitagawa, H.**, Occurrence of two different endorphins in the salmon pituitary, *Biochem. Biophys. Res. Commun.*, 92, 1278, 1980.
90. **Niall, H. D.**, The evolution of peptide hormones, *Ann. Rev. Physiol.*, 44, 615, 1982.
91. **Seeburg, P. H.**, The human growth hormone gene family: nucleotide sequences show recent divergence and predict a new polypeptide hormone, *DNA*, 1, 239, 1982.
92. **Martens, G. J. M., Terwel, D., and Bussemakers, M.**, unpublished data, 1986.
93. **Wilcox, J. N., Gee, C. E., and Roberts, J. L.**, In situ cDNA:mRNA hybridization:development of a technique to measure mRNA levels in individual cells, *Methods Enzymol.*, 124, 510, 1986.
94. **Roberts, J. L., Eberwine, H., and Gee, C. E.**, Analysis of POMC gene expression by transcription assay and *in situ* hybridization histochemistry, in *Cold Spring Harbour Symposium on β-Quantitative Biology*, 28, 385, 1983.
95. **Gurdon, J. B. and Melton, D. A.**, Gene transfer in amphibian eggs and oocytes, *Annu. Rev. Genet.*, 15, 189, 1981.
96. **Norris, K., Norris, F., Christiansen, F., and Fiil, N.**, Efficient site-directed mutagenesis by simultaneous use of two primers, *Nucleic Acids Res.*, 11, 503, 1983.
97. **Lubbe, A. and Schaffner, W.**, Tissue-specific gene expression, *Trends Neurosc.*, 3, 100, 1985.

Chapter 7

REGULATION OF PRO-OPIOMELANOCORTIN (POMC) BIOSYNTHESIS IN THE AMPHIBIAN AND MOUSE PITUITARY INTERMEDIATE LOBE

Y. Peng Loh, Stela Elkabes, and Brenda Myers

TABLE OF CONTENTS

I. Introduction 86

II. Regulation of POMC Biosynthesis at the Transcriptional and Translational Levels 86

    A. POMC Biosynthesis in the Amphibian Intermediate Pituitary Under Different Physiological Paradigms 86
        1. POMC mRNA Levels and POMC Synthesis During Background Adaptation 86
        2. The Dopamine Receptor and Changes in POMC Synthesis with Dopamine and 8-Bromo-cAMP Treatment In Vitro 90
    B. POMC Biosynthesis in the Mouse Intermediate Pituitary During Salt Loading 93
        1. Changes in POMC mRNA Levels, POMC Synthesis, and α-MSH Levels in the Peripheral Blood 85

III. Regulation of Post–translational Processing of POMC and POMC-Derived Products 95
    A. The Enzymology and Regulation of Proteolytic Processing of POMC 95
    B. Acetylation of POMC-Derived Peptides 96
    C. Amidation of POMC-Derived Peptides 97

IV. Conclusions 97

Acknowledgments 98

References 99

I. INTRODUCTION

α-Melanotropin (α-MSH) and β-endorphin are synthesized from a common prohormone, pro-opiomelanocortin (POMC).[1-5] This prohormone is present both in the intermediate and anterior lobe of the pituitary.[6] The amino acid structure of POMC has been derived from the cDNA sequence for several species including bovine, human, rat, mouse and amphibian (*Xenopus laevis* frog).[7-11] The structure reveals pairs of basic residues flanking the hormones to be cleaved (see Figure 1). These pairs of basic residues represent cleavage signals for prohormone processing.[12] The primary sequence of POMC appears to be generally well conserved among the species, one difference being the absence of a pair of basic residues within the β-lipotropin portion of the molecule in rodents and, hence, the lack of a processing site for the production of β-MSH.[9,10] The α-MSH sequence appears to be identical in all the mammalian species, but the first amino acid, serine, in the mammalian sequence, is replaced by an alanine in the *X. laevis* sequence.[11] POMC is differentially processed in the anterior and intermediate lobes of the pituitary. In the anterior lobe, the processing pathway for the prohormone terminates with the production of ACTH, β-lipotropin and some β-endorphin, while in the intermediate lobe, the final products are α-MSH, β-endorphin$_{1-31}$ and β-endorphin$_{1-27}$ (Figure 2). (For more details, see Chapter 3 by Dores, this Volume).

Several mechanisms may exist for the regulation of the synthesis of POMC-derived peptides. These include regulation at the transcriptional level; at the post-transcriptional level (e.g., POMC mRNA stability); translational level, and post-translational level (i.e., the processing of POMC). In this chapter, we will examine the regulation of POMC and α-MSH biosynthesis at these different levels in the intermediate lobe.

Two model systems were chosen for our studies: the amphibian (*Xenopus laevis* frog) and mouse intermediate lobe. Regulation of POMC biosynthesis was examined in the intermediate lobe of these animals under various experimental conditions, both in vivo and in vitro. The in vivo experimental perturbation employed was background adaptation for the frog, since the role of α-MSH in the control of skin color is well documented in amphibians.[13] For the mouse, the effect of salt loading on POMC biosynthesis in the intermediate lobe was studied in view of the implications of the osmoregulatory role of α-MSH in aldosterone secretion.[14] In vitro studies aimed at leading to the understanding of the mechanism of coupling of extracellular signalling (e.g., by a neurotransmitter acting at the receptor level) to the intracellular event of regulating POMC gene expression in the intermediate lobe cell, will also be discussed.

II. REGULATION OF POMC BIOSYNTHESIS AT THE TRANSCRIPTIONAL AND TRANSLATIONAL LEVELS

A. POMC Biosynthesis in the Amphibian Intermediate Lobe Pituitary Under Different Physiological Paradigms

1. POMC mRNA Levels and POMC Synthesis During Background Adaptation

During black-background adaptation of the amphibian (*Xenopus laevis* frog), the melanophore index (a measure of melanocyte expansion) of the skin increased to 5 as compared to white-background-adapted animals, which had punctuate melanocytes and a melanophore index of 1. This increase in melanophore index (i.e., melanocyte expansion and skin darkening) is mediated by an increase in secretion of the hormone, α-MSH, from the intermediate lobe of the pituitary.[13,15] Concomitantly, there is an increase of α-MSH levels in the intermediate lobe, presumably reflecting an increase in synthesis of this hormone.[16] Indeed, studies on the biosynthesis of POMC (the prohormone for α-MSH) showed that when white-background-adapted animals were transferred to a black background for 7 days, the incorporation of [^{3}H]arginine into POMC increased 5-fold[17] (Figure 3). Conversely, when black-

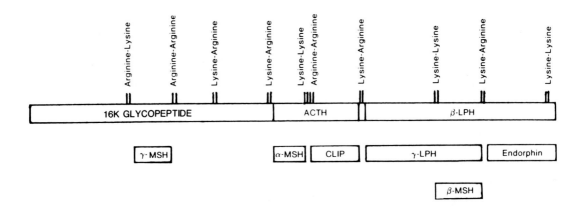

FIGURE 1. Diagrammatic representation of human pro-opiomelanocortin structure showing the pairs of basic residues.

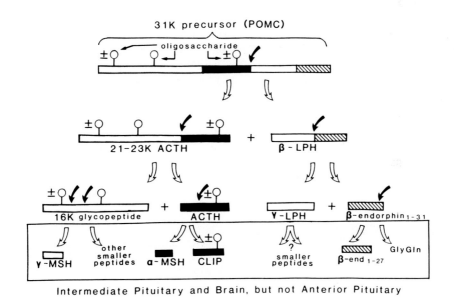

FIGURE 2. Pathway of processing of pro-opiomelanocortin (POMC). The black arrows indicate cleavage sites at pairs of basic amino acids to yield the various products.

background-adapted animals were transferred to a white background for 7 days, the incorporation of [^{3}H]arginine into POMC decreased by 76% (Figure 4). This decrease in POMC synthesis probably reflects the decreased demand for α-MSH during white-background adaptation. Thus, POMC synthesis appears to be coordinately coupled to the level of secretion of α-MSH during background adaptation.

To determine if POMC synthesis is regulated at the transcriptional level, changes in POMC mRNA levels were measured. Due to the small amount of tissue available in the

88 *The Melanotropic Peptides*

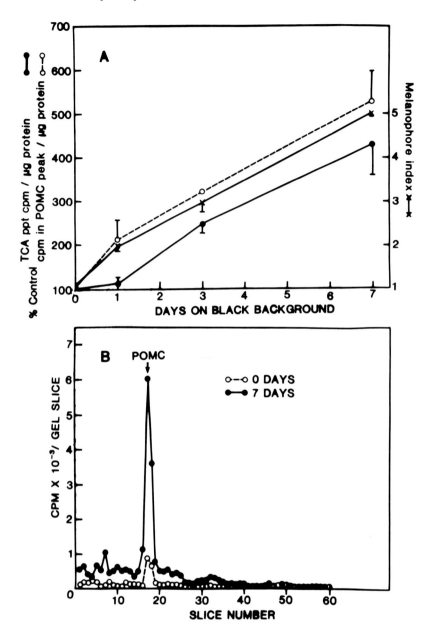

FIGURE 3. Effect of transferring white-background-adapted animals to black background. (A) Animals were sacrificed at various times after black-background adaptation, and the NILs were pulse labeled for 1 hr in [^{3}H]arginine. The left ordinate shows cpm of trichloroacetic acid (TCA) precipitated (ppt) per µg of protein (●—●) and cpm in the POMC peak per µg of protein (○---○) in the NILs, expressed as a percent of the value from white-background-adapted animals just prior to transfer to a black background (control). The control values for cpm of trichloroacetic acid precipitated per µg of protein and cpm in POMC peak per µg of protein were 913 ± 134 and 98 ± 12, respectively, and they were arbitrarily set to 100%. The right ordinate shows the melanophore index of the skin (×—×) recorded at the time of dissection of the animals. Results are expressed as the mean ± S.E. (n = 3). (B) acid-urea gel profiles showing [^{3}H]arginine-labeled POMC from animals that had been transferred to black background, at day 0 (○---○) and day 7 (●—●). (From Loh Y.P., et al., *J. Biol. Chem.*, 260, 8956, 1985, With permission.)

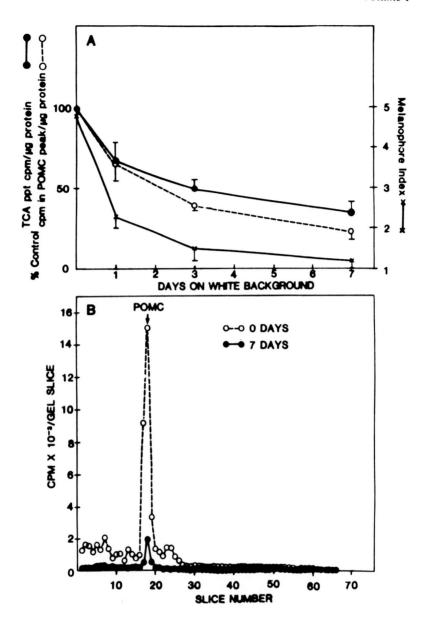

FIGURE 4. Effect of transferring black-background-adapted animals to white background. (A) Animals were killed at various times after white-background adaptation, and the NILs were pulse-labeled for 1 hr in [^{3}H]arginine. The left ordinate shows cpm of trichloroacetic acid (TCA) precipitated (ppt) per μg of protein (●—●) and cpm in the POMC peak per μg of protein (○---○) in the NILs, expressed as a percent of the value from black background-adapted animals just prior to transfer to a white background (control). The control values for cpm of trichloroacetic acid precipitated per μg of protein and cpm in POMC peak per μg of protein were 3176 ± 370 and 494 ± 39, respectively, and they were arbitrarily set to 100%. The right ordinate shows the melanophore index of the skin (×—×) recorded at the time of dissection of the animals. The results were expressed as the mean ± S.E. (n = 3). (B) Acid urea gel profiles showing [^{3}H]arginine-labeled POMC from animals that had been transferred to white ackground at day 0 (○---○) and day 7 (●—●). (From Loh Y. P., et al., *J. Biol. Chem.*, 260, 8956, 1985, With permission.)

intermediate lobe, the approach used was quantitative *in situ* hybridization. Animals were adapted to a white or black background and their pituitaries removed, frozen immediately and sectioned. The sections were put on slides and then processed for hybridization using a [^{35}S]-labeled synthetic (48 mer) probe complimentary to the α-MSH region of the frog (*X. laevis*) POMC mRNA. This POMC probe has been verified by Northern gel analysis to specifically hybridize to the ~1300 nucleotide *X. laevis* POMC mRNA reported by Martens, Civelli, and Herbert.[11] After the hybridization and washes, the sections were opposed to cover slips which were previously dipped in emulsion (Kodak® NTB-3). Slides were exposed for 3 weeks, developed, and the grain density scanned using an image analysis system. The probe copies/μ^3 were then calculated from the reflectance using brain paste standards of known radioactivity which were exposed for the same period of time. Figure 5 shows bright and dark field views of a section from the intermediate lobe of a black-background-adapted animal. The grains were intensely localized to the intermediate lobe and there were essentially no grains over the neural lobe. On the basis of probe copies/μ^3, POMC mRNA levels in the intermediate lobe were found to be 4-fold greater in black- vs. white-background-adapted animals. Thus, POMC biosynthesis appears to be regulated both at the transcriptional and translational levels during background adaptation.

2. The Dopamine Receptor and Changes in POMC Synthesis with Dopamine and 8-Bromo-cAMP Treatment In Vitro

To further understand the regulation of POMC synthesis from the physiological to the cellular level, it is necessary to utilize an in vitro system. We have, therefore, examined the regulation of POMC synthesis in organ-cultured *X. laevis* neurointermediate lobes.[17] Morphological studies have shown that the *X. laevis* intermediate lobe is innervated by dopamine fibers[18] and numerous studies have shown that α-MSH secretion from this lobe is under inhibitory control by dopamine.[16,19-21] Moreover, De Volcanes and Weatherhead have demonstrated that selective destruction of the dopamine fibers resulted in prolonged darkening of the frog and the loss of ability to adapt to a white background.[22] These observations suggest that dopamine may be the extracellular signal that triggers a series of cellular events leading to the regulation of POMC synthesis in the intermediate lobe of the frog during background adaptation.

We further tested this hypothesis by examining the effect of dopamine on POMC biosynthesis in organ-cultured neurointermediate lobes from black- and white-background-adapted animals.[17] POMC synthesis was assayed by the incorporation of [^{3}H]Arg into POMC after 1 hr of pulse labeling. Dopamine (50 μM) treatment of organ-cultured neurointermediate lobes (NILs) from black-background-adapted animals, which were actively synthesizing POMC, resulted in a decrease of POMC synthesis to 24.6% of control after 7 days.[17] In the absence of dopamine, these NILs after 7 days in culture, maintained their synthesis of POMC at control levels (day 0). Conversely, when NILs from white-background-adapted animals were treated with dopamine for 3 days, POMC synthesis continued to be suppressed, maintaining the low level of synthesis seen at day 0. However, in the absence of dopamine, POMC synthesis rose to 700% of control, after 7 days in culture.[17] Treatment of lobes with actinomycin D, which blocks RNA synthesis, prevented this rise in POMC synthesis.[74] These data indicate that dopamine not only regulates secretion of α-MSH, but also POMC biosynthesis at the transcriptional and translational levels in the frog intermediate lobe, and further support an inhibitory role of this neurotransmitter in regulating POMC synthesis during background adaptation of the animal.

In order to dissect the mechanism of action of dopamine in mediating the regulation of POMC synthesis, the dopamine receptor was characterized.[17] Dopamine receptors may be classified based upon the result of receptor stimulation on adenylate cyclase activity. Stimulation of D-1 receptors is known to result in enhanced activity of adenylate cyclase (and

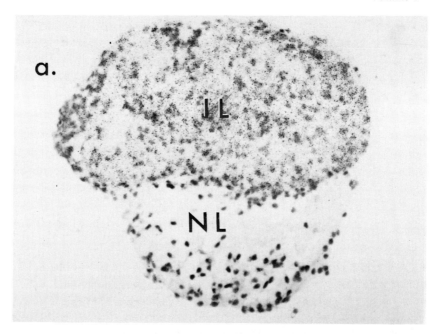
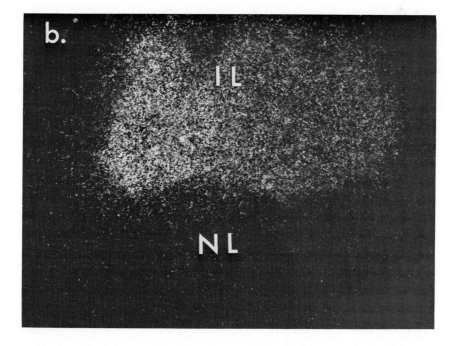

FIGURE 5. Bright field (a), and dark field (b) views of a cross-section of the *Xenopus laevis* intermediate lobe after *in situ* hybridization with a [^{35}S]-labeled POMC oligonucleotide probe. Note the localization of the silver grains to the intermediate lobe only. IL, intermediate lobe; NL, neural lobe.

Table 1
EFFECT OF DOPAMINE RECEPTOR LIGANDS ON IMMUNOREACTIVE α-MSH RELEASE

Treatment	IR α-MSH released[a]
	ng/NIL·180 min
Control	11.7 ± 1.0
LY-171555 (30 nM)	1.2 ± 0.3[b]
LY-171555 (30 nM) + SCH-23390 (1 μM)	0.9 ± 0.02[b]
SCH-23390 (1 μM)	11.9 ± 3.3
SKF-38393 (1 μM)	11.4 ± 2.1

[a] Values reported are the mean ± S.E., n = 4.
[b] $p < 0.001$ vs. control using the Student's t test.

enhanced cAMP accumulation in the cell), whereas stimulation of D-2 receptors produces a diminution of adenylate cyclase activity.[23] Recently, the first generation of selective ligands has become available to study each class of dopamine receptor. LY-171555 and YM-09151-2 are selective D-2 receptor agonists and antagonists, respectively, whereas SKF-38393 and SCH-23390 are selective D-2 receptor agonists and antagonists, respectively. The effects of those agents on immunoreactive α-MSH secretion was used to characterize the X. laevis intermediate lobe (IL) dopamine receptor. LY-171555 inhibited α-MSH release from the IL in a dose-dependent manner.[17] The effect of LY-171555 was reversed in a dose-dependent manner by YM-09151-2. SKF-38393 or SCH-23390 did not alter α-MSH release. In addition, SCH-23390 did not affect LY-171555-induced inhibition of α-MSH secretion (Table 1). Based upon this pharmacological analysis, the dopamine receptor in the X. laevis IL belongs to the D-2 category, as does the dopamine receptor in the rat IL. Further support that the X. laevis dopamine receptor belongs to the D-2 category is derived from studies on cAMP accumulation in this tissue. Dopamine alone caused a decrease in IL cAMP content and significantly inhibited forskolin-stimulated cAMP accumulation in the IL. This effect of dopamine was reversed by the selective D-2 antagonist, YM-09151-2. These studies suggest that the D-2 receptor in X. laevis is coupled to adenylate cyclase in an inhibitory manner, as is the rat IL D-2 receptor.[24] Based on the model for the adenylate cyclase complex, the dopamine receptor in the frog IL may be coupled to the inhibitory subunit (N_i) of the complex, as has been hypothesized for the dopamine receptor in the rat IL.[25]

The nature of the dopamine receptor in the frog IL suggests that cAMP is a second messenger in regulating POMC synthesis. Our recent data[17] (Table 2) showing reversal of the inhibition of POMC synthesis by dopamine with the addition of 8-Br-cAMP to the culture medium, support such a role for cAMP, and indicate that such changes in the intracellular level of cAMP can modulate POMC synthesis in the IL. This is further supported by the finding that while 8-Br-cAMP can reverse the dopamine inhibition of POMC synthesis, it did not reverse dopamine inhibition of secretion. 8-Br-cAMP also had no effect on stimulating secretion of α-MSH in the IL. Thus, cAMP probably acts cytoplasmically or in the nucleus, rather than on secretion per se in stimulating POMC synthesis.

We propose from the above studies that during white-background adaptation of X. laevis, dopamine is secreted from the nerve terminals impinging on the intermediate lobe. The dopamine then acts by stimulating an IL D-2 receptor, which results in a diminution of adenylate cyclase activity and subsequently lowers intracellular levels of cAMP. This, in turn, results in decreased POMC synthesis. During black-background adaptation, reduction of dopamine levels via the inhibition of secretion from the dopaminergic nerve terminals results in a disinhibition of adenylate cyclase, elevation of intracellular content, and stim-

Table 2
EFFECT OF DOPAMINE AND 8-Br-cAMP ON POMC SYNTHESIS

	Control[a]	
	Nonimmunoprecipitate (%)	Immunoprecipitate[b] (%)
Minus dopamine, minus 8-Br-cAMP	298 ± 46 (4)	264
Plus 8-Br-cAMP (1 mM)	274 ± 15 (4)[c]	248
Plus dopamine (10 μM)	101 ± 11 (4)[d]	100
Plus dopamine (10 μM), plus 8-Br-cAMP (1 mM)	220 ± 23 (5)[c]	205

[a] NILs from white-background-adapted animals were cultured for 3 days in medium 199 without drugs or containing either dopamine or 8-Br-cAMP, or both. NILs were then pulse-labeled in [^{3}H]arginine for 1 hr. The [^{3}H]arginine cpm in the POMC peak per μg of protein present in the cultured lobes was expressed as a percent of the control. The control value equals the cpm in the POMC peak per μg protein in the freshly dissected NILs from white-background-adapted animals and was arbitrarily set to 100%. The cpm in POMC per μg protein corresponding to 100% = 115 ± 11. Values reported are the mean ± S.E. The number of experiments is shown in parentheses.
[b] An aliquot of the homogenized NIL from each experiment in the group was pooled and immunoprecipitated with ACTH antiserum. The immunoprecipitated POMC peak was identified by gel electrophoresis and expressed as cpm in the immunoprecipitate POMC peak per μg of protein.
[c] Not significant vs. lobes incubated without dopamine and 8-Br-cAMP.
[d] p <0.01 vs. lobes incubated without dopamine and 8-Br-cAMP using the Student's t test.

ulation of POMC synthesis. Such a mechanism of action of dopamine on POMC synthesis has been proposed for the rat intermediate lobe.[26-28]

How cAMP acts to trigger synthesis of POMC is unclear at the present time. It is known that the activation of some protein kinases is cAMP dependent.[29,30] Perhaps the phosphorylation of specific proteins may be involved in the regulation of translation and transcription of the POMC mRNA. Studies on the POMC gene in the rat suggest that there are consensus sequences upstream from the transcription site which are cAMP-responsive elements.[31] cAMP may activate the phosphorylation of a *trans* factor which binds directly to the POMC gene to regulate its expression. Future studies at the genomic level should shed more light on the regulation of expression of the POMC gene by cAMP.

B. POMC Biosynthesis in the Mouse Intermediate Pituitary During Salt Loading
1. Changes in POMC mRNA Levels, POMC Synthesis, and α-MSH Levels in Peripheral Blood

The sequence of the mouse, human, bovine, and rat POMC genes have been characterized.[32-35] The mouse genome contains two POMC-related gene sequences; α-POMC and β-POMC genes which are located in different chromosomes.[32] The α-POMC gene in the mouse is very similar to the single POMC gene found in human, bovine, and rat genomes. The sequence of the mouse β-POMC gene is quite different from that of the α-POMC gene. It has a translation stop signal in place of the first amino acid in β-endorphin (Tyr) and it appears not to be expressed, suggesting that it is a pseudogene.[32]

In mammals, POMC mRNA has been detected by Northern blot analysis in the anterior and neurointermediate lobes of the pituitary, as well as the hypothalamus, amygdala, and cerebral cortex of the rat.[36,37] The POMC mRNA in the pituitary and in the brain varies in size as well as in quantity.[36,37] POMC mRNA in the pituitary is 1150 bases long, while that in the brain is only 1050 bases. The POMC mRNA levels in the neurointermediate lobe are 20 times higher than in the anterior lobe which, in turn, are one or two orders of magnitude higher than those in brain tissues.[37] The anterior and intermediate lobes of the pituitary gland exhibit tissue-specific regulation of POMC gene expression. In the anterior lobe, expression of POMC is inhibited by glucocorticoids released by the adrenal cortex and stimulated by CRF.[38-42] In the intermediate lobe, the POMC gene expression is unaffected by glucocorticoids, but is inhibited by dopamine.[43]

In mammals, the function of ACTH from the anterior lobe, is well studied, but the role of α-MSH from the intermediate lobe is less well defined. Very early studies suggested a possible involvement of the intermediate lobe in osmoregulation. This was indicated by changes in the content of α-MSH and the histology of the intermediate lobe in rats, after ingestion of hypertonic saline.[44] In addition, it was shown that immunoreactive plasma β-endorphin levels were decreased in salt-loaded rats, resulting in alterations in non-opiate β-endorphin binding sites in the kidney and the adrenal gland.[45] Such observations suggested that ingestion of hypertonic saline may have effects on POMC synthesis, in parallel to its effects on secretion of POMC-derived peptides from the intermediate lobe. We, therefore, investigated the effects of hypertonic saline ingestion on the POMC mRNA levels and POMC synthesis in the intermediate lobe of mouse pituitary. Parallel studies were carried out in the anterior lobe to determine whether similar changes occur in both lobes.

For these studies, mice were salt-loaded for various days by substituting their drinking water with 2% NaCl. The animals were then sacrificed and the neurointermediate lobes removed for RNA extraction or POMC biosynthesis studies. For analysis of POMC mRNA, total RNA was extracted by the guanidinium thiocyanate procedure, followed by density gradient centrifugation, using a CsTFA-EDTA gradient.[46] The RNA preparation was then analyzed for POMC mRNA by the Northern blot technique, using a mouse [^{32}P]-labeled 946 bp POMC probe (a gift from Dr. E. Herbert) for hybridization. The amount of [^{32}P]-labeled probe that hybridized to the POMC mRNA was determined by scanning the autoradiographs and was expressed as density per μg rRNA. To determine the amount of rRNA in the samples, the photonegatives of the ethidium-bromide-stained gels were scanned. The density over the 28S RNA band was measured and read against a standard curve of calf liver rRNA. Figure 6 shows that after two days of salt-loading, there was a decrease in POMC mRNA levels in the intermediate lobe to 50.0 ± 4.3% of control, followed by an increase to 193.7 ± 5.2% above control, 9 days after salt-loading. In contrast, the POMC mRNA levels in the anterior lobe did not show a decrease after two days of salt-loading. Thus the fall in POMC mRNA level appears to be unique to the intermediate lobe. Analysis of α-MSH in the peripheral blood showed a 25% decrease in α-MSH levels as compared to control after 2 days, but returned to control levels after 9 days. The decrease in POMC mRNA probably reflects a feedback due to decrease in secretion of α-MSH. Similarly, analysis of [^{3}H]Arg incorporation into POMC in the intermediate lobe also showed a decrease after two days salt-loading and then returned to control levels after 9 days. Thus, there appears to be close coupling between a decrease in POMC mRNA levels and translation of POMC mRNA, and a decrease in α-MSH secretion 2 days after salt-loading. The increase in POMC mRNA above control at 9 days may be due to an overcompensation in the restoration of POMC mRNA levels in the cell when the α-MSH secretion returned to control levels. The changes in POMC mRNA levels may be the result of a change in transcription rate or mRNA degradation. Interestingly, at 9 days after salt treatment, [^{3}H]Arg incorporation into POMC was not above control, suggesting another point of regulation of POMC biosynthesis, at the translational level.

The physiological significance of the decrease in α-MSH secretion, POMC mRNA levels, and translation of POMC mRNA in the intermediate lobe may be related to the need to inhibit aldosterone secretion.[14] Previous work has indicated that α-MSH stimulates the secretion of aldosterone from the adrenal cortex.[14] Since it is well documented that this hormone is involved in Na$^+$ retention in the kidneys,[47] it seems reasonable that under conditions of salt-loading, α-MSH secretion should decrease, leading to a shut-off of POMC synthesis. The anterior lobe, however, does not synthesize α-MSH and therefore, POMC

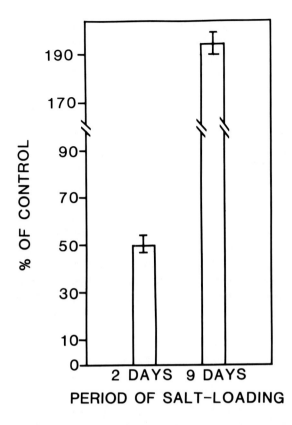

FIGURE 6. Northern gel analysis of POMC mRNA levels in the mouse intermediate lobe during salt-loading, expressed as a percent of control. Error bars show the S.E.M. (n = 4).

synthesis in that lobe was not inhibited during salt loading. Studies are now in progress to determine the extracellular factors that may trigger the inhibition of secretion of α-MSH and POMC synthesis in the intermediate lobe during salt-loading. One possible candidate is arginine vasopressin, which is greatly increased in the blood during salt-loading.[48] The hormone may either act directly on the intermediate lobe or exert its effect via the stimulation of the catecholaminergic systems that project to the intermediate lobe.

III. REGULATION OF POST–TRANSLATIONAL PROCESSING OF POMC AND POMC-DERIVED PRODUCTS

The post–translational processing of POMC to various peptides found in the intermediate lobe, involves several steps. The first is a series of proteolytic events leading to the generation of cleaved peptides. These proteolytic steps include cleavages at paired basic residues, followed by the trimming off of extended basic residues from the C- and N-termini of the cleaved peptides. These peptides are then further modified at the N-terminus by acetylation and/or at the C-terminus by amidation to yield the final products which are secreted. The terminal modifications are very important, for example, for α-MSH to achieve full melanotropic potency, it has to be both acetylated and amidated at the N- and C-termini, respectively.[49] Elucidating the enzymes involved in these post-translational modification steps would facilitate the understanding of the regulation of these events.

A. The Enzymology and Regulation of Proteolytic Processing of POMC

POMC is processed by a cascade of proteolytic steps beginning with a cleavage at a Lys-

Arg pair between the ACTH and β-lipotropin sequence (see Figure 1). This is then followed by the cleavage of the 16K glycopeptide and β-lipotropin, again at Lys-Arg residues to yield ACTH and β-endorphin, respectively.[7,50] Recently, an enzyme has been purified from bovine intermediate lobe secretory vesicles that can accomplish these cleavages. The enzyme, named pro-opiomelanocortin-converting enzyme (PCE), has been characterized as an aspartyl protease.[51] It has an acidic pH optimum and is a 70,000 mol. wt. glycoprotein. PCE is highly specific for paired basic residues. The enzyme cleaves POMC, as well as the intermediates, 21 to 23K ACTH and β-lipotropin to yield ACTH and β-endorphin, respectively. However, the enzyme does not cleave the Lys-Lys-Arg-Arg residues of ACTH to yield α-MSH,[52] nor will it cleave the Lys-Lys pair at the C-terminal portion of β-endorphin to yield truncated forms of β-endorphin found in the intermediate lobe.[52] In vitro studies on PCE suggest that the enzyme concentration is important in the regulation of the processing of POMC. Depending on the enzyme concentration used, human β-lipotropin was found to be either fully processed to β-endorphin and β-MSH, or to β-endorphin and γ-lipotropin[52] (see Figure 1). Thus, theoretically, by regulating the enzyme to substrate concentration within a secretory vesicle where processing occurs, it may be possible to attenuate the degree of processing and, hence, the end-products generated. The differential processing of β-lipotropin in the intermediate and anterior lobes may be under such a regulatory mechanism.

Two other proteases, an aminopeptidase B-like enzyme and a carboxypeptidase B-like enzyme, are necessary for the removal of the N- and C-terminal-extended basic residues, respectively, from the cleaved peptides following PCE action. A carboxypeptidase B-like enzyme has been found in secretory vesicles of pituitary and chromaffin cells.[53,54] This enzyme has been purified and characterized as a metalloprotease, which is stimulated by Co^{++}. The enzyme has recently been cloned and the amino acid sequence derived from the cDNA sequence. An aminopeptidase B-like enzyme has been detected in secretory vesicles of pituitary and shown to be a metalloprotease which is also stimulated by Co^{++}. (For reviews of the proteolytic prohormone processing enzymes, see References 55,56). However, it seems unlikely that these exopeptidases will play a key role in rate-limiting the production of α-MSH and β-endorphin.

To determine if the overall rate of POMC processing is altered under conditions where POMC synthesis is changed, pulse-chase experiments were carried out. With background adaptation of the frog, dopamine treatment of the frog intermediate lobe and salt-loading of the mouse, no difference in the rate of processing of POMC coupled to changes in the synthesis of the prohormone (see Section II) was detected. It would appear, therefore, that the attenuation of the rate of proteolytic processing of POMC is not a mechanism used for regulating α-MSH and β-endorphin production in the intermediate lobe, at least under these experimental conditions.

B. Acetylation of POMC-Derived Peptides

Acetylation of β-endorphin renders this peptide inactive as an opiate. Yet much of the β-endorphin in the intermediate lobe is in the acetylated form.[57] It is possible that the acetylated form has a function, yet to be elucidated, in another target tissue other than the nervous system. Alternatively, the acetylation step may be looked upon as a regulatory step to inactivate the peptide as an opiate, while at the same time activating the α-MSH molecules to its full potency as a melanotropic hormone, in the case of amphibians. In rodents, the intermediate lobe α-MSH is acetylated and diacetylated, although the physiological significance of the mono- and di-acetylation is not known.[58] Interestingly in the rat brain, neither the β-endorphin nor α-MSH is acetylated (i.e., desacetyl α-MSH is the major form of α-MSH).[59,60]

Recently, an acetylation enzyme of ~75,000 mol. wt. and pH optimum of 6.5 to 7.0 has been detected in intermediate lobe secretory vesicles that acetylates both β-endorphin at the tyrosine position, and α-MSH at a serine position.[61,63] The enzyme requires acetyl Co A as a co-factor. Whether this enzyme can be readily activated or inactivated to alter acetylation within the secretory vesicle under different physiological states of the cell, is not known. There is a report that in the rat intermediate lobe cells in culture, acetylation of β-endorphin is modulated by the neurotransmitter, dopamine.[64] However, this modulation may be due to changes in the acetylation enzyme activity, or prolonged exposure of β-endorphin to the acetylation enzyme in the vesicle as a result of decreased secretion in the presence of dopamine. In the brain, the expression of the acetylation enzyme gene may be repressed, or the enzyme activity inhibited, giving rise to nonacetylated POMC-derived peptides. Further studies on the enzyme are necessary to evaluate these different possibilities.

Interestingly, an example exists in *Xenopus laevis* where there is a pathway in which acetylation of desacetyl α-MSH is coupled to secretion of the hormone.[65,66] In this case, it is possible that the acetylation enzyme is inactive in the secretory vesicle (which has an acidic internal pH) and is activated only upon exposure to a neutral pH environment during exocytosis. If this hypothesis is correct, then the combination of selective cellular compartmentalization and pH dependence of the enzyme may be a mechanism for regulating post-translational processing events.

C. Amidation of POMC-Derived Peptides

Several POMC-derived peptides are amidated, α-MSH, the joining peptide[67] and γ-MSH (see Figure 2). Although the function of amidation of these peptides is not known, amidation probably represents an important step for biological activity since numerous peptides are amidated.[67] It has been found that peptides whose structure ends with a glycine just before a pair of basic residues at the C-terminus, is generally amidated. An amidation enzyme named peptidyl-glycine α-amidating monooxygenase (PAM) has been purified from bovine pituitary.[68] This enzyme requires the C-terminal glycine of a peptide as the NH_2 donor, and requires ascorbate, Cu^{++}, and molecular oxygen as co-factors for enzymatic activity.[69,70] Deficiencies in any one of these co-factors could lead to a decrease in PAM activity and a decrease in amidation. For example, it has been reported that the lack of ascorbate in the medium of cultured intermediate lobe cells results in a lack of amidation of POMC-derived peptides, and this phenomenon can be reversed by the addition of ascorbate.[71] At present, it is not known if the amidation step is regulated under different physiological conditions of the cell, or that amidation occurs whenever the appropriate peptide structure and enzyme are present together.

IV. CONCLUSIONS

The studies we have presented on the biosynthesis of POMC in the intermediate lobe have yielded much insight into the coupling of transcription, translation, and processing of POMC with secretion of products in this tissue. In the frog intermediate lobe, transcription and translation of POMC mRNA and proteolytic processing of the prohormone appear to be "on-line", and closely linked to secretion of the POMC-derived products. This statement is borne out by the data showing that black- and white-background adaptation, resulting in an increase or decrease of secretion of α-MSH, respectively, is closely paralleled with a change in POMC mRNA levels and incorporation of [^{3}H]arginine into POMC in the same direction. In vitro studies also showed similar regulation of POMC synthesis by dopamine, an inhibitor of α-MSH release. Moreover, these studies revealed that once POMC is synthesized, it is rapidly processed and there appears to be no regulation at the processing level. This is perhaps expected, since there is very little prohormone stored in the lobe.[75] Thus, the regulatory mechanism for POMC biosynthesis in the frog intermediate lobe seems to be a direct feedback from secretion to transcription. Negative feedback on POMC synthesis displayed by the frog intermediate lobe is mediated by dopamine at the receptor level. The signal resulting from dopamine binding to the receptor is transduced to the genome via negative coupling to adenylate cyclase and lowering of cellular cAMP levels, which then result in a shut-off POMC synthesis. An increase in cAMP is then required to reactivate POMC gene expression.

The regulation of POMC biosynthesis in the mammalian intermediate lobe shows great similarities to the frog. However, there are some subtle differences. In the mouse, with salt-loading for 2 days which resulted in a decrease of α-MSH secretion, there was a parallel decrease in transcription and translation reflected by a decrease in POMC mRNA levels and [^{3}H]arginine incorporation into POMC. After 9 days salt-loading, α-MSH secretion returned to normal, but POMC mRNA increased to twice control level, suggesting that there was feedback between transcription and secretion, except that there was an over compensation in the regulatory mechanism. However, after 9 days salt-loading, translation of POMC mRNA was at control levels, indicating that translation was not simply an "on-line" process, yielding a quantitatively parallel change with transcription as in the frog. Thus, secretion can feedback at the transcriptional and translational levels differentially in the mammal. This may be due to lower rates of secretion of POMC products than in the amphibian; and the greater complexity in the regulation of secretion, which necessitates better fine-tuning in the regulation of POMC biosynthesis in the mammalian intermediate lobe. As in the frog, the lack of storage of high levels of POMC suggests that regulation of α-MSH synthesis at the processing level is also insignificant in this mammalian tissue.

Another level of regulation of α-MSH content within the intermediate lobe cell which heretofore have not been eluded to, is degradation. Previous studies have revealed that dopamine treatment of neurointermediate lobes from dark-background-adapted frogs resulted in the rapid degradation of newly synthesized POMC-derived peptides within 6 hr of synthesis.[72] Therefore, in a situation where transcription and translation is at a maximum and secretion is inhibited, the cell degrades 76% of its newly synthesized products. Stimulation of degradation may be the rapid mechanism for coping with overproduction, whereas supression of POMC gene expression may be for long-term attenuation.

It is hoped that the studies described in this chapter have given some insight into how an intermediate lobe cell couples secretion to the synthesis of α-MSH. The regulatory pattern of POMC synthesis seen in the intermediate lobe cell may be extended to other peptide-hormone-synthesizing endocrine cells or peptidergic neurons. For example, prolactin synthesis in the mamotroph appears to be regulated by dopamine and cAMP in a very similar manner.[73] It should also be appreciated that to fully understand the cell-biological dynamics of POMC metabolism under any given perturbation (e.g., alcohol treatment, stress), it is necessary to conduct studies at the level of transcription, translation, post-translational processing, degradation, and secretion. Future work on POMC synthesis in the intermediate lobe should reveal more of how second messengers, such as cAMP, mediate the trigger for POMC gene expression at the genomic level, how translation of the POMC mRNA is regulated, and how post-translational modification (e.g., acetylation) of POMC-derived peptides may be modulated.

ACKNOWLEDGMENTS

The authors thank Drs. Tom Zoeller, Mark Goldman, and Michael Lang for their contributions to various aspects of this work described in this chapter. We also thank Mr. Baldwin Wong for technical assistance, Mr. Douglas Shen for doing the artwork, and Ms. Jodi Hiltbrand and Mrs. Maxine Shaefer for typing the manuscript.

REFERENCES

1. **Mains, R. E., Eipper, B. A., and Ling, N.**, Common precursor to corticotropin and endorphins, *Proc. Natl. Acad. Sci. U. S. A.*, 74, 3014, 1977.
2. **Roberts, J. L. and Herbert, E.**, Characterization of a common precursor to corticotropin and β-lipotropin: cell free synthesis of the precursor and identification of corticotropin peptides in the molecule, *Proc. Natl. Acad. Sci. U.S.A.*, 74, 5300, 1977.
3. **Loh, Y. P.**, Immunological evidence for two common precursors to corticotropins, endorphins, and melanotropin in the neurointermediate lobe of the toad pituitary, *Proc. Natl. Acad. Sci. U. S. A.*, 76, 797, 1979.
4. **Crine, P., Seidah, N. G., Heanotte, L., and Chrétien, M.**, Two large glycoprotein fragments related to the NH_2 terminal part of the adrenocorticotropin β-lipotropin precursors are the end products of the maturation process in the rat pars intermedia, *Can. J. Biochem.*, 58, 1318, 1980.
5. **Martens, G. J. M., Biermans, P. P. J., Jenks, B. G., and van Overbeeke, A. P.**, Biosynthesis of two structurally different pro-opiomelanocortins in the pars intermedia of the amphibian pituitary gland, *J. Biochem.*, 126, 17, 1982.
6. **Crine, P., Gianoulakis, C., Seidah, N. G., Gossard, F., Pezella, P. D., Lis, M., and Chrétien, M.**, Biosynthesis of β-endorphin from β-lipotropin, and a larger molecular weight precursor in rat pars intermedia, *Proc. Natl. Acad. Sci. U.S.A.*, 75, 4719, 1978.
7. **Nakanishi, S., Inoue, A., Kita, T., Nakamura, M., Chang, A. C. Y., Cohen, S. N., and Numa, S.**, Nucleotide sequence of cloned cDNA for bovine corticotropin-β-lipotropin precursor, *Nature (London)*, 1278, 423, 1979.
8. **Chang, A. C. Y., Cochet, M., and Cohen, S. N.**, Structural organization of human genomic DNA encoding the pro-opiomelanocortin peptide, *Proc. Natl. Acad. Sci., U.S.A.*, 77, 4890, 1980.
9. **Drouin, J. and Goodman, H. M.**, Most of the coding region of rat ACTH/β-LPH precursor gene lacks intervening sequences, *Nature (London)*, 288, 610, 1980.
10. **Uhler, M. and Herbert, E.**, Complete amino acid sequence of mouse proopiomelanocortin derived from the nucleotide sequence of pro-opiomelanocortin, cDNA, *J. Biol. Chem.*, 258, 257, 1983.
11. **Martens, G. J. M., Civelli, O., and Herbert, E.**, Nucleotide sequence of cloned cDNA for pro-opiomelanocortin in the amphibian, Xenopus laevis, *J. Biol. Chem.*, 260, 13685, 1985.
12. **Docherty, K. and Steiner, D.**, Post-translational proteolysis in polypeptide hormone biosynthesis, *Ann. Rev. Physiol.*, 44, 625, 1982.
13. **Hadley, M. E., Heward, C. B., Hruby, Y. J., Sawyer, T. K., and Yang, Y. C. S.**, Biological actions of melanocyte-stimulating hormone, in *Peptides of the Pars Intermedia*, Ciba Foundation Symp. 81, Pitman Medical, London, 1981, 244.
14. **Vinson, G. P., Whitehouse, B. J., Dell, A., Etienne, T., and Morris, H. R.**, Characterization of an adrenal zona glomerulosa stimulating component of posterior pituitary extracts as α-MSH, *Nature (London)*, 284, 484, 1980.
15. **Wilson, J. F. and Morgan, M. A.**, α-Melanotropin-like substances in the pituitary and plasma of Xenopus laevis in relation to color change responses, *Gen. Comp. Endocrinol.*, 38, 172, 1979.
16. **Loh, Y. P., Li, A., Gritsh, H. A., and Eskay, R. L.**, Immunoreactive α-melanotropin and β-endorphin in the toad pars intermedia. Dissocation in storage, secretion and subcellular localization, *Life Sci.*, 29, 1599, 1981.
17. **Loh, Y. P., Myers, B., Wong, B., Parish, D. C., Lang, M., and Goldman, M. E.**, Regulation of pro-opiomelanocortin synthesis by dopamine and cAMP in the amphibian pituitary intermediate lobe, *J. Biol. Chem.*, 260, 8956, 1985.
18. **Terlou, M. and Van Kooten, H.**, Microspectrofluorometric identification of formaldehyde induced fluorescence in hypothalamic nuclei of Xenopus laevis tadpoles, *Z. Zellforsch, Mikrosk. Anat.*, 147, 529, 1974.
19. **Bower, A., Hadley, M. E., and Hruby, Y. J.**, Biogenic amines and control of melanophore stimulating hormone release, *Science*, 184, 70, 1974.
20. **Jenks, B. G. and Van Overbeeke, A. P.**, Biosynthesis and release of neurointermediate lobe peptides in the aquatic toad, Xenopus laevis, adapted to black background, *Comp. Biochem. Physiol.*, 66, 71, 1980.
21. **Loh, Y. P. and Gainer, H.**, Biosynthesis, processing and control of release of melanotropic peptides in the neurointermediate lobe of Xenopus laevis, *J. Gen. Physiol.*, 70, 35, 1977.
22. **De Volcanes, B. and Weatherhead, B.**, Stereological analysis of the effects of 6-hydroxydopamine on the ultrastructure of the pars intermedia of the pituitary of Xenopus laevis, *Gen. Comp. Endocrinol.*, 28, 205, 1976.
23. **Kebabian, J. W. and Calne, D. G.**, Multiple receptors for dopamine, *Nature (London)*, 277, 93, 1979.
24. **Cote, T. E., Grewe, C. W., Tsuruta, K., Stoof, J. C., Eskay, R. L., and Kebabian, J. W.**, D-2 dopamine receptor-mediated inhibition of adenylate cyclase activity in the intermediate lobe of the rat pituitary gland requires guanosine 5'-triphosphate, *Endocrinology*, 110, 1957, 1982.

25. **Rodbell, M.,** The role of hormone receptors and GTP regulatory proteins in membrane transduction, *Nature (London),* 284, 17, 1980.
26. **Chen, C. L. C., Dionne, F. T., and Roberts, J. L.,** Regulation of the pro-opiomelanocortin mRNA levels in rat pituitary by dopaminergic compounds, *Proc. Natl. Acad. Sci. U.S.A.,* 80, 2211, 1983.
27. **Cote, T. E., Felder, R., Kebabian, J. W., Sekura, R. D., Reisine, T., and Affolter, H. U.,** D-2 Dopamine receptor-mediated inhibition of pro-opiomelanocortin synthesis in rat intermediate lobe, *J. Biol. Chem.,* 261, 4555, 1986.
28. **Beaulieu, M., Felder, R., and Kebabian, J. W.,** D-2 Dopaminergic agonists and adenosine $3',5'$-monophosphate directly regulate the synthesis of α-melanocyte-stimulating hormone-like peptides by cultured rat melanotroph, *Endocrinology,* 118, 1032, 1982.
29. **Walaas, S. I., Aswad, D. W., and Greengard, P.,** A dopamine- and cAMP-regulated phosphoprotein enriched in dopamine-innervated brain regions, *Nature (London),* 301, 69, 1983.
30. **Huganir, R. L. and Greengard, P.,** cAMP-dependent protein kinase phosphorylates the nicotinic acetylcholine receptor, *Proc. Natl. Acad. Sci. U.S.A.,* 80, 1130, 1983.
31. **Herbert, E., Comb, H., Thomas, G., Liston, D., Douglass, J., Thorne, B., and Martin, M.,** Tissue specific regulation of expression of opioid peptides, *DNA,* 5, 68, 1986.
32. **Uhler, H., Herbert, E., D'Eustrachio, P., and Ruddle, F. D.,** The mouse genome contains two nonallelic pro-opiomelanocortin genes, *J. Biol. Chem.,* 258, 9444, 1983.
33. **Whitfeld, P. L., Seeburg, P. H., and Shine, J.,** The human pro-opiomelanocortin gene: organization, sequence and interspersion with repetitive DNA, *DNA,* 1, 133, 1982.
34. **Nakanishi, S., Teranishi, Y., Watanabe, Y., Notake, M., Noda, M., Kakidani, H., Jingami, H., and Numa, S.,** Isolation of characterization of the bovine corticotropin/β-lipotropin precursor gene, *Eur. J. Biochem.,* 115, 429, 1981.
35. **Drouin, J., Chamberland, M., Charron, J., Jeannotte, L., and Nemer, M.,** Structure of the rat pro-opiomelanocortin (POMC) gene, *FEBS Lett.,* 193, 354, 1985.
36. **Civelli, O., Birnberg, N., and Herbert, E.,** Detection and quantitation of pro-opiomelanocortin mRNA in pituitary and brain tissues from different species, *J. Biol. Chem.,* 257, 6783, 1982.
37. **Herbert, E., Birnberg, N., Civelli, O., Lissitzky, J. C., Uhler, M., and Durrin, L.,** Regulation of genetic expression of pro-opiomelanocortin in pituitary and extra pituitary tissues of mouse and rat in *Regulatory Peptides: from Molecular Biology to Function,* Costa, E. and Trabucchi, M., Eds., New York, 1982, 9.
38. **Roberts, J. L., Budarf, M. L., Baxter, J. D., and Herbert, E.,** Selective reduction of pro ACTH/endorphin proteins and mRNA by glucocorticoids in pituitary tumor cells, *Biochemistry,* 18, 4907, 1979.
39. **Schachter, B. S., Johnson, L. K., Baxter, J. D., and Roberts, J. L.,** Differential regulation by glucocorticoids of pro-opiomelanocortin mRNA levels in the anterior and intermediate lobes of the rat pituitary, *Endocrinology,* 110, 1442, 1982.
40. **Vale, W., Spiess, J., Rivier, C., and Rivier, J.,** Characterization of a 41 residue ovine hypothalamic peptide that stimulates secretion of corticotropin and β-endorphin, *Science,* 213, 1394, 1981.
41. **Herbert, E., Allen, R. G., and Paquette, T. L.,** Reversal of dexamethasone inhibition of adrenocorticotropin release in a mouse pituitary tumor cell line either by growing cells in the absence of dexamethasone or by addition of hypothalamic extract, *Endocrinology,* 102, 218, 1978.
42. **Gagner, J. P. and Drouin, J.,** Opposite regulation of pro-opiomelanocortin gene transcription by glucocorticoids and CRH, *Molec. Cell. Endocrinology,* 40, 25, 1985.
43. **Rosa, P. A., Policastro, P., and Herbert, E.,** A cellular basis for the differences in regulation of synthesis and secretion of ACTH/endorphin peptides in anterior and intermediate lobes of the pituitary, *J. Exp. Biol.,* 89, 215, 1979.
44. **Howe, A. and Thody, A. J.,** The effect of ingestion of hypertonic saline on the melanocyte-stimulating hormone content and histology of the pars intermedia of the rat pituitary, *J. Endocrinology,* 46, 201, 1970.
45. **Dave, J. R., Rubinstein, N., and Eskay, R. L.,** Evidence that β-endorphin binds to specific receptors in rat peripheral tissues and stimulates the adenylate cyclase-adenosine $3',5'$-monophosphate system, *Endocrinology,* 117, 1389, 1985.
46. **Affolter, H. U. and Reisine, T.,** Corticotropin releasing factor increases pro-opiomelanocortin messenger RNA in mouse anterior pituitary, *J. Biol. Chem.,* 260, 15477, 1985.
47. **Martin, R. F., Jones, W. J., and Hayslett, J. P.,** Animal model to study the effect of adrenal hormones on epithelial function, *Kidney Intl.,* 24, 386, 1983.
48. **Sladek, C. D. and Knigge, K. M.,** Osmotic control of vasopressin release by rat hypothalamo-neurohypophyseal explants in organ culture, *Endocrinology,* 101, 1834, 1977.
49. **Schwyzer, R. and Eberle, A.,** On the molecular mechanisms of α-MSH receptor interactions, in *Frontiers of Hormone Research,* Van Wimersma Greidanus, Tj. B., Ed., S. Karger, Basel, Switzerland, 1977, 18.
50. **Eipper, B. A. and Mains, R. E.,** Structure and biosynthesis of pro-adrenocorticotropin/endorphin and related peptides, *Endocr. Rev.,* 1, 1, 1980.

51. **Loh, Y. P., Parish, D. C., and Tuteja, R.,** Purification and characterization of a paired basic residue-specific pro-opiomelanocortin converting enzyme from bovine pituitary intermediate lobe secretory vesicles, *J. Biol. Chem.*, 260, 7194, 1985.
52. **Loh, Y. P.,** Kinetic studies on the processing of human β-lipotropin by bovine pituitary intermediate lobe pro-opiomelanocortin-converting enzyme, *J. Biol. Chem.*, 261, 11949, 1986.
53. **Fricker, L. D. and Snyder, S. H.,** Enkephalin convertase: purification and characterization of a specific enkephalin-synthesizing carboxypeptidase localized to adrenal chromaffin granules, *Proc. Natl. Acad. Sci. U.S.A.*, 79, 3886, 1982.
54. **Hook, V. Y. H. and Loh, Y. P.,** Carboxypeptidase B-like converting enzyme activity in secretory granules of rat pituitary, *Proc. Natl. Acad. Sci. U.S.A.*, 81, 2777, 1984.
55. **Loh, Y. P., Brownstein, M. J., and Gainer, H.,** Proteolysis in neuropeptide processing and other neural functions, *Annu. Rev. Neurosci.*, 7, 189, 1984.
56. **Loh, Y. P.,** Peptide precursor processing enzymes within secretory vesicles, *Ann. N.Y. Acad. Sci.*, 493, 392, 1987.
57. **Zakarian, S. and Smyth, D.,** Distribution in active and inactive forms of endorphins in rat pituitary and brain, *Proc. Natl. Acad. Sci. U.S.A.*, 76, 5972, 1979.
58. **Goldman, M. E., Beaulieu, M., Kebabian, J. W., and Eskay, R. L.,** α-Melanocyte stimulating hormone-like peptides in the intermediate lobe of the rat pituitary gland: characterization of content and release *in vitro*, *Endocrinology*, 112, 435, 1983.
59. **Smyth, D. G.,** β-endorphin and related peptides, *Br. Med. Bull.*, 39, 25, 1983.
60. **Loh, Y. P., Eskay, R. L., and Brownstein, M.,** α-MSH-like peptides in rat brain — identification and changes in level during development, *Biochem. Biophys. Res. Comm.*, 94, 916, 1980.
61. **Glembotski, C. C.,** Acetylation of α-melanotropin and β-endorphin in the rat intermediate pituitary, *J. Biol. Chem.*, 257, 10493, 1982.
62. **Chappell, M. C., Loh, Y. P., and O'Donohue, T. L.,** Evidence for an opiomelanocortin acetyltransferase in the rat pituitary neurointermediate lobe, *Peptides*, 3, 405, 1982.
63. **Chappell, M. C., O'Donohue, T. L., Millington, W. R., and Kempner, E. S.,** The size of enzymes acetylating α-melanocyte-stimulating hormone and β-endorphin, *J. Biol. Chem.*, 261, 1088, 1986.
64. **Ham, J. and Smyth, D. G.,** Regulation of bioactive β-endorphin processing in rat pars intermedia, *FEBS Lett.*, 175, 407, 1984.
65. **Martens, G. J. M., Jenks, B. G., and van Overbeeke, A. P.,** N-α-acetylation is linked to α-MSH release from pars intermedia of the amphibian pituitary gland, *Nature (London)*, 294, 558, 1981.
66. **Goldman, M. E. and Loh, Y. P.,** Intracellular acetylation of desacetyl α-MSH in the *Xenopus laevis* neurointermediate lobe, *Peptides*, 5, 1129, 1985.
67. **Mains, R. E., Eipper, B. A., Glembotski, C. C., and Dores, R. M.,** Strategies for the synthesis of bioactive peptides, *Trends Neurosci.*, 6, 229, 1983.
68. **Murthy, A., Mains, R. E., and Eipper, B. A.,** Purification and characterization of peptidylglycine α-amidating monooxygenase from bovine neurointermediate pituitary, *J. Biol. Chem.*, 261, 1915, 1986.
69. **Bradbury, A. F., Finnie, M. D. A., and Smyth, D. G.,** Mechanism of C-terminal amide formation by pituitary enzymes, *Nature (London)*, 298, 686, 1982.
70. **Bradbury, A. F. and Smyth, D. G.,** Amidation of synthetic peptides by a pituitary enzyme: specificity and mechanism of the reaction, in *Peptides*, Walter de Gruyter, Berlin, 1983, 381.
71. **Glembotski, C. C.,** The α-amidation of α-melanocyte stimulating hormone in intermediate pituitary requires ascorbic acid, *J. Biol. Chem.*, 259, 13041, 1984.
72. **Loh, Y. P. and Jenks, B. G.,** Evidence for two different turnover pools of adrenocorticotropin, α-melanocyte-stimulating hormone, and endorphin-related peptides released by the frog pituitary neurointermediate lobe, *Endocrinology*, 109, 54, 1981.
73. **Maurer, R. A.,** Adenosine, 3',5'-monophosphate derivatives increase prolactin synthesis and prolactin messenger ribonucleic acid levels in ergocryptine-treated pituitary cells, *Endocrinology*, 110, 1957, 1982.
74. **Loh, Y. P.,** unpublished data.
75. **Loh, Y. P.,** unpublished data.

Chapter 8

PRO-OPIOMELANOCORTIN IN THE AMPHIBIAN PARS INTERMEDIA: A NEUROENDOCRINE MODEL SYSTEM

Bruce G. Jenks, B. M. L. (Lidy) Verburg-Van Kemenade, and Gerard J. M. Martens

TABLE OF CONTENTS

I. Introduction 104
    A. Background Adaptation in Amphibians: a Neuroendocrine Reflex 104
    B. The Pars Intermedia: a Model System for Studying POMC 105
    C. Aim and Scope of this Review 106

II. Biosynthesis and Processing of Amphibian POMC 106
    A. Biosynthesis of POMC during Background Adaptations 107
    B. Processing of Amphibian POMC 109
    C. Acetylation of Amphibian MSH 112

III. Secretion of POMC-Derived Peptides 115
    A. Characteristics of Spontaneous Release 115
    B. Multiple Factors Regulating Secretion from Amphibian Melanotrophs .. 116
    C. Coordinate vs. Noncoordinate Regulation of Peptide Release 117

IV. Concluding Remarks 119

Acknowledgments 119

References 120

I. INTRODUCTION

A major secretory product of the pars intermedia of the pituitary gland is the melanotropic peptide alpha-melanophore-stimulating hormone (α-MSH). The role of this hormone in the process of background adaptation in amphibians has, in the past, been extensively reviewed.[1-4] However, there has been a recent development which makes a reexamination of this subject of value, namely the discovery of the precursor for α-MSH, the multifunctional protein pro-opiomelanocortin (POMC). This introduction will first give a short account of the role of the pars intermedia during background adaptation in amphibians and then outline why this tissue, in view of its physiological function, can be considered a good model system for the study of POMC-producing cells.

A. Background Adaptation in Amphibians: a Neuroendocrine Reflex

Most amphibians possess the remarkable ability to alter the color of their skin by altering the distribution of pigment in pigment-containing cells (chromatophores) of their integument. The dermal melanophores are most directly involved in the rapid changes of pigment distribution associated with background adaptation. The melanophores contain the black pigment, melanin, within pigment-filled granules, the melanosomes. In fully white-background-adapted animals, these granules are aggregated around the nuclei of the cells (melanophore index, MI = 1, see Figure 1), while in animals adapted to a black background, the melanosomes are dispersed throughout the cells (MI = 5), and consequently the skin takes on a dark coloration (reviewed by Bagnara and Hadley[1]). Early experiments showed the involvement of the pituitary gland in the regulation of pigment distribution. Removal of this gland in amphibians leads to a permanent paling of the skin.[5,6] The source of the factor stimulating dispersion proved to be the pars intermedia[7,8] and hence, the hormone was called Intermedin, later to be replaced by the now more familiar name melanophore-stimulating hormone, MSH. The full significance of these early observations became clear through a series of classical experiments by Hogben and co-workers (reviewed by Waring[2]). These investigators showed that in the South African clawed toad, *Xenopus laevis*, a determining factor in the regulation of pigment dispersion in dermal melanophores is the fraction of overhead light which is reflected from the background, thus impinging on the upper retina (Figure 1). They proposed that this environmental cue ultimately regulates hormone secretion from the pituitary gland.

Background adaptation in amphibians is an example of a neuroendocrine reflex. Reception of the environmental signal (color of background) involves differential stimulation of dorsal and ventral parts of the retina which is likely followed by central processing of this neuronal input. Precisely how the retina discriminates the light sensations and which pathways are involved in the central processing of this information is largely unknown. That the neuronal information originating from the eyes ultimately reaches the pars intermedia, has been shown from extracellular recordings made within this gland.[9,10] In the coupling of the neuronal input to the secretion of MSH there is little doubt that the hypothalamus plays a significant role. Early experiments by Etkin[11] showed that the hypothalamus exerts primarily an inhibitory influence, and a number of studies have been devoted to the identification of hypothalamic MSH release-inhibiting factor(s). Many of these early studies concerned the possible involvement of an aminergic fiber system innervating the pars intermedia of amphibians[12-15] and, indeed, catecholamines, including adrenalin and dopamine, were found to be potent inhibitors of MSH secretion.[16-19] From these studies evolved the classical view of regulation of the intermediate lobe melanotroph cell, with central integration of neuronal input leading to an activation or inactivation of an inhibitory catecholaminergic system innervating the pars intermedia (Figure 1). Presumably, in an animal on a black background, integration of neuronal inputs from the eyes leads to an inhibition of the aminergic neurons

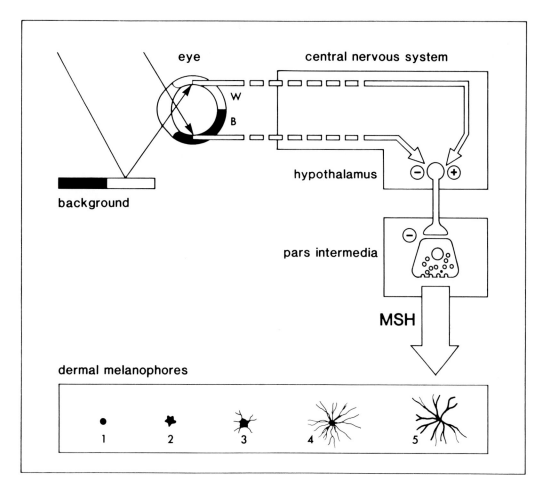

FIGURE 1. The neuroendocrine reflex regulating pigment dispersion in dermal melanophores of amphibians during the process of background adaptation. In this classic view, the neuronal input concerning color of background is centrally integrated, leading to an activation or inactivation of an inhibitory hypothalamic neuronal system innervating the melanotroph cells of the pars intermedia. Secretion of MSH from this gland leads to a dispersion of pigment within the melanophores and a consequent darkening of the animal (see text for discussion).

innervating the pars intermedia. Consequently, there is secretion of MSH from the intermediate lobe melanotrophs and thus expansion of pigment in dermal melanophores. It was already appreciated in the early 1970s however, that the regulation of secretion of MSH from the melanotroph was probably much more complex, and a number of potential regulatory neurotransmitters were indicated.[20-23] It is now evident, that the pars intermedia is regulated by multiple factors, both classical neurotransmitters and neuropeptides, which can either stimulate or inhibit MSH secretion. Attention will be given to some of these recent findings later in this review. This topic is more extensively considered by Tonon, M. C., et al., (Chapter 9, this Volume).

B. The Pars Intermedia: a Model System for Studying POMC

POMC was first characterized in a mouse pituitary tumor-cell line,[24,25] but it was soon shown to be produced by a number of different tissues including endocrine cells of the pituitary gland and within neuronal networks of the central nervous system (CNS). Tissue-specific processing of the precursor, which includes proteolytic cleavage and other post-translational events such as N-terminal acetylations, can generate a number of different peptide hormones or neuropeptides.[24-28] Among the potential products of processing are adrenocorticotropic hormone (ACTH), α-MSH, β-MSH, γ-MSH, β-lipotropic hormone (β-LPH), and the endogenous opiate, β-endorphin. Knowledge of the regulatory mechanisms involved in the generation and release of these diverse biological signals will contribute to an understanding of neuronal and neuroendocrine communication. The heterogeneous nature of neuronal tissue of the CNS makes this tissue impractical for many of the in vitro techniques required to study regulation at the cellular and molecular levels. Tissue heterogeneity has been less of a problem in studies with the ACTH-producing cells of the anterior pituitary gland, and progress is being made in defining mechanisms regulating these cells in response to stress.[29,30] In theory, the pars intermedia of the pituitary gland should be a good tissue to study the function of POMC cells, in that it constitutes an almost homogeneous population of POMC-producing cells. Also, processing of the precursor in the pars intermedia is very similar to that occurring in the CNS, and, therefore, it might act as a model for central POMC neuronal systems.[27] A major drawback of the mammalian pars intermedia is that its function is not well understood, and thus physiological manipulations to activate or inactivate these POMC cells are difficult. The melanotroph cells of the amphibian pars intermedia, however, have a well-defined neuroendocrine function, namely the production and secretion of MSH during adaptation of the animal to a black background. Ultrastructural studies have been particularly useful in defining the physiological functioning of these cells.[31-36] The melanotrophs of a fully white-background-adapted animal are extremely rich in secretory granules, but the biosynthetic apparatus of these cells is undeveloped. Such features are typical for inactive storage-cells. These same cells in black-background-adapted amphibians are characterized by a well-developed endoplasmic reticulum and a proliferated Golgi complex, indications of biosynthetically active cells. The biosynthetic and secretory activity of these cells can thus be manipulated simply by changing the color of the background, an attractive feature in studies concerning the mechanisms regulating these activities. Also, because the physiological function of the pars intermedia of amphibians is known, the physiological significance of experimental findings can be considered. In view of the multiple neuronal afferents involved in the regulation of this gland, an understanding of mechanisms utilized by the melanotroph in integrating the neuronal input may also prove useful for extrapolation to POMC-systems present in the CNS.

C. Aim and Scope of this Review

The physiological significance of a POMC cell will be determined largely by the nature of its secretory signal, considered both qualitatively (i.e., which of the potential POMC-derived peptides are present) and quantitatively (i.e., the amount of peptide(s) secreted). A number of factors can be involved in determining the ultimate secretory signal, including the rate of biosynthesis of the precursor, the direction of proteolytic processing of the precursor, possible post-translational modifications of the peptides, and finally, the rate of exocytosis. Regulation of the composition of the secretory signal would be expected to involve, at least in theory, regulation of these factors. This review will consider the biosynthesis, processing and release of POMC-derived peptides in the amphibian intermediate lobe melanotroph. It will address, where possible, questions concerning the regulation of these events, and the physiological significance of this regulation.

II. BIOSYNTHESIS AND PROCESSING OF AMPHIBIAN POMC

The presence of a precursor to α-MSH in amphibians was first demonstrated in the species *Xenopus laevis*.[37] Subsequently, immunological evidence was presented suggesting the biosynthesis of two forms of the precursor in the pars intermedia of this species.[38] This suggestion was confirmed by the isolation of two prohormones and, from the tryptic mapping characteristics of these proteins, it was concluded that *Xenopus* produces two sequentially different forms of POMC.[39] This same study revealed that both forms of the prohormone are glycosylated and that this glycosylation concerns a single site within each prohormone, namely within their γ-MSH region. The POMC protein of a second amphibian species, *Rana ridibunda*[40] has now been partially characterized. This species synthesizes only one form of POMC, which also proved to be glycosylated at a single site, situated in the γ-MSH region.

A. Biosynthesis of POMC During Background Adaptations

As outlined in the introduction, there are numerous ultrastructural studies indicating that the melanotrophs of black-background-adapted amphibians are biosynthetically extremely active cells, while in white-background-adapted animals these cells have the appearance of less active storage cells. Concerning the events occurring within the melanotrophs during background adaptation (i.e., as discerned from background-transfer experiments), our discussion must, of necessity, be restricted largely to the species *Xenopus laevis*. It is only with this species that detailed electron microscopical and biosynthetic studies have been conducted to follow changes occurring in the pars intermedia during physiological adaptations. For this species a number of biosynthetic studies have shown that the in vitro incorporation of radioactive amino acids by neurointermediate lobes obtained from black-background-adapted animals, is up to 10-fold higher than that displayed by this same tissue obtained from white-background-adapted animals.[37,41-43] Analysis of the newly synthesized material by either electrophoretic methods[37] or HPLC,[39,44,45] has shown that the high biosynthetic activity of tissue from black-background-adapted animals concerns biosynthesis of POMC and POMC-derived peptides, such as the melanotropins (see the following section concerning processing of POMC). The conclusion drawn from these in vitro studies, namely that the intermediate lobe melanotrophs of black-background-adapted animals have a high rate of POMC biosynthesis, while those of white-background-adapted animals have a low level of prohormone synthesis, has been confirmed in in vivo experiments where labeled amino acids were administered to living animals.[46]

After transferring fully black-background-adapted *Xenopus laevis* to a white background, the biosynthetic activity of the intermediate lobe remains at a high level for several days following transfer.[42,43,47] Thus, the decline in POMC biosynthesis occurs only very slowly during adaptation to a white background. During this adaptation, the melanophore index of dermal melanophores drops significantly within the first hour on white background and, with respect to this index, the animal is fully adapted to its new background within one day.[42,43,47,48] It has been shown that there is a high correlation between plasma concentration of α-MSH and melanophore index readings in *Xenopus*.[49] Therefore, in sharp contrast to the slow inhibition of POMC biosynthesis, there is apparently a very rapid inhibition of secretion of POMC-derived peptides from the intermediate lobe during white-background adaptation. Presumably, the sustained biosynthetic activity during the first few days on a white background replenishes stores of hormone which become depleted on a black background. This assumption is supported by ultrastructural data,[31] showing a build-up of mature secretory granules in the melanotrophs during white-background adaptation. Also, in vivo labeling experiments have shown that radiolabeled POMC-derived peptides accumulate in the intermediate lobe when the animal is transferred to a white background.[46]

An intriguing question for endocrinologists concerns the relationship between biosynthesis of a hormone and its release. Some hormones, notably the steroid hormones, are synthesized on demand. Most peptide-producing endocrine cells, however, display some capacity to store their secretory products. It is clear that the amphibian melanotroph possesses a mechanism to ensure an adequate storage of hormone before the biosynthetic apparatus is dismantled (i.e., through protracted biosynthetic activity following cessation of hormone demand). Treatment of amphibian neurointermediate lobes with dopamine, which is very likely a

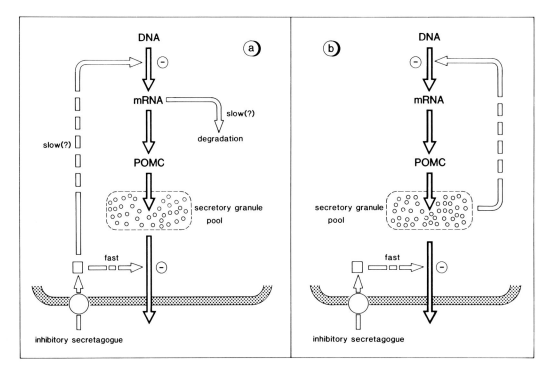

FIGURE 2. Possible mechanisms for the inhibition of POMC biosynthesis in the intermediate lobe of amphibians. (a) Mechanism whereby there is a direct effect of inhibitory secretagogue on both synthesis and secretion. The inhibitory secretagogue activates or inhibits an intracellular 2nd messenger system (circle and square) which is directly involved in regulation of both peptide release and POMC-biosynthesis. The observation that the storage pool of secretory peptides is replenished before synthesis terminates could be accounted for by either a slow inhibition of transcription or a slow degradation of POMC mRNA. (b) Mechanism whereby the inhibitory secretagogue has a direct and rapid effect on secretion, but only indirectly affects POMC biosynthesis. In this mechanism, an internal feedback system involving the size of the storage pool of POMC-derived peptides regulates biosynthesis.

physiologically important MSH-release inhibiting factor, leads to a very rapid inhibition of MSH secretion.[50-53] In superfusion experiments, complete inhibition of the secretory process can be achieved within 10 to 15 min.[51,53] In addition, there is evidence that this neurotransmitter may have a very slow effect on biosynthetic processes within these cells. While a 6 hr treatment of the neurointermediate lobe of the frog *Rana ridibunda* with dopamine had no effect on the biosynthesis of POMC and POMC-related peptides,[52] a more extended treatment of *Xenopus* tissue with dopamine (1 day) was shown to lead to an inhibition of prohormone biosynthesis.[47] While it is too early to comment on the precise nature of the relationship between the regulation of biosynthesis and release in the amphibian melanotroph, several possibilities are worth considering (Figure 2). There could be a direct relationship between regulation of synthesis and secretion, the only difference being the speed with which the intracellular mechanisms activated (or inactivated) by the secretagogue can lead to inhibition of the two processes (Figure 2a). An alternative is a more indirect action of the secretagogue on synthesis, such as an internal feedback mechanism emanating from the accumulating secretory material (or storage granules). In this case, inhibition of synthesis would only be a consequence of the primary action of the secretagogue, namely the inhibition of release (Figure 2b). Measurements of the rate of POMC gene transcription and of the half-life of POMC mRNA, will be required to test such models. With the introduction of recombinant DNA techniques probes are now available to quantify levels of mRNA, and these probes can also be utilized to measure the turn-over of the message, such as in in vitro nuclear transcription run-on assays.[54] It is worth mentioning at this point that the results of studies with mammalian intermediate lobes also indicate that there is a relationship between regulation of synthesis and secretion. In the rat, dopamine is also an extremely rapid and potent MSH-release inhibiting factor,[55,56] and it has been shown that treatment with dopamine receptor agonists leads to a decrease in POMC mRNA levels in the pars intermedia melanotrophs[57] and a decrease in the level of POMC biosynthesis.[58,59] Relatively long treatment with the receptor agonists was required to demonstrate these latter effects (1 to 4 days), which again raises questions concerning the precise nature of the relationship between regulation of synthesis and secretion.

After transferring *Xenopus* from a white to a black background, the response of the dermal melanophores has been shown to be extremely rapid, reaching the fully black-background-adapted state of dispersion (index 4 to 5) within a few hours.[2,42,43,48] Clearly, activation of the secretory process for release of MSH from the intermediate lobe is a rapid event. The capacity of this tissue for in vitro incorporation of radioactive amino acids shows a significant increase within 8 to 16 hr after transferring the animal to a black background.[42,43] There is also ultrastructural evidence indicating a very rapid activation of the biosynthetic process within the melanotrophs.[31] It should be stressed, however, that despite this apparently rapid activation of the biosynthetic process after transferring the animal to a black background, many days of black-background adaptation are required for the pars intermedia to acquire its full biosynthetic capacity.[42,43,47] Therefore, it seems reasonable to conclude that the animal relies on its stores of MSH for both the initial adaptation to a black background, and for the maintenance of pigment dispersion during the period that the biosynthetic capacity of the melanotrophs is being developed. This conclusion finds support in electron microscopical studies,[31,32] showing a rapid reduction in the number of storage granules within the intermediate lobe melanotrophs during adaptation of the animal to a black background. In keeping with these findings, radioimmunoassays and bioassays show that in the tissue there is a decrease in the levels of MSH during adaptation to a black background.[42,48,60]

B. Processing of Amphibian POMC

High-Performance Liquid Chromatography (HPLC) has proven to be a most appropriate method for the separation and structural characterization of the amphibian POMC-derived peptides. By combining HPLC with a number of other techniques including immunoprecipitations, selective amino acid incorporations, bioassays, radioimmunoassays, and tryptic and chymotryptic mapping techniques, many of the peptides synthesized by *Xenopus* neurointermediate lobes, have been characterized, and the pathways for their biosynthesis elucidated by the use of pulse-chase experiments.[44-46] Figure 3 summarizes the processing of POMC in the intermediate lobe melanotrophs of *Xenopus laevis*. Biosynthesis begins with production of the two prohormones, which are synthesized in approximately equal amounts. HPLC does not separate these two prohormones and SDS-gel electrophoresis was required for their complete resolution and isolation.[39] Following biosynthesis of the two prohormones, they are rapidly cleaved. In line with the presence of two sequentially different prohormones, HPLC resolved two endorphins and two CLIP peptides among the products appearing during chase incubations.[45] The rapid appearance of these peptides, and the 17K/18K-products, during the chase-period indicates that larger cleavage products of processing, such as β-LPH and the N-terminal ACTH-containing fragments of the prohormone molecules (represented by dotted lines in Figure 3), may only be transitory intermediates in the processing of the prohormones. Also identified among the HPLC-resolved peptides was des-N α-acetyl-α-MSH (desAc-α-MSH), while almost no acetylated form of this peptide was found in the tissue.[45,61] The fact that the acetylated peptide was found to be a major product in the incubation medium, however, leads to the proposal that N-terminal acetylation of MSH in melanotrophs of *Xenopus laevis* is associated with the secretory process.[61] This topic receives

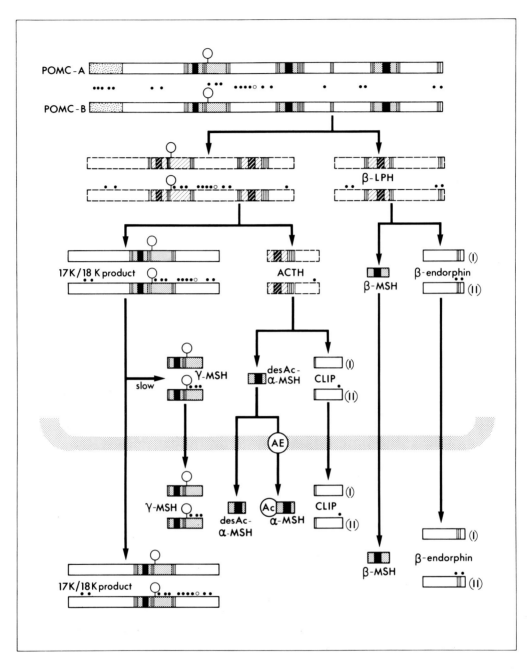

FIGURE 3. Schematic representation of the processing of POMC by intermediate lobe melanotrophs of the amphibian *Xenopus laevis*. The structures of the two prohormones are deduced from nucleotide sequences of POMC cDNA clones. Locations of amino acid differences between these structures are indicated by filled dots and the single amino acid deletion in POMC-A by an open dot. Positions of Asn-linked glycosylations are shown by open circles. Sites where two or more basic amino acids occur (potential proteolytic cleavage sites) are indicated by open vertical bars. Products indicated in dotted lines have not been detected in *Xenopus* intermediate lobes and may represent short-lived intermediates in the processing event. The pattern of processing is based on pulse-chase analyses combined with structural characterization of the products formed. The acetylation enzyme for the conversion of desAc-α-MSH to α-MSH is depicted to be associated with the cell membrane for convenience only. The cellular location of this enzyme in amphibian melanotrophs is unknown (see text for discussion).

more detailed attention in the next section. The basic pattern of processing, elucidated from the in vitro pulse-chase experiments (Figure 3), has been confirmed by in vivo labeling experiments.[46] In general, pairs of basic amino acids have been found to be potential recognition sites for the proteolytic enzymes involved in precursor processing.[62,63] The terminal products of POMC processing identified in biosynthetic studies are consistent with the locations of the dibasic cleavage sites within the prohormones (Figure 3), as deduced from the structure of *Xenopus* POMC cDNA.[64] In view of the presence of cleavage sites flanking the β-MSH moiety within the precursors, it was surprising that β-MSH was not identified in the HPLC analysis of the newly synthesized peptides. To examine this further, experiments have now been conducted which have utilized the structural information obtained from the sequence of *Xenopus* POMC cDNA, and have involved selective amino acid incorporation studies combined with tryptic and chymotryptic mapping of radiolabeled peptides. These studies have shown that β-MSH is indeed a biosynthetic product of the intermediate lobe tissue, that it is rapidly synthesized, and that in the HPLC analysis, this peptide coelutes with γ-MSH (Verburg-van Kemenade, B.M.L. and Jenks, B.[131] unpublished observations). The in vitro pulse-chase analysis shows that following their biosynthesis the various newly synthesized radioactive peptides are rapidly and spontaneously released to the incubation medium. Discussion of this event is reserved for later in this review.

Autoradiographic analysis at the ultrastructural level indicates that during pulse-chase experiments with neurointermediate lobes of *Xenopus*, the newly synthesized material follows the expected intracellular pathways for secretory peptides, namely synthesis in the rough endoplasmic reticulum, transfer to the Golgi apparatus, and finally packaging into secretory granules.[65,66] A comparison of the time-course of these intracellular events with chromatographic data concerning the time-course for processing of the prohormones, indicates that cleavage of the prohormones to form the POMC-derived peptides probably takes place in the secretory granules.[66] In studies with mammalian tissue, partially purified enzyme preparation have indeed been isolated from secretory granule fractions of melanotrophs, and these enzymes shown to be capable of cleaving POMC.[67,68]

Analysis of biosynthesis in the melanotrophs of another amphibian species, *Rana ridibunda*, shows that processing of POMC proceeds in a manner similar to that described for *Xenopus*.[69] The expected high molecular weight intermediates of processing, such as β-LPH and the large N-terminal ACTH-containing fragment of POMC, could not be detected.[40,69] Therefore, it was proposed that once the processing enzymes or enzyme complex begins to cleave an individual prohormone molecule, this cleavage proceeds rapidly to yield desAc-α-MSH, CLIP, endorphin, and 18K-fragment. Acetylated forms of MSH were found only in the incubation medium,[69,70] and it was again concluded that the acetylation reaction is associated with the secretory process (see next section). The N-terminal 18K-fragment of the prohormone, containing the glycosylated γ-MSH moiety, proved to be a long-lived intermediate in POMC processing. A large amount of intact 18K-fragment was found to be secreted into the incubation medium during pulse-chase experiments, and it was therefore concluded that at least some of this product could be considered a terminal product of processing of POMC in amphibian intermediate lobe melanotrophs.

The processing of POMC in intermediate lobes of amphibians is essentially the same as that described for the mammalian pars intermedia,[24-26,71] with the exception of the post-translational acetylation reaction to produce α-MSH and, possibly, acetylated forms of endorphins (see next section). As mentioned in the introduction, POMC is produced by a number of different tissues (e.g., intermediate lobe and anterior lobe of the pituitary gland, neuronal networks of the CNS), and within these tissues processing of the precursor protein is apparently tissue specific. Concerning the regulation of proteolysis of the precursor to form secretory peptides, studies with the intermediate lobe of *Xenopus laevis* have led to the concept that the carbohydrate group on the prohormone could play an important role in determining the direction of proteolytic processing.[72,73] Similar studies, conducted with a corticotropin-producing mouse tumor-cell line,[74,75] with mouse intermediate lobe tissue,[76] and with intermediate lobes of another amphibian species, *Rana ridibunda*,[40] have failed to support this concept. Therefore, it appears unlikely that glycosylation of POMC is of general importance in regulating the direction of proteolysis. It should be mentioned, however, that in most species the γ-MSH moiety of the prohormone is glycosylated and, as discussed above, the N-terminal glycoslyated fragment of the prohormone is only slowly processed in amphibian intermediate lobe tissue to give rise to γ-MSH. Possibly, the carbohydrate on this fragment hinders access of proteolytic enzymes to the dibasic cleavage sites. This could have important consequences concerning the peptide content of the secretory signal of the melanotroph, to be discussed later.

The differences in POMC processing among different types of tissues (i.e., production of α-MSH in intermediate lobe melanotrophs and ACTH in anterior lobe corticotrophs) will undoubtedly reflect, at least in part, the presence of different enzymes or enzyme-systems within the POMC-cells of these tissues. Presumbaly, events during embryonic cell differentiation lead to differences in the expression of genes coding for the processing enzymes. The important question remains whether regulatory mechanisms exist within an individual fully differentiated POMC cell, such that the direction of processing or other post-translational events can be altered, thus giving rise to different peptide profiles in the secretory signal. This possibility would greatly enhance the versatility of POMC-synthesizing cells in regulatory processes. In mammalian studies, this question has received considerable attention, particularly the possibility that the N-terminal acetylation of MSH or the endorphins could be an important regulatory point in determining the biological potency of secretory signals from POMC-cells.[27,28,77-83] As outlined in the next section, studies with amphibian intermediate lobe tissue may make an important contribution in furthering this concept.

C. Acetylation of Amphibian MSH

The discussion of acetylation reactions in amphibian melanotrophs will involve a comparison with the situation in mammals. Therefore, a brief summary will first be given of the major findings from mammalian studies. The reader is referred to the Chapter by Dores (Chapter 3, this volume) for more details concerning these mammalian studies.

Biosynthetic studies with mammalian intermediate lobe melanotrophs have shown that both desAc-α-MSH and β-endorphin are rapidly acetylated following their biosynthesis.[84-86]

For MSH, this acetylation can involve potential acetylation sites on its N-terminal serine, and, indeed, in addition to α-MSH also N,O,-diacetyl-α-MSH has been shown to be a biosynthetic product in the intermediate lobe of a number of mammals.[87-89] For β-endorphin, it has been shown that following the acetylation of its N-terminal tyrosine, this peptide undergoes a very slow proteolysis of its C-terminal region in mammalian melanotrophs to give rise to acetylated peptides of 27 amino acids or shorter in place of the 31 amino acid parent molecule.[80,81,90-93] The acetylation reaction is an example of a tissue-specific process (both MSH and endorphin can be found in their nonacetylated forms in other POMC-tissues), and it is believed that the same enzyme is responsible for acetylation of both peptides in the intermediate lobe.[28,94-97] These N-terminal acetylations are clearly of physiological importance in that they affect the biological activity of the peptides concerned; α-MSH is approximately 10 times more potent in stimulating dermal melanophores than the nonacetylated form of the peptide,[98-100] while the acetylation of β-endorphin completely eliminates the opiate activity of this peptide.[80,101-103] In view of the above, it is important to consider whether these peptides are secreted in their acetylated or nonacetylated forms.

There is very little known concerning post-translational modification of endorphins in amphibian intermediate lobe tissue. Biosynthetic studies have offered no evidence for an N-terminal acetylation of these peptides,[45,69] although a thorough analysis of their acetylation status has not been conducted. It should be mentioned that despite the view that the same enzyme is responsible for the acetylation of both MSH and endorphin, these acetylations are not necessarily concomitant events. The endorphin of the pars intermedia of the reptile, *Anolis carolinensis,* is reported to be nonacetylated, while the MSH of this species is in the acetylated form.[104] In the intermediate lobe of the fetal mouse, β-endorphin undergoes an N-terminal acetylation, while the acetylation of MSH begins only following birth, and it takes a number of days before the level of acetylation has reached the level found in adult lobes.[88]

The acetylation of MSH in the amphibian intermediate lobe has received considerable attention, which is not surprising in view of the physiological function of this peptide in amphibians. In vitro pulse-chase studies with neurointermediate lobes of both the toad, *Xenopus laevis,*[45] and the frog, *Rana ridibunda,*[69] have revealed that following its biosynthesis, the newly synthesized desAc-α-MSH remains in the tissue almost entirely in the nonacetylated form. This is in sharp contrast to the situation in rat and mouse intermediate lobes where acetylation of the newly synthesized desAc-α-MSH has gone essentially to completion within 8 hr of chase-incubation.[76,84,88] In the frog, there was still no detectable radioactive α-MSH associated with the tissue following 16 hr of chase-incubation.[69] In *Xenopus* intermediate lobes, desAc-α-MSH always remains the major tissue-form of the peptide, accounting for over 70% of newly synthesized MSH.[45,61,105,106] Analysis of incubation medium from pulse-chase experiments with amphibian tissue, however, shows that acetylated forms of MSH make a major contribution to the profile of radioactive peptides secreted into the medium. For *Xenopus,* only one form of acetylated peptide was found, which is not surprising in view of the fact that N-terminal serine normally found in α-MSH is substituted by an alanine in *Xenopus* MSH.[64,107] Alanine has only one potential acetylation site, namely its free amino-group. The acetylated form of *Xenopus* MSH can account for 50 to 100% of the newly synthesized MSH in the medium.[53,61,105,108] The acetylated peptides found in the chase-incubation media from experiments with intermediate lobes of *Rana ridibunda,* were identified as the mono-acetylated and di-acetylated forms of MSH.[52] These acetylated peptides account for approximately 50% of the radioactive α-MSH-related peptides in the medium.[52,69,70] The observation that the acetylated forms of MSH are always major products in the medium of incubated intermediate lobes, while they are minor products or are not detected at all in the tissue, leads to the conclusion that, at least in these two species of amphibians, the acetylation of MSH is associated with the secretory process. Both acetylated and nonacetylated forms of MSH are found in incubation media, indicating that both forms of the peptide can be considered potential secretory products of the intermediate lobe.

To measure the steady-state levels of desAc-α-MSH and α-MSH, studies have been conducted where HPLC analysis is combined with radioimmunoassays. The findings support the conclusions drawn from the biosynthetic studies with amphibian intermediate lobes. For *Xenopus,* the major immunoactive form of MSH associated with the tissue is desAc-α-MSH (accounting for approximately 75% of the immunoactivity) while in the medium of incubated lobes the contribution of α-MSH to the immunoactivity is over 50% of the total.[105] In studies with *Rana ridibunda,* only immunoactive desAc-α-MSH is detected in intermediate lobe tissue extracts while the acetylated forms of this peptide accounts for about one-half of the MSH immunoactivity released into the incubation media.[70] In sharp contrast to these results, analysis of neurointermediate lobes of mammals show that almost all immunoreactive MSH in the melanotrophs are in acetylated forms,[87-89] a finding in keeping with the rapid intracellular acetylation occurring in this tissue.

In mammals the enzyme for the acetylation of MSH and β-endorphin has been shown to be associated with the intracellular secretory granules.[28,67,95,109,110] Nothing is known concerning the cellular location of the acetylating enzyme in amphibians, and it is only for convenience that this enzyme is shown to be associated with the plasma membrane in models such as that depicted in Figure 3. From a mechanistic point of view, one could speculate that association of the enzyme with secretory granules, which probably have a low internal pH,[68,110] might be more appropriate. The enzyme could be inactive or display only low activity at low pH, and become highly active during secretion because of the electrochemical changes that occur within secretory granules during the process of exocytosis. Of possible relevance to this discussion, however, is a recent observation that cell suspensions of *Xenopus* intermediate lobe melanotrophs completely lack the ability to acetylate MSH.[105] This is the only biosynthetic abnormality associated with the isolated cells, when compared to intact tissue. These results might suggest that in the preparation of cell suspensions membrane structures essential for acetylation are disrupted, or alternatively, that the extracellular microenvironment experienced by melanotrophs in intact tissue is important for the acetylation process. Interestingly, in the rat and mouse, where acetylation is strictly an intracellular event, cell-culture techniques with intermediate lobe melanotrophs are routinely used,[84,85,90,111] and there has never been a report of an impairment of the acetylation system with isolated cells. This again illustrates that the acetylation process in the melanotrophs of amphibians is very different from that occurring in mammalian melanotrophs.

At this point, it should be mentioned that there is a different view concerning the process of acetylation of MSH in *Xenopus* melanotrophs. Goldman and Loh,[112] with data which do not differ substantially from those presented above, conclude that the major route for acetylation of MSH in *Xenopus* is an intracellular route analogous to that reported for mammals. They too, find that desAc-α-MSH is the major immunoactive form of MSH in the tissue and, to account for the enhanced level of α-MSH to the medium, they evoke the concept of independent ''pools'' of secretory peptides within *Xenopus* melanotrophs. In this theory a preferential release of peptides from a more mature secretory ''pool'', which would contain mostly the acetylated form of MSH, would account for the enhanced level of α-MSH found in the incubation medium. In our view, the association of some acetylated MSH with the intermediate lobe tissue in *Xenopus* might indicate the presence of some acetylated peptide in the extracellular space, and/or reflect a low basal activity of the acetylation enzyme, although a role for independent secretory ''pools'' can not be completely ruled out. One point worth mentioning is that acceptance of the above ''pool'' theory would also necessitate also accepting that *Xenopus* and *Rana* have developed different strategies for the regulation of the acetylation of their MSH. *Rana* has no acetylated MSH in its melanotrophs[40,52,69,70] and thus it would presumably lack this mature ''pool'' of secretory peptides to draw upon for release of acetylated MSH. The results of Goldman and Loh indicate that in vitro inhibition of secretion with dopamine leads to an accumulation of acetylated peptide in the tissue. While a rapid accumulation of acetylated peptide in release-inhibited lobes would certainly argue for a major role for intracellular acetylation, the only data presently available on this question indicate that this is not the case. An in vivo labeling experiment conducted with *Xenopus,* shows that after 10 days on a white background (where secretion is under tonic inhibition), the major form of MSH in the intermediate lobe is still desAc-α-MSH.[46] These results indicate that intracellular acetylation within the storage granules of *Xenopus* melanotrophs is at best an extremely slow process, which can hardly be considered a major route for acetylation, and in no way can be considered analogous to the rapid acetylation reaction observed in mammalian melanotrophs.

There is evidence from biosynthetic studies with intermediate lobes of *Rana ridibunda,* that the acetylation of MSH may be a regulated event. It was found that treating the lobes with the neurotransmitter dopamine leads not only to an inhibition of secretion, but also to a dose-dependent inhibition of the secretion-associated acetylation of MSH.[52] At very high levels of dopamine the secretory signal was virtually devoid of α-MSH but desAc-α-MSH was still detected in the incubation medium. If this same phenomenon occurs in vivo, then it would imply that the secretory signal of the pars intermedia of a black-background-adapted animal is rich in the highly melanotropic acetylated form of MSH, while the secretory signal of a white-background-adapted animal, the pars intermedia of which is under dopaminergic inhibition, would be the less melanotropic nonacetylated peptide. Therefore, this may be the first example of a POMC cell altering the profile of its secretory output in response to different physiological circumstances.

III. SECRETION OF POMC-DERIVED PEPTIDES

A. Characteristics of Spontaneous Release

During pulse-chase experiments with amphibian neurointermediate lobes, the newly synthesized POMC-derived peptides are spontaneously released.[37,45,69,113,114] One characteristic of this release is that there is usually a lag-perod of 1 to 1.5 hours following the pulse-incubation before radioactive peptides begin to appear in chase-incubation media. This lag-period corresponds to a close approximation to the time required for radioactive secretory granules to become associated with the cell membrane[66] and probably reflects the time constraints for the intracellular biosynthesis, transport, and processing of POMC in preparation for secretion. The spontaneous appearance of the newly synthesized peptides in the media is in agreement with the concept that hypothalamic regulation of the pars intermedia is primarily inhibitory.[11-19] That it reflects a true secretory process, and is not simply a leakage of material from the cells, is demonstrated by the fact that this release can be inhibited by dopamine[37,113,114] and other MSH-release-inhibiting factors.[108]

Chromatographic analysis of the newly synthesized peptides released to the incubation medium during chase incubations allows for a complete survey of the secretory signal, as opposed to immunoassays with which secretion of only selected peptides is measured. There is one inherent drawback in such an analysis, namely that secretion of only the newly synthesized population of POMC-derived peptides is measured, the behavior of which may or may not reflect steady-state conditions. This may not be a serious problem, since the melanotropes of black-background-adapted amphibians are in a partially depleted state with respect to stores of hormones, as judged from both electron-microscopical analysis[31,32,66] and assays for MSH immuno- and bioactivity.[42,48,60] These cells would presumably rely very heavily on newly synthesized peptides for their secretory signal, and thus analysis of these peptides would, to a close approximation, reflect overall events associated with secretion. Probably reflecting this reliance on newly synthesized peptides, the release of these peptides from amphibian intermediate lobes is very rapid. In a typical pulse-chase experiment with *Xenopus* intermediate lobes, almost all radioactive POMC-derived peptides are released within 5 hr of terminating the pulse incubation.[45,114]

Analysis of the newly synthesized peptides released from intermediate lobes of both *Xenopus*[37,45,114] and *Rana*[40,52,70] show that all peptides, including even a small amount of POMC, are released (i.e., concomitant release). The apparent secretion of a precursor could result from some prohormone becoming associated with a constitutive pathway, described for secretory cells of mammals.[115] Alternatively, it could be a characteristic of in vitro incubated lobes, the cells of which are completely devoid of hypothalamic regulation and may thus release peptides at a rate higher than that occurring in the living animal (i.e., there is insufficient time for processing of all the precursor). A quantitative analysis of the rate of release of individual POMC-derived peptides has proven to be very difficult because of the dynamic nature of the event (such an analysis would have to consider both the rate of biosynthesis and rate of release of each peptide). The only noteworthy disjunction to the apparent concomitant nature of the secretory process is the relatively low level of desAc-α-MSH in the medium, a phenomenon which reflects the secretion-associated acetylation of this peptide discussed earlier. Altogether, analysis of the spontaneous release of newly synthesized peptides gives the impression that these POMC-derived peptides are a concomitantly released group of peptides.

The release of immunoreactive α-MSH from superfused neurointermediate lobes of *Rana ridibunda*[116-119] and *Xenopus laevis*,[53,106,108,120,121] has been extensively analyzed. The radioimmunoassay used in these studies is directed towards the C-terminal region of the peptide,[122] and thus both the nonacetylated and acetylated forms of the peptide are measured. Analysis of secretion from neurointermediate lobes of black-background-adapted *Xenopus* shows that spontaneous release of MSH is usually maintained at a relatively constant rate over many hours of superfusion. In contrast, lobes of white-background-adapted animals have difficulty in maintaining in vitro secretion of immunoactive α-MSH. Secretion from these latter lobes often begins at very high levels, but falls during the course of superfusion giving rise to declining base-lines in spontaneous secretion.[106] Analysis of these lobes following superfusion shows that many of them have depleted their stores of immunoactive MSH. The difference in the behavior of lobes from white- and black-background-adapted animals in in vitro superfusion very likely reflects the difference in the biosynthetic capacity of their melanotrophs. The melanotroph of the white-background-adapted animal, in contrast to that of a black-background-adapted animal, is biosynthetically inactive, and is thus incapable of maintaining the high rate of in vitro secretion.

In addition to the α-MSH related peptides, endorphins are the only other POMC-derived peptides of amphibians analyzed using radioimmunoassays.[108,120,123] All these studies involved the use of an assay for mammalian β-endorphin, and in at least two of them,[108,120] it was found that *Xenopus* intermediate-lobe extracts did not give parallel dilution curves with the standard curve of the assay. Therefore, the results of these latter studies can not be considered quantitative. In analyzing the immunoactive endorphin(s) released during these superfusion experiments, however, all superfusion fractions were measured at the same dilution and they thus give a good evaluation of the relative rate of secretion of endorphins. The immunoactive endorphin was found to be spontaneously released from the lobes and, with the limited data available, it can be concluded that the rate of this basal secretion closely parallels that of immunoactive MSH, again indicating a concomitant release of POMC-derived peptides. In view of the sequence differences between many of the amphibian POMC-peptides and their mammalian counterpart,[64,107] a careful selection of radioimmunoassay systems will have to be made in applying mammalian assays to quantify amphibian peptides.

B. Multiple Factors Regulating Secretion from Amphibian Melanotrophs

Experiments involving superfusion techniques have been used with neurointermediate lobes of *Rana ridibunda* and *Xenopus laevis* to expand on earlier studies concerning the nature of hypothalamic regulation of MSH secretion. Analysis of MSH secretion during superfusion has a number of advantages over methods involving static incubations of neurointermediate lobes. Superfusion allows for an analysis of the time-course of MSH release from individual lobes and thus the basal rate of secretion can be established. Discrete pulses of potential regulatory factors can be introduced during the superfusion and their effect on basal secretion determined. Thus, both control and experimental data are collected from the same lobe, an obvious advantage over static incubation methods which usually involve comparisons being made between control and secretagogue-treated tissue.[50,112,123] Recent investigations have often included immunocytochemical analysis to establish the morphological basis for regulation by the potential secretagogue. This topic is extensively reviewed by Tonon et al. (Chapter 9, this Volume), and only those facts most pertinent to our further discussions concerning the regulation of the peptide content of the secretory signal will be given here.

The pars intermedia of amphibians is clearly regulated by multiple factors, both classical neurotransmitters and neuropeptides. These factors reach the pars intermedia through either direct innervation of the gland, or in some cases through diffusion from nerves terminating in the pars nervosa. While some species differences have been found in both the nature of the innervation of the neurointermediate lobes and in the effect of secretagogues on MSH release, it can be said that both *Rana* and *Xenopus* have similar strategies in the regulation of MSH secretion. Both species possess catecholamine receptors, and dopamine inhibits secretion from both species.[51,53] The intermediate lobe of both species is innervated by a GABAergic system and these tissues possess both $GABA_A$ and $GABA_B$ receptors.[105,108,118,124,125] In *Xenopus*, GABA can induce inhibition through either receptor type.[105,125] For *Rana*, however, while $GABA_B$ receptors are involved with inhibition of secretion, the $GABA_A$ receptor mechanism gives stimulation of MSH secretion.[118] The tripeptide thyrotropin-releasing hormone (TRH) stimulates secretion[117,121] and the more recently discovered neuropeptide-Y (NPY) inhibits the secretory process of both species.[119,120] The observation that many of the regulatory factors act directly on isolated melanotrophs of *Xenopus*,[105] (cell suspensions) indicates that these cells constitute a major center of neuroendocrine integration. Recent studies, moreover, indicate that more than one intracellular second messenger system is involved in the integration of the diverse neuronal signals. Cyclic-AMP is known to stimulate secretion of MSH from *Xenopus* melanotrophs,[50,106] and it has now been established that the $GABA_B$ receptor mechanism inhibiting MSH secretion does so through an inhibition of the adenylate cyclase system.[124] $GABA_A$ receptor agonists, however, have no inhibitory effect on cyclic-AMP production by the melanotrophs,[124] and thus the inhibitory action of these substances cannot be attributed to a direct or indirect action on the adenylate cyclase system.

C. Coordinate vs. Noncoordinate Regulation of Peptide Release

The significance of the complex neuronal systems involved in the regulation of secretion from amphibian melanotrophs, is not understood. These cells have, in addition, complex biosynthetic pathways which have, through the multifunctional precursor POMC, the potential to generate diverse biological signals. Therefore, one wonders whether at least some of the regulation may concern the qualitative nature of the secretory signal. The secretagogue could, in theory at least, alter the secretory profile by affecting either the direction of processing of the precursor or the selection of peptides for secretion. At present, however, there is no evidence that regulatory factors can influence the direction of processing of POMC, although research in this area is still in a very preliminary stage. Treatment of *Xenopus* neurointermediate lobe tissue with dopamine,[114] GABA,[108] or NPY,[120] has no apparent effect on the profile of newly synthesized POMC-peptides cleaved from the precursor. The degree of inhibition exerted on the secretion of each of the POMC-derived peptides appeared to be approximately equal for a given factor, and in this sense they can be considered to exert a coordinate effect on secretion. These studies involved only a single concentration of the respective secretagogue and concerned only a single experiment in each case. Therefore, any subtle effects on secretory profiles could have been overlooked. A more extensive examination of the effects of dopamine have been conducted with the species *Rana ridibunda*.[52] Dopamine had no significant effect on the cleavage of the precursor and, in an analysis of secreted peptides, it was shown that the degree of inhibition of release exerted by any given concentration of dopamine is the same on each of the POMC-derived peptides. This latter observation was true over a wide range of dopamine concentrations and involved all POMC-derived peptides except those related to α-MSH, analysis of which was complicated by the fact that dopamine also affected their acetylation. Altogether it can be said that there is reasonable but by no means extensive evidence for a coordinate regulation of secretion of POMC-derived peptides from amphibian melanotrophs.

The possibility of noncoordinate regulation of the secretion of POMC-derived peptides would imply the presence of populations of intracellular secretory granules which differ in both their peptide content (i.e., their profile of POMC-derived peptides) and in the mechanisms involved in the regulation of their secretion. These independent secretory "pools" could be sequestered within the same cell (compartmentalization) or be represented in two different cell-types within the same tissue. There have been three reports concerning the possible presence of secretory "pools" of POMC-derived peptides in *Xenopus,* one of which dealth with acetylated and nonacetylated forms of α-MSH,[112] discussed in detail in a previous section. Another report described the possibility that a selective intracellular degradation of POMC-derived peptides could lead to qualitative differences in the secretory signal of the malanotroph, namely its content of the peptide β-LPH.[126] The third reports an α-MSH-enriched population of secretory granules.[123] In our opinion, there is, in each case, other possible explanations to account for the experimental data and, without going into detail, we therefore decline to conclude that the presence of independent secretory "pools" of POMC-derived peptides is firmly established.

Despite the lack of conclusive evidence for the presence of independently regulated secretory pools of peptides, the question remains extremely relevant, given the potential of POMC to produce peptides with different biological activities. In the rat, secretory pools have been established for two different hormone-producing cells, namely the cells which produce prolactin[127] and parathyroid hormone, PTH.[128,129] In both cases, the pools concern newly synthesized peptides vs. peptides sequestered in a mature storage compartment, although it is not known whether a single cell contains both kinds of pools or if the different pools correspond to subcategories of a general cell type.[130] For PTH-producing cells, the two pools have been shown to be independently regulated; low Ca^{2+} stimulates release of newly synthesized PTH, whereas dibutryl cyclic-AMP and -adrenergic receptor agonists stimulate secretion of hormone from the mature secretory pool.[128] Findings from earlier ultrastructural studies of amphibian melanotrophs may give some indication for an amorphological basis for the presence of "peptide-pools" in these cells,[32] namely the marked differences observed in the structure of mature secretory granules of white-background-adapted animals (large fibrous granules) in comparison to granules of black-background-adapted animals (smaller and electron dense). In this regard it is perhaps relevant to relate that cyclic-AMP analogues have been found to be very effective in stimulating secretion of MSH from neurointermediate lobes of white- but not of black-background-adapted *Xenopus*.[106] While the reason for this is not clear, these results raise the possibility of cyclic-AMP-dependent and cyclic-AMP-independent secretory pathways in amphibian melanotrophs. Another difference between melanotrophs of white- and black-background-adapted animals is their response to the neuropeptide, TRH. This peptide is effective in stimulating MSH release from melanotrophs of only white-background-adapted *Xenopus laevis*.[121] The analysis of secretion from melanotrophs of fully white- and fully black-background-adapted amphibians may ultimately be instrumental in characterizing independently regulated secretory pools within POMC cells.

The existence of newly synthesized vs. mature secretory granules could have important implications for the peptide composition of the secretory signal. For example, as mentioned earlier, the conversion of the 18K-product to γ-MSH is a very slow process and the 18K-product is a major secretory product from neurointermediate lobes of black-background-adapted animals. One might expect that by slowing down or inhibiting the secretory process, such as by placing the animal on a white background, the intracellular conversion process could go to completion, and thus γ-MSH would become a major peptide of the mature secretory granule of the white-background-adapted animal. The potential for regulation through such a mechanism is nicely illustrated in studies with intermediate lobe melanotrophs of the rat.[77,78] It has been found that after several days in culture these cells display an accelarated rate of release of immunoreactive endorphins, when compared to fresh cell preparations. Chromatographic analysis of these peptides revealed that the relative proportion of nonacetylated forms of endorphin is considerably enhanced over that released from fresh cells. In that acetylation of endorphins in the rat takes a number of hours to be completed,[85,90] it seems safe to assume that, with the accelarated rate of secretion from the cultured cells, there was simply insufficient time for the acetylation of the peptides to occur prior to their release. This assumption finds support in the observation that treating the cells with dopamine not only leads to a slower rate of secretion, but also causes an increase in the contribution of acetylated peptides to the secretory profile. While these results might reflect an in vitro artifact, they nonetheless give an indication for a possible mechanism regulating the secretory signal of POMC cells. The fact that the biosynthetic and secretory activity of the melanotrophs of the amphibian intermediate lobe can be physiologically manipulated (i.e., change in color of background) may prove important in establishing examples of physiological mechanisms, analogous to that described above, for the manipulation of the secretory signal from POMC-producing cells.

IV. CONCLUDING REMARKS

A central theme in our analysis of the amphibian melanotroph cell has been the question, can the profile of POMC-derived peptides in the secretory signal of this cell be regulated, or, is the qualitative output of this cell irreversibly programmed during embryonic differentiation? There is evidence that, in general, the release of POMC-derived peptides from this cell can be described as concomitant, and that regulatory factors exert a coordinate effect on their release. There are, nonetheless, already indications for exceptions to this generalization. It is interesting and perhaps not coincidental, in view of the physiological function of this cell, that one of the potential sites for regulating the peptide content of the secretory signal concerns the acetylation of MSH. This regulation is, moreover, in concordance with the known physiological function, showing that the secretory product of the cell of a black-background-adapted animal is the highly melanotropic acetylated form of MSH and indicating that this same cell of the white-background-adapted animal may release the less melanotropic nonacetylated peptide. The melanotroph cell of the amphibian pars intermedia is very complex, considering both its potential for producing diverse biological signals and the complex nature of the hypothalamic system regulating its secretory output. Further analysis of the physiological functioning of this cell should help in both determining the potential plasticity of a POMC cell to generate diverse secretory signals and to delineate the mechanisms involved in regulating these signals. Such studies may also help clarify why complex integrative mechanisms have evolved at the level of the intermediate lobe melanotroph for a seemingly simple neuroendocrine reflex, namely that of background adaptation.

ACKNOWLEDGMENTS

We thank Professor A. P. van Overbeeke for his dedication and leadership in the development of this research and Dr. Hubert Vaudry for a very inspiring collaboration. We also thank H. V. for critically reading the present manuscript. Research grants from the Netherlands Organization for the Advancement of Pure Research (ZWO) and from the European Economic Community (EEC) are also gratefully acknowledged.

REFERENCES

1. **Bagnara, J. T. and Hadley, M. E.**, *Chromatophores and Color Change: The Comparative Physiology of Animal Pigmentation*, Prentice Hall, Englewood Cliffs, New Jersey, 1973.
2. **Waring, H.**, *Color Change Mechanisms of Cold-Blooded Vertebrates*, Academic Press, New York, 1963.
3. **Terlou, M., Goos, H. J. Th., and van Oordt, P. G. W. J.**, Hypothalamic regulation of pars intermedia function in amphibians, *Fortschritte Zoologie*, 140, 117, 1974.
4. **Goos, H. J. Th. and Terlou, M.**, Hypothalamic control of MSH secretion in lower vertebrates, in *Frontiers of Hormone Research, Vol. 4*, van Wimersma Greidanus, Tj. B., Ed., S. Karger, Basel, Switzerland, 1977, 51.
5. **Smith, P. E.**, Experimental ablation of the hypophysis in the frog embryo, *Science*, 44, 280, 1916.
6. **Allen, B. M.**, The results of extirpation of the anterior lobe of hypophysis and of the thyroid of *Rana pipiens* larvae, *Science*, 44, 755, 1916.
7. **Atwell, W. J.**, On the nature of pigmentation changes following hypophysectomy in the frog larva, *Science*, 49, 48, 1919.
8. **Swingle, W. W.**, The relation of the pars intermedia of the hypophysis to pigmentation changes in the anuran larvae. *J. Exp. Zool.*, 34, 119, 1921.
9. **Oshima, K. and Gorbman, A.**, Evidence for a double-innervated secretory unit in the anuran pars intermedia. I. Electrophysiological studies, *Gen. Comp. Endocrinol.*, 13, 98, 1969.
10. **Dawson, D. C. and Ralph, C. L.**, Neural control of the amphibian pars intermedia: electrical response evoked by illumination of the lateral eyes, *Gen. Comp. Endocrinol.*, 16, 611, 1971.
11. **Etkin, W.**, Hypothalamic inhibition of pars intermedia activity in the frog, *Gen. Comp. Endocrinol.*, 1, 148, 1962.
12. **Jorgensen, C. B. and Larsen, O. J.**, Neuroendocrine mechanisms in lower vertebrates, in *Neuroendocrinology, Vol. II*, Martini, L. and Ganong, W. F., Eds., Academic Press, New York, 1967, 485.
13. **Enemar, A. and Falk, B.**, On the presence of adrenergic nerves in the pars intermedia of the frog, *Rana temporaria*, *Gen. Comp. Endocrinol.*, 5, 577, 1965.
14. **Enemar, A., Falk, B., and Iturriza, F. C.**, Adrenergic nerves in the pars intermedia of the toad, *Bufo arenarum*, *Z. Zellforsch.*, 77, 325, 1967.
15. **Terlou, M. and Ploemacher, R. E.**, The distribution of monoamines in the tel-, di-, and mesencephalon of *Xenopus laevis* tadpoles, with special reference to the hypothalamo-hypophysial system, *Z. Zellforsch.*, 137, 521, 1973.
16. **Iturriza, F. C.**, Monoamines and control of the pars intermedia of the toad pituitary, *Gen. Comp. Endocrinol.*, 6, 19, 1966.
17. **Iturriza, F. C.**, Further evidence for the blocking effect of catecholamines on the secretion of melanocyte-stimulating hormone in toads, *Gen. Comp. Endocrinol.*, 12, 417, 1969.
18. **Goos, H. J. Th.**, Hypothalamic control of the pars intermedia in *Xenopus laevis* tadpoles, *Z. Zellforsch.*, 97, 118, 1969.
19. **Bower, A., Hadley, M. E., and Hruby, V. J.**, Biogenic amines and control of melanophore stimulating hormone release, *Science*, 184, 70, 1974.
20. **Hadley, M. E., Hruby, V. J., and Bower, A.**, Cellular mechanisms controlling melanophore stimulating hormone release, *Gen. Comp. Endocrinol.*, 26, 24, 1975.
21. **Hadley, M. E., Bower, A., and Hruby, V. J.**, Regulation of melanophore stimulating hormone (MSH) release, *Yale J. Biol. Med.*, 46, 602, 1973.
22. **Hadley, M. E. and Bagnara, J. T.**, Regulation of release and mechanism of action of MSH, *Am. Zool.*, 15, (Suppl.1), 81, 1975.
23. **Hadley, M. E., Davis, M. D., and Morgan, C. M.**, Cellular control of melanocyte-stimulating hormone secretion, in *Frontiers of Hormone Research, Vol. 4*, van Wimersma Greidanus, Ed., S. Karger, Basel, Switzerland, 1977, 94.
24. **Eipper, B. A. and Mains, R. E.**, Structure and biosynthesis of pro-adrenocorticotropin/endorphin and related peptides, *Endocr. Rev.*, 1, 1, 1980.
25. **Herbert, E.**, Discovery of pro-opiomelanocortin: a cellular polyprotein, *Trends Biochem. Sci.*, 6, 184, 1981.
26. **Jenks, B. G., Leenders, H. J., Verburg-van Kemenade, B. M. L., Tonon, M. C., and Vaudry, H.**, Strategies in the regulation of secretory signals from proopiomelanocortin-producing cells, in *Neuroendocrine Molecular Biology*, Fink, G., Harmar, A. J., and McKerns, Ed., Plenum Press, New York, 1986, 281.
27. **O'Donohue, T. L. and Dorsa, D. M.**, The opiomelanotropinergic neuronal and endocrine systems, *Peptides*, 3, 353, 1982.
28. **O'Donohue, T. L.**, Opiomelanotropin acetyltransferase regulates actions of opiomelanotropinergic neurons, in *Integrative Neurohumoral Mechanisms*, Endroczi, E., Ed., Elsevier Scientific Publishers, Amsterdam, 1983, 295.

29. **Lowry, P. J., Estivariz, F. E., Silas, L., Linton, E. A., McLean, C., and Crombe, K.,** The case for pro-γ-MSH as the adrenal growth factor, *Endocr. Res.,* 10, 243, 1984.
30. **Schachter, B. S., Johnson, L. K., Baxter, J. D., and Roberts, J. L.,** Differential regulation by glucocorticoids of proopiomelanocortin mRNA levels in the anterior and intermediate lobes of the rat pituitary, *Endocrinology,* 106, 1442, 1982.
31. **Weatherhead, B. and Whur, P.,** Quantification of the ultrastructural changes in the "melanocyte-stimulating hormone cell" of the pars intermedia of the pituitary of *Xenopus laevis,* produced by change of background colour, *J. Endocrinol.,* 53, 303, 1972.
32. **Hopkins, C. R.,** Studies on secretory activity in the pars intermedia of *Xenopus laevis.* 1. Fine structural changes related to the onset of secretory activity in vivo, *Tissue Cell,* 2, 59, 1970.
33. **Cohen, A. G.,** Observations on the pars intermedia of *Xenopus laevis, Nature (London),* 215, 55, 1967.
34. **Imai, K.,** Color change and pituitary function, in *Biology of the Normal and Abnormal Melanocyte,* Kawamura, T., Fitzpatrick, T. B., and Seiji, M., Eds., University Park Press, London, 1971, 17.
35. **Pehlemann, F. W.,** Ultrastructure and innervation of the pars intermedia of the pituitary of *Xenopus laevis, Gen. Comp. Endocrinol.,* 9, 481, 1967.
36. **Saland, L. C.,** Ultrastructure of the frog pars intermedia in relation to the hypothalamic control of hormone release, *Neuroendocrinology,* 3, 72, 1968.
37. **Loh, Y. P. and Gainer, H.,** Biosynthesis, processing, and control of release of melanotropic peptides in the neurointermediate lobe of *Xenopus laevis, J. Gen. Physiol.,* 70, 37, 1977.
38. **Loh, Y. B.,** Immunological evidence for two common precursors to corticotropins, endorphins, and melanotropin in the neurointermediate lobe of the toad, *Proc. Natl. Acad. Sci. U.S.A.,* 76, 796, 1979.
39. **Martens, G. J. M., Biermans, P. P. J., Jenks, B. G., and van Overbeeke, A. P.,** Biosynthesis of two structurally different pro-opiomelano-cortins in the pars intermedia of the amphibian pituitary gland, *Eur. J. Biochem.,* 126, 17, 1982.
40. **Vaudry, H., Jenks, B. G., Verburg-van Kemenade, B. M. L., and Tonon, M. C.,** Effects of tunicamycin on biosynthesis, processing and release of proopiomelanocortin-derived peptides in the intermediate lobe of the frog *Rana ridibunda, Peptides,* 7, 163, 1986.
41. **Thornton, V. F.,** The effect of change of background color on the in vitro incorporation of labeled amino acids into the intermediate lobe of the pituitary in *Xenopus laevis, Gen. Comp. Endocrinol.,* 22, 250, 1974.
42. **Jenks, B. G., van Overbeeke, A. P., and McStay, B. F.,** Synthesis, storage, and release of MSH in the pars intermedia of the pituitary gland of *Xenopus laevis* during background adaptation, *Can. J. Zool.,* 55, 922, 1977.
43. **Whur, P. and Weatherhead, B.,** Rates of incorporation of ^{3}H-leucine into protein of the pars intermedia of the pituitary in the amphibian *Xenopus laevis* after change of background color, *J. Endocrinol.,* 51, 521, 1971.
44. **Martens, G. J. M., Jenks, B. G., and van Overbeeke, A. P.,** Analysis of peptide biosynthesis in the neurointermediate lobe of *Xenopus laevis* using high-performance liquid chromatography: occurrence of small bioactive products, *Comp. Biochem. Physiol.,* 67B, 493, 1980.
45. **Martens, G. J. M., Jenks, B. G., and van Overbeeke, A. P.,** Biosynthesis of pairs of peptides related to melanotropin, corticotropin and endorphin in the pars intermedia of the amphibian pituitary gland, *Eur. J. Biochem.,* 122, 1, 1982.
46. **Martens, G. J. M., Soeterik, F., Jenks, B. G., and van Overbeeke, A. P.,** In vivo biosynthesis of melanotropins and related peptides in the pars intermedia of *Xenopus laevis, Gen. Comp. Endocrinol.,* 49, 73, 1983.
47. **Loh, Y. P., Myers, B., Wong, B., Parish, D. C., Lang, M., and Goldman, M. E.,** Regulation of proopiomelanocortin synthesis by dopamine and cAMP in the amphibian pituitary intermediate lobe, *J. Biol. Chem.,* 260, 8956, 1985.
48. **Thornton, V. F.,** The effect of change of background color on the melanocyte-stimulating hormone content of the pituitary of *Xenopus laevis, Gen. Comp. Endocrinol.,* 17, 554, 1971.
49. **Wilson, J. F. and Morgan, M. A.,** α-melanotropin-like substances in the pituitary and plasma of *Xenopus laevis* in relation to color change responses, *Gen. Comp. Endocrinol.,* 38, 172, 1979.
50. **Jenks, B. G.,** Control of MSH synthesis and release in the aquatic toad, *Xenopus laevis,* in *Frontiers of Hormone Research, Vol. 4,* van Wimersma Greidanus, Tj. B., Ed., S. Karger, Basel, Switzerland, 1977, 63.
51. **Tonon, M. C., Leroux, P., Stoeckel, M. E., Jegou, S., Pelletier, G., and Vaudry, H.,** Catecholaminergic control of α-melanocyte-stimulating hormone (α-MSH) release by frog neurointermediate lobe in vitro: evidence for direct stimulation of α-MSH release by thyrotropin-releasing hormone, *Endocrinology,* 112, 133, 1984.
52. **Jenks, B. G., Verburg-van Kemenade, B. M. L., Tonon, M. C., and Vaudry, H.,** Regulation of biosynthesis and release of pars intermedia peptides in *Rana ridibunda:* dopamine affects both acetylation and release of α-MSH, *Peptides,* 6, 913, 1985.

53. **Verburg-van Kemenade, B. M. L., Jenks, B. G., Tonon, M. C., and Vaudry, H.,** Characteristics of receptors for dopamine in the pars intermedia of the amphibian pituitary, *Neuroendocrinology,* 44, 446, 1986.
54. **Roberts, J. L., Eberwine, H., and Gee, C. E.,** Analysis of POMC gene expression by transcription assay and *in situ* hybridization histochemistry, in *Cold Spring Harbor Symp. Quant. Biol. Vol. 2,* 8, 1983, 385.
55. **Stroll, G., Martin, R., and Voigt, K. H.,** Control of peptide release from cells of the intermediate lobe of the rat pituitary, *Cell Tiss. Res.,* 236, 561, 1984.
56. **Cote, T. E., Eskay, R. L., Frey, E. A., Grewe, C. W., Munemura, M., Stoff, J. C., Tsuruta, K., and Kebabian, J. W.,** Biochemical and physiological studies of the beta-adrenoceptor and the D-2 dopamine receptor in the intermediate lobe of the rat pituitary gland: a review, *Neuroendocrinology,* 35, 217, 1982.
57. **Chen, C. L. C., Dionne, F. T., and Roberts, J. L.,** Regulation of the proopiomelanocortin mRNA levels in the rat pituitary by dopaminergic compounds. *Proc. Natl. Acad. Sci. U.S.A.,* 80, 2211, 1983.
58. **Cote, T. E., Felder, R., Kebabian, J. W., Sekura, R. D., Reisine, T., and Affolter, H. U.,** D-2 dopamine receptor-mediated inhibition of pro-opiomelanocortin synthesis in rat intermediate lobe, *J. Biol. Chem.,* 261, 4555, 1986.
59. **Beaulieu, M., Felder, R., and Kebabian, J. W.,** D-2 dopaminergic agonists and adenosine 3′-5′-monophosphate directly regulate the synthesis of α-melanocyte-stimulating hormone-like peptides by cultured rat melanotrophs, *Endocrinology,* 118, 1032, 1986.
60. **Burgers, A. C. J., Imai, K., and van Oordt, G. J.,** The amount of melanophore-stimulating hormone in single pituitary glands of *Xenopus laevis* kept under various conditions, *Gen. Comp. Endocrinol.,* 3, 53, 1963.
61. **Martens, G. J. M., Jenks, B. G., and van Overbeeke, A. P.,** N-α-acetylation is linked to α-MSH release from pars intermedia of the amphibian pituitary gland, *Nature (London),* 294, 558, 1981.
62. **Steiner, D. F., Patzelt, C., Chan, S. J., Quinn, P. S., Tager, H. S., Nielsen, D., Lernmark, A., Noyes, B. E., Agarwal, K. E., Gabbay, K. H., and Rubenstein, A. H.,** Formation of biologically active peptides, *Proc. Roy. Soc. London,* 210, 45, 1980.
63. **Douglass, J., Civelli, O., and Herbert, E.,** Polyprotein gene expression: generation of diversity of neuroendocrine peptides, *Annu. Rev. Biochem.,* 53, 665, 1984.
64. **Martens, G. J. M., Civelli, C., and Herbert, E.,** Nucleotide sequence of cloned cDNA for pro-opiomelanocortin in the amphibian *Xenopus laevis, J. Biol. Chem.,* 260, 13685, 1985.
65. **Hopkins, C. R.,** The biosynthesis, intracellular transport, and packaging of melanocyte-stimulating peptides in the amphibian pars intermedia. *J. Cell Biol.,* 53, 642, 1972.
66. **Jenks, B. G., Martens, G. J. M., van Helden, H. P. M., and van Overbeeke, A. P.,** Biosynthesis and release of melanotropins and related peptides by the pars intermedia in *Xenopus laevis,* in *Current Trends in Comparative Endocrinology,* Lofts, B. and Holmes, W. N., Eds., Hong Kong University Press, Hong Kong, 1985, 149.
67. **Loh, Y. P. and Gainer, H.,** Characterization of pro-opiocortin converting activity in purified secretory granules from rat pituitary intermediate lobe, *Proc. Natl. Acad. Sci. U.S.A.,* 79, 100, 1982.
68. **Chang, T. L., Gainer, H., Russell, J. T., and Loh, Y. P.,** Pro-opiocortin converting enzyme activity in bovine neurosecretory granules, *Endocrinology,* 111, 1607, 1982.
69. **Vaudry, H., Jenks, B. G., and van Overbeeke, A. P.,** Biosynthesis, processing and release of pro-opiomelanocortin related peptides in the intermediate lobe of the pituitary gland of the frog *(Rana ridibunda), Peptides,* 5, 905, 1984.
70. **Vaudry, H., Jenks, B. G., and van Overbeeke, A. P.,** The frog pars intermedia contains only the non-acetylated form of α-MSH: acetylation to generate α-MSH occurs during the release process, *Life Sci.,* 33, 97, 1983.
71. **Lazure, C., Seidah, N. G., Pelaprat, D., and Chrétien, M.,** Proteases and posttranslational processing of prohormones: a review, *Can. J. Biochem. Cell Biol.,* 61, 501, 1983.
72. **Loh, Y. P. and Gainer, H.,** The role of glycosylation on the biosynthesis, degradation and secretion of the ACTH-β-lipotropin common precursor and its peptide products, *FEBS Lett.,* 96, 269, 1978.
73. **Loh, Y. P. and Gainer, H.,** The role of the carbohydrate in the stabilization, processing, and packaging of the glycosylated adrenocorticotropin-endorphin common precursor in toad pituitaries, *Endocrinology,* 105, 474, 1979.
74. **Budarf, M. L. and Herbert, E.,** Effect of tunicamycin on the synthesis, processing and secretion of pro-opiomelanocortin peptides in mouse pituitary cells, *J. Biol. Chem.,* 257, 10128, 1982.
75. **Phillips, M. A., Budarf, M. L., and Herbert, E.,** Glycosylation events in the processing and secretion of pro-ACTH-endorphin in mouse pituitary tumor cells, *Biochemistry,* 20, 1666, 1981.
76. **Jenks, B. G., Ederveen, J. H. M., Feyen, J. H. M., and van Overbeeke, A. P.,** The functional significance of glycosylation of proopiomelanocortin in melanotrophs of the mouse pituitary gland, *J. Endocrinol.,* 107, 365, 1985.
77. **Ham, J., McFarthing, K. G., Toogood, C. I. A., and Smyth, D. G.,** Influence of dopaminergic agents on β-endorphin processing in rat pars intermedia, *Biochem. Soc. Transcr.,* 12, 927, 1984.

78. **Ham, J. and Smyth, D. G.**, Regulation of bioactive β-endorphin processing in rat pars intermedia, *FEBS Lett.*, 175, 407, 1984.
79. **O'Donohue, T. L., Handelmann, G. E., Chanconas, T., Miller, T. S., and Jacobowitz, D. M.**, Evidence that N-acetylation regulates the behavioral activity of α-MSH in the rat and human central nervous systems, *Peptides*, 2, 333, 1981.
80. **Smyth, D. G., Massey, D. E., Zakarian, S., and Finnie, M. D.**, Endorphins are stored in biologically active and inactive forms: Isolation of alpha-N-acetyl peptides, *Nature (London)*, 279, 252, 1979.
81. **Chrétien, M., Seidah, N. G., and Dennis, M.**, Processing of precursor polyproteins in rat brain: regional differences in acetylation of POMC peptides, in *Central and Peripheral Endorphins: Basic and Clinical Aspects*, Muller, E. E., and Genazzani, A. R., Eds., Raven Press, New York, 1984, 27.
82. **Burbach, J. P. H., van Toll, H. H. M., Wiegant, V. M., van Ooijen, R. A., and Maes, R. A. A.**, Identification of N-α-acetyl-α-endorphin and N-α-acetyl-γ-endorphin isolated from the neurointermediate lobe of the rat pituitary gland, *J. Biol. Chem.*, 260, 6663, 1985.
83. **Wiegant, V. M., Verhoef, J., Burbach, J. P. H., van Amerongen, A., Gaffori, O., Sitsen, J. M. A., and de Wied, D.**, N-α-acetyl-γ-endorphin is an endogenous non-opiate neuropeptide with biological activity, *Life Sci.*, 36, 2277, 1985.
84. **Mains, R. E., and Eipper, B. A.**, Synthesis and secretion of corticotropins, melanotropins and endorphins by rat intermediate pituitary cells, *J. Biol. Chem.*, 254, 7885, 1979.
85. **Mains, R. E., and Eipper, B. A.**, Differences in the post-translational processing of β-endorphin in rat anterior and intermediate pituitary, *J. Biol. Chem.*, 256, 5683, 1981.
86. **Jenks, B. G., van Daal, J. H. H. N., Scharenberg, J. G. M., Martens, G. J. M., and van Overbeeke, A. P.**, Biosynthesis of pro-opiomelanocortin-derived peptides in the mouse neurointermediate lobe, *J. Endocrinol.*, 98, 19, 1983.
87. **Rudman, D., Chawla, R. K., and Hollins, B. M.**, N,O-diacetyl-serine α-melanocyte-stimulating hormone, a naturally occurring melanotropic peptide, *J. Biol. Chem.*, 254, 10102, 1979.
88. **Leenders, H. J., Janssens, J. J. W., Theunissen, H. J. M., Jenks, B. G., and van Overbeeke, A. P.**, Acetylation of melanocyte-stimulating hormone and β-endorphin in the pars intermedia of the perinatal pituitary gland in the mouse, *Neuroendocrinology*, 43, 166, 1986.
89. **Browne, C. A., Bennett, H. P. J., and Solomon, S.**, Isolation and characterization of corticotropin- and melanotropin-related peptides from the neurointermediate lobe of the rat pituitary by reversed-phase liquid chromatography, *Biochemistry (U.S.A.)*, 20, 4538, 1981.
90. **Eipper, B. A. and Mains, R. E.**, Further analysis of post-translational processing of β-endorphin in rat intermediate lobe, *J. Biol. Chem.*, 256, 5689, 1981.
91. **Liotta, A. S., Yamaguchi, H., and Krieger, D. T.**, Biosynthesis and release of β-endorphin, N-acetyl-β-endorphin, β-endorphin-(1-27) and N-acetyl-β-endorphin-(1-27)-like peptides by rat pituitary neurointermediate lobe: β-endorphin is not further processed by anterior lobe, *J. Neurosci.*, 1, 585, 1981.
92. **Akil, H., Ueda, Y., Lin, H. L., and Watson, S. J.**, A sensitive coupled HPLC/RIA technique for separation of endorphins: multiple forms of β-endorphin in the pituitary intermediate versus anterior lobe, *Neuropeptides*, 1, 429, 1981.
93. **Autelitano, D. J., Smith, A. I., Lolait, S. J., and Funder, J. W.**, Dopaminergic agents differentially alter β-endorphin processing patterns in the rat pituitary neurointermediate lobe, *Neurosci. Lett.*, 59, 141, 1985.
94. **Barnea, A. and Cho, G.**, Acetylation of adrenocorticotropin and β-endorphin by hypothalamic and pituitary acetyltransferases, *Neuroendocrinology*, 37, 434, 1983.
95. **Chappell, M. C., Loh, Y. P., and O'Donohue, T. L.**, Evidence for an opiomelanotropin acetyltransferase in the rat pituitary intermediate lobe, *Peptides*, 3, 405, 1982.
96. **Glembotski, C. C.**, Characterization of the peptide acetyltransferase activity in bovine and rat intermediate pituitaries responsible for the acetylation of β-endorphin and α-melanotropin, *J. Biol. Chem.*, 257, 10501, 1982.
97. **Chappell, M. C., O'Donohue, T. L., Millington, W. R., and Kempner, E. S.**, The size of enzymes acetylating α-melanocyte-stimulating hormone and β-endorphin, *J. Biol. Chem.*, 261, 1088, 1968.
98. **Guttmann, S. T. and Boissonnas, R. A.**, Influence of the N-terminal extremity of α-MSH on the melanophore stimulating activity of this hormone, *Experientia*, 17, 265, 1961.
99. **Rudman, D., Hollins, B. M., Kutner, M. H., and Moffitt, S. D.**, Three types of α-melanocyte-stimulating hormone: bioactivities and half-lives, *Am. J. Physiol.*, 245, 47E, 1983.
100. **Eberle, A. N.**, Structure and chemistry of peptide hormones of the intermediate lobe, in *Peptides of the Pars Intermedia*, Evered, D. and Lawrensen, G., Eds., Pitman Medical, Summit, New Jersey, 1981, 13.
101. **Akil, H. Young, E., Watson, S. J., and Coy, D. H.**, Opiate binding properties of naturally occurring N- and C-terminus modified beta-endorphins, *Peptides*, 2, 289, 1981.

102. Li, C. H., Tseng, L. F., Jibson, M. D., Hammonds, R. G., Yamashiro, D., and Zaoral, M., β-endorphin (1-27): acetylation of α-amino groups enhances immunoreactivity but diminishes analgesic and receptor binding activities with no change in circular dichroism spectra, *Biochem. Biophys. Res. Commun.*, 97, 932, 1980.
103. Deakin, J. F. W., Dostrovsky, J. O., and Smyth, D. G., Influence of N-terminal acetylation and C-terminal proteolysis on the analgesic activity of β-endorphin, *Biochem. J.*, 189, 501, 1980.
104. Dores, R. M., Further characterization of the major forms of reptile beta-endorphin, *Peptides*, 4, 897, 1983.
105. Verburg-van Kemenade, B. M. L., Jenks, B. G., and Driessen, A. G. H., GABA and dopamine act directly on melanotropes of *Xenopus* to inhibit MSH secretion, *Brain Res. Bull.*, 17, 697, 1986.
106. Verburg-van Kemenade, B. M. L., Jenks, B. G., and van Overbeeke, A. P., Regulation of melanotropin release from the pars intermedia of the amphibian *Xenopus laevis:* evaluation of the involvement of serotonergic, cholinergic or adrenergic receptor mechanisms. *Gen. Comp. Endocrinol.*, 63, 471, 1986.
107. Martens, G. J. M., Expression of two proopiomelanocortin genes in the pituitary gland of *Xenopus laevis:* complete structures of the two preprohormones, *Nucleic Acid Res.*, 14, 3791, 1986.
108. Verburg-van Kemenade, B. M. L., Tappaz, M., Paut, L., and Jenks, B. G., GABAergic regulation of melanocyte-stimulating hormone secretion from the pars intermedia of *Xenopus laevis:* Immunocytochemical and physiological evidence, *Endocrinology*, 118, 260, 1986.
109. Glembotski, C. C., Acetylation of α-melanotropin and β-endorphin in the rat intermediate pituitary: subcellular localization, *J. Biol. Chem.*, 257, 10493, 1982.
110. Gainer, H., Russell, J. T., and Loh, Y. P., The enzymology and intracellular organization of peptide precursor processing: the secretory vesicle hypothesis, *Neuroendocrinology*, 40, 171, 1985.
111. Eipper, B. A. and Mains, R. E., Existence of a glycosylated precursor to ACTH and endorphin in the anterior and intermediate lobes of the rat pituitary, *J. Supramolec. Struct.*, 8, 247, 1978.
112. Goldman, M. E. and Loh, Y. P., Intracellular acetylation of desacetyl-α-MSH in the *Xenopus laevis* neurointermediate lobe, *Peptides*, 5, 1129, 1984.
113. Jenks, B. G. and van Overbeeke, A. P., Biosynthesis and release of neurointermediate lobe peptides in the aquatic toad, *Xenopus laevis*, adapted to black background, *Comp. Biochem. Physiol.*, 66C, 71, 1980.
114. Martens, G. J. M., Jenks, B. G., and van Overbeeke, A. P., Microsuperfusion of neurointermediate lobes of *Xenopus laevis:* concomitant and coordinately controlled release of newly synthesized peptides, *Comp. Biochem. Physiol.*, 69C, 75, 1981.
115. Gumbiner, B. and Kelly, R. B., Two distinct intracellular pathways transport secretory and membrane glycoproteins to the surface of pituitary tumor cells, *Cell*, 28, 51, 1982.
116. Tonon, M. C., Leroux, P., Jenks, B. G., Gouteux, S., Guy, J., Pelletier, G., and Vaudry, H., The intermediate lobe of the amphibian pituitary: An endocrine gland which gives rise to multiple hormonal peptides and is regulated by multiple control mechanisms, *Ann. Endocrinol.*, 46, 69, 1985.
117. Tonon, M. C., Leroux, P., Leboulenger, F., Delarue, C., Jegou, S., and Vaudry, H., Thyrotropin-releasing hormone stimulates the release of melanotropin from frog neurointermediate lobes in vitro, *Life Sci.*, 26, 869, 1980.
118. Adjeroud, S., Tonon, M-C., Lamacz, M., Leneveu, E., Stoekel, M. E., Tappaz, M. L., Cazin, L., Danger, J. M., Bernard, C., and Vaudry, H., GABAergic control of α-melanocyte-stimulating hormone (α-MSH) release by frog neurointermediate lobe in vitro, *Brain Res. Bull.*, 17, 717, 1986.
119. Danger, J. M., Leboulenger, F., Guy, J., Tonon, M-C., Benyamina, M., Martel, J. C., Saint-Pierre, S., Pelletier, G., and Vaudry, H., Neuropeptide-Y in the intermediate lobe of the frog pituitary acts as an α-MSH-release inhibiting factor, *Life Sci.*, 39, 1183, 1986.
120. Verburg-van Kemenade, B. M. L., Jenks, B. G., Danger, J. M., Vaudry, H., Pelletier, G., and Saint-Pierre, S., A NPY-like peptide may function as MSH-release inhibiting factor in *Xenopus laevis*, *Peptides*, 8, 61, 1987.
121. Verburg-van Kemenade, B. M. L., Jenks, B. G., Visser, T., Tonon, M-C., and Vaudry, H., Assessment of TRH as a potential MSH release stimulating factor in *Xenopus laevis*, *Peptides*, 8, 69, 1987.
122. Vaudry, H., Tonon, M. C., Delarue, C., Vaillant, R., and Kraicer, J., Biological and radioimmunological evidence for melanocyte stimulating hormone (MSH) of extrapituitary origin in the rat brain, *Neuroendocrinology*, 27, 9, 1978.
123. Loh, Y. P., Li, A., Gritsch, H. A., and Eskay, R. L., Immunoreactive α-MSH and β-endorphin in the toad pars intermedia: dissociation in storage, secretion and subcellular localization, *Life Sci.*, 29, 1599, 1981.
124. Verburg-van Kemenade, B. M. L., Jenks, B. G., and Houben, A. G. M. J., Regulation of cyclic-AMP synthesis in amphibian melanotrope cells by catecholamine and GABA receptors, *Life Sci.*, 40, 1859, 1987.
125. Verburg-van Kemenade, B. M. L., Jenks, B. G., Lenssen, F. J. A., and Vaudry, H., Characterization of GABA receptors in the neurointermediate lobe of the amphibian, *Xenopus laevis*, *Endocrinology*, 120, 622, 1987.

126. **Loh, Y. P. and Jenks, B. G.**, Evidence for two different pools of adrenocorticotropin, α-melanocyte-stimulating hormone, and endorphin-related peptides released by the frog pituitary neurointermediate lobe, *Endocrinology*, 109, 54, 1981.
127. **Dannies, P. S. and Rudnick, M. S.**, Prolactin: multiple intracellular processing routes plus several potential mechanisms for regulation, *Biochem. Pharmacol.*, 23, 2845, 1984.
128. **Morrissey, J. J. and Cohn, D. V.**, Secretion and degradation of parathyroid hormone as a function of intracellular maturation of hormone pools, *J. Cell Biol.*, 83, 521, 1979.
129. **Hanley, D. A. and Wellings, P. G.**, Dopamine-stimulated parathyroid hormone release in vitro: further evidence for a two-pool model of parathyroid hormone secretion, *Can. J. Physiol. Pharmacol.*, 63, 1139, 1985.
130. **Bienkowski, R. S.**, Intracellular degradation of newly synthesized secretory proteins, *Biochem. J.*, 214, 1, 1983.

Chapter 9

MULTIHORMONAL CONTROL OF MELANOTROPIN SECRETION IN COLD-BLOODED VERTEBRATES

Marie-Christine Tonon, Jean-Michel Danger, Marek Lamacz, Philippe Leroux, Saida Adjeroud, Ann C. Andersen, Lidy Verburg-van Kemenade, Bruce G. Jenks, Georges Pelletier, Lise Stoeckel, Arlette Burlet, Gotfryd Kupryszewski, and Hubert Vaudry

TABLE OF CONTENTS

I. Introduction 128

II. Control of the Pars Intermedia by Biogenic Amines 128
    A. Dopamine 128
        1. Anatomical Basis 129
        2. Control of the Pars Intermedia Secretion by Dopamine 131
    B. Noradrenaline 133
        1. Anatomical Basis 133
        2. Control of the Pars Intermedia Secretion by Noradrenaline 134
    C. GABA 134
        1. Anatomical Basis 135
        2. Control of the Pars Intermedia Secretion by GABA 136
    D. Serotonin 138
        1. Anatomical Basis 138
        2. Control of the Pars Intermedia Secretion by Serotonin 139
    E. Other Neurotransmitters 140
        1. Anatomical Basis 140
        2. Control of the Pars Intermedia Secretion by Other Neurotransmitters 140

III. Control of the Pars Intermedia by Regulatory Peptides 141
    A. Thyrotropin-Releasing Hormone 141
        1. Anatomical Basis 141
        2. Control of the Pars Intermedia Secretion by TRH 141
    B. Neuropeptide Tyrosine 146
        1. Anatomical Basis 146
        2. Control of the Pars Intermedia Secretion by NPY 148
    C. Corticotropin-Releasing Factor 150
        1. Anatomical Basis 150
        2. Control of the Pars Intermedia Secretion by CRF 150
    D. Neurohypophyseal Peptides 151
        1. Anatomical Basis 151
        2. Control of the Pars Intermedia Secretion by Neurohypophyseal Peptides 151

IV. Concluding Remarks 152

Acknowledgments 153

References 153

I. INTRODUCTION

The pars intermedia of the pituitary has been described as a "compact, almost avascular structure exhibiting an abundant and diffuse innervation".[1] The sparse vascularity and the rich innervation of this tissue is unusual in an endocrine gland and these structural peculiarities raise important histophysiological questions as to the access of the nutrients and the evacuation of the secretory products released by the endocrine cells. The most intriguing problem, however, concerns the mechanism of control of the pars intermedia.

The intermediate lobe of the pituitary synthesizes a multifunctional glycoprotein, called pro-opiomelanocortin (POMC), which gives rise, through specific processing to a variety of bioactive peptides.[2-5] The post-translational processing of POMC has been the subject of a great number of studies performed in mammals (Loh et al., Chapter 7, this Volume), lizards (Dores, Chapter 3, this Volume), and two anuran species (Jenks et al., Chapter 9, this Volume). Several biosynthetic products elaborated and released by intermediate lobe cells exhibit melanophore-expending activity (Jenks et al., Chapter 9, this Volume). In fish, amphibians, and some reptiles, these melanotropins are considered to play a major role in the adaptation of skin color to changes in environmental illumination.[6] Thus, rapid and delicate regulation of the secretory activity of the pars intermedia is of prime importance in cold-blooded vertebrates.

In all vertebrates examined so far, with the exception of lizards, the pars intermedia is profusely innervated by fibers coursing between the parenchymal cells. In mammals, this innervation mainly comprises aminergic fibers originating from the hypothalamus.[7] The nature of the transmitters contained in these nerve terminals and the extent of their penetration varies considerably between species.[1,8] In some mammalian species, such as the rabbit, neurosecretory-like fibers containing typical peptidergic neurosecretory granules have been described.[9,10] In lower vertebrates, the intermediate lobe is abundantly innervated by aminergic and peptidergic terminals, and our knowledge about the involvement of these potential neuroregulators in the control of pars intermedia secretion has markedly increased during the last decade.

In such submammalian vertebrates as amphibians, the intermediate lobe of the pituitary appears to be a most complex neuroendocrine transducer system, in that this endocrine gland produces multiple hormonal peptides and undergoes multi-factorial regulation. The purpose of this chapter is to assess, from a comparative point of view, the present state of our knowledge concerning the aminergic and peptidergic factors involved in the regulation of the pars intermedia.

II. CONTROL OF THE PARS INTERMEDIA BY BIOGENIC AMINES

A. Dopamine

The first evidence that the pars intermedia is under hypothalamic inhibitory control was provided by Etkin[11] who showed that ectopic transplantation of the pituitary in the tadpole induces excessive activity of the pars intermedia, as indicated by the intense pigmentation of the host animal. The same darkening response can be obtained in normal tadpoles by disrupting the hypothalamic-pituitary connections;[12,13] hyperpigmentation is always associated with hypertrophy of pars intermedia cells and reduction of the cellular MSH content.[14,15] In the adult *Bufo arenarum*, transplantation of the pars intermedia into the kidney or the eye results in maximal darkening.[16] Similar experiments have been performed in many other species,[17-19] leading to the concept that in all vertebrate classes the activity of the pars intermedia is under inhibitory regulation by the hypothalamus. Evidence that catecholamines are involved in the negative control of intermediate lobe secretion was first presented by Stoppani et al.,[20,21] who showed that administration of sympathomimetic drugs cause melanophore concentration in *Bufo arenarum*. Inhibitors of monoamine-oxidase, such as lysergic acid diethylamide, block the secretory activity of pars intermedia cells in *Xenopus laevis*[22] and *Bufo arenarum*.[23] Direct evidence that dopamine is one of the hypothalamic factors inhibiting MSH secretion was provided by Hadley's group[24] in the rat. Later, it became evident that dopamine is the major MSH-release inhibiting factor in all vertebrate species[25-27] except leporidae.[28]

1. Anatomical Basis

A direct innervation of the parenchymal cells of the pars intermedia was reported by Ramon y Cajal[29] as early as 1894 and the existence of nerve fibers penetrating the pars intermedia was subsequently described in a number of mammalian species (reviewed by Wingstrand[30]). In mammals, most of these fibers contain aminergic[7,31,32] and cholinergic[33,34] neurotransmitters, although neurosecretory terminals have occasionally been described.[1,9,10,35-37] The distribution of aminergic neurons in the brain and pituitary of all vertebrate classes has been determined by the histofluorescence technique of Falck and Hillarp.[38] Subsequently, antibodies against the enzymes controlling the biosynthesis of catecholamines, tyrosine hydroxylase and dopamine β-hydroxylase have become available.[39-42] These antisera have enabled the localization of catecholaminergic neurons in the central nervous system at both the optic and electron microscopic levels.[43,44] Autoradiography after in vivo administration of dopamine has been also applied successfully in the study of the distribution of the noradrenaline[45,46] and dopamine[47,48] neuronal systems. At the electron microscopic level noradrenergic and dopaminergic endings can also be identified after intraventricular injection of the sympatholytic agent 6-hydroxydopamine (6-OHDA). More recently, antibodies have been raised against dopamine[49] and have been used for the immunocytochemical localization of dopamine-containing structures within the brain and pituitary of various species.[50,51]

In mammals, the major ascending dopamine neuronal system originates in the ventral part of the mesencephalon, essentially in the pars compacta of the substancia nigra.[38,52,53] Dopaminergic perikarya, originally identified as A8, A9, and A10 project either to the caudate nucleus, putamen and nucleus accumbens (nigro-striatal system) or to the olfactory tubercle, septum and amygdala, frontal cingulate and entorhinal cortex (mesocortical system).[52] Other groups of dopaminergic neurons are packed in hypothalamic nuclei (areas designated A11, A12, A13, and A14). The incerto-hypothalamic system[54] is composed of areas A11, A13, and A14 which send out intrahypothalamic or spinal cord projections,[55] whereas the arcuate and periventricular dopaminergic perikarya belonging to the A12 region send projections to the median eminence (tuberoinfundibular system) and to the intermediate lobe of the pituitary (tuberohypophyseal system).[56] In amphibians, the distribution of catecholaminergic neurons has been extensively studied by Terlou,[57] Prasada-Rao,[58] and Goos.[59] In the telencephalon, catecholamine-containing perikarya are located in the olfactory bulb, and send widespread projections into the brain, in particular to the nucleus accumbens and nucleus lateralis septi.[57] In the hypothalamus, catecholamines are present in neurons bordering the preoptic recess organ (PRO),[58] in the paraventricular organ (PVO),[59] and in the dorsal infundibular nucleus (DIN).[57] In the PRO, catecholaminergic neurons contain only dopamine. They send projections caudally across the preoptic region which pass around the optic chiasma and extend posteriorly along the ventral part of the infundibulum towards the median eminence. Although Prasada-Rao and Hartwig[58] called this dopaminergic nerve bundle the "preoptic recess organ-hypophysial tract" they could not determine whether this tract actually terminates in the pars intermedia of *Rana temporaria*. In the PVO, most aminergic perikarya contain dopamine; however, some of them are presumably serotoninergic.[58] Fibers originating from the PVO project both towards the preoptic area where various peptidergic hypophysiotropic cell bodies are located[60-66] and dorsally towards the thalamus and telencephalon.[57] A dense tract of fibers emerging from the PVO extends ventrally towards the infundibulum and terminates

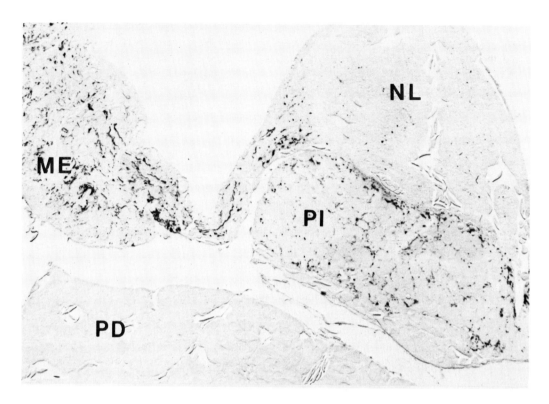

FIGURE 1. Immunohistochemical localization of tyrosine hydroxylase (TH) in the frog pituitary. Immunoreactive fibers are observed in the median eminence (ME). Numerous TH positive fibers are distributed throughout the intermediate lobe (IL) while, in the adjacent neural lobe (NL), labeled fibers are restricted to the border facing the pars intermedia. The distal lobe (DL) is totally devoid of specific staining.

in the median eminence. Many processes originating from the PVO course inside the pituitary stalk and penetrate the pars intermedia. The DIN is a paired nucleus, located at the posterior lateral end of the PVO, which contains dopaminergic and serotoninergic perikarya.[58] The exact projections of these neurons are not known and it cannot be excluded that axons originating from the DIN join the PRO-hypophyseal tract and participate in the dopaminergic innervation of the intermediate lobe.[67]

The presence of catecholaminergic fibers among the parenchymal cells of the pars intermedia is well documented. Iturriza,[68] the first to study the ultrastructure of the amphibian pars intermedia, found nerve fibers containing synaptic vesicles in close contact with the endocrine cells of the toad *Bufo arenarum*. Other electron-microscopical studies have confirmed the presence of aminergic-containing terminals in the pars intermedia of various amphibians: *Xenopus laevis*,[69-71] *Rana esculenta*,[72,73] *Rana pipiens*,[74] *Rana nigromaculata*,[75] *Hyla regilla*,[76] and *Rana ridibunda* (Figure 1). Some of these studies have been carried out by combining radioautography and electron microscopy.[71-73] Using the fluorescence microspectrofluorimetric techinque, the existence of a dense aminergic innervation in the intermediate lobe of several anuran species was confirmed.[57,58,77-80] Autoradiographic and histofluorescence studies indicate that both dopaminergic and noradrenergic fibers exist in the amphibian pars intermedia.[58,71,77] However, direct immunohistochemical (i.e., using dopamine antibodies) or biochemical evidence (i.e., HPLC-electrochemical detection of dopamine and noradrenaline in intermediate lobe extracts) for the presence of both catecholaminergic fibers, is lacking in amphibians. In contrast, dopamine innervation has been demonstrated in the fish pituitary using radioautography[81,82] and immunocytochemistry at the electron-microscopic level.[51] Lesion studies indicate that the sources of catecholaminergic innervation in the neurointermediate lobe of the goldfish are the *nucleus recessus lateralis* and the *nucleus posterioris paraventricularis* of the PVO,[83] but not the anterior ventral preoptic region, although dopaminergic neurons of this region project to the adenohypophysis.[84,85] In the rat, mouse, and pig the innervation of the intermediate lobe of the pituitary is essentially dopaminergic in nature, as revealed by histofluorescent,[86,87] radioautographic,[88] and biochemical techniques.[89,90] Occasionally, noradrenergic nerve fibers are observed in the intermediate lobe.[91] It must be noted, however, that in the rabbit, antiserum against TH does not reveal any axons in the intermediate lobe of the hypophysis.[28]

2. Control the of Pars Intermedia Secretion by Dopamine

In fish, pharmacological studies indicate that dopaminergic fibers, which make synaptic contacts with pars intermedia cells, mediate the inhibitory control of the hypothalamus.[92-95] In the toad, *Bufo arenarum*,. Iturriza[96] showed that administration of reserpine and α-methyl-*m*-tyrosine, which deplete the monoaminergic fibers, both induce a marked increase in the melanophore index. The same author found that monoamine oxydase inhibitors produce bleaching of the skin in dark-background-adapted toads.[96] Similarly, in vivo treatment of *Xenopus* tadpoles with reserpine simultaneously causes a disappearance of catecholamines in the caudal hypothalamus and melanophore dispersion.[97] In *Bufo arenarum*, transplantation of the neurointermediate lobe into the anterior chamber of the eye induces skin darkening; injection of dopamine into the eye 15 days after the operation results in a marked reduction of the melanophore index, indicating that catecholamine mimics the hypothalamic control of the pars intermedia.[16] However, in vivo manipulations of the catecholaminergic systems must be interpreted with a degree of caution since catecholamine can induce melanin concentration by acting directly at the level of the melanophore.[98-100] In mammals, electrothermic lesions of the ventromedian hypothalamus, where dopaminergic perikarya of the tubero-hypophyseal tract are located, cause a reduction in pituitary α-MSH content[101], and an increase in plasma α-MSH levels.[102] Treatment of rats with haloperidol or pimozide significantly stimulates α-MSH secretion in vivo,[101,103,104] while administration of the dopaminergic agonists apomorphine or 2-Br-α-ergocryptine (CB154) causes a marked reduction of circulating α-MSH levels.[104] In agouti mice, pituitary pars intermedia grafted in adult yellow mice produces an intense coat darkening of newly grown hair; administration of bromoergocryptine or apomorphine to the operated animals prevents the darkening effect.[105] In the lizard, *Anolis carolinesis*, pimozide causes skin darkening in normal animals but the dopaminergic antagonist has no effect in hypophysectomized animals.[106] However, such in vivo experiments do determine whether catecholaminergic neurons exert a direct effect on pituitary melanotrophs or whether they act indirectly via other aminergic or peptidergic neuronal systems. Therefore, in vitro experiments have markedly contributed to the understanding of the role of dopamine in the control of pars intermedia secretion.

Bower, Hadley, and Hruby[24] first demonstrated the ability of dopamine to inhibit the release of melanotropin secretion from the intermediate lobe of amphibians and mammals in vitro. Their pioneer work was confirmed in many mammalian species, with the exception of the rabbit,[28] using either perifused neurointermediate lobes[107-110] or intermediate lobe cells in primary culture.[111-113] In both systems, dopamine and its agonists induce a marked inhibition of α-MSH secretion. Pharmacological characterization of dopamine receptors has been performed in rat[112,114,115] and bovine intermediate lobes.[116] According to Munemura et al.,[111] the dopamine receptor in the intermediate lobe can be assigned to the category designated D2 (based on the classification of Kebabian and Calne[117]) i.e., negatively coupled to the adenylate cyclase system.[111,112,115] Using incubations of amphibian intermediate lobes, several groups have shown that dopamine blocks α-MSH secretion.[24,118-122] In *Rana ridibunda*, apomorphine and CB154 are more potent than dopamine in inhibitng α-MSH secretion (Figures 2 and 3).[120,123] In *Xenopus laevis* dopaminergic receptors have been carefully

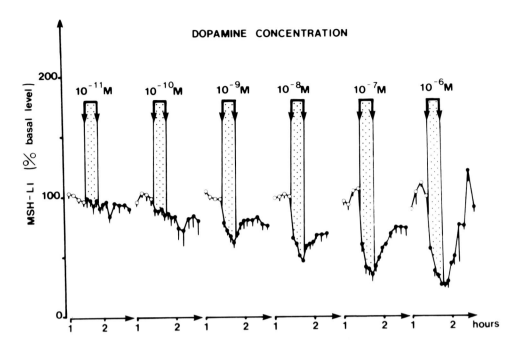

FIGURE 2. Effect of increasing concentrations of dopamine on the release of α-MSH by perifused frog neurointermediate lobes (NIL). Administration of dopamine (10^{-11} to 10^{-6} M) for 20 min periods, induced a dose-related inhibition of α-MSH release. Each profile represents the mean ($\pm$ S.E.M.) of at least three independent perifusion experiments. The reference level of α-MSH (100% basal level) was calculated for each experiment as the mean α-MSH secretion rate during 30 min (four consecutive fractions, ○–○) just preceding infusion of dopamine. (From Adjeroud, S., et al., *Gen. Comp. Endocrinol.*, 64, 1986, in press. With permission.)

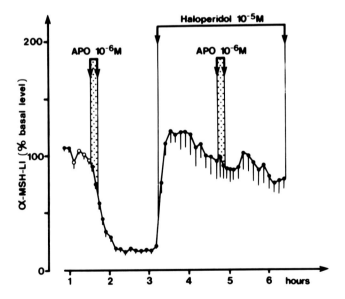

FIGURE 3. Effect of apomorphine in the absence or presence of a dopaminergic antagonist, haloperidol on the release of α-MSH by perifused frog NIL. Apomorphine (APO), a dopaminergic agonist, administered for 10 min at a concentration of 10^{-6} M, induced a marked inhibition of the basal secretion of α-MSH. The effect of APO was reversed by the infusion of haloperidol (10^{-5} M). See legend to Figure 2 for additional information. (From Tonon, M.C., et al., *Endocrinology*, 112, 133, 1983. With permission.)

characterized by administering selective agonists and antagonists to perifused neurointermediate lobes.[122] It should be mentioned that the perifusion technique used by several groups to investigate dopaminergic control of frog pars intermedia cells[120-123] has several advantages over static incubations. This method avoids accumulation of the secreted peptides and thus eliminates a possible self-inhibition of MSH release by high ambient concentration of the hormone.[124] The perifusion technique is also particularly appropriate for the study of rapid changes in hormone secretion and makes it possible to investigate biphasic responses of the endocrine cells to exogenous factors. Finally, this approach provides valuable information regarding the kinetics of the responses of the glands to various secretagogues. In *Xenopus laevis*, receptors for dopamine display both analogies and discrepancies with those described in mammals. First, it would appear that the receptor involved is not only a dopamine receptor, but also displays a high sensitivity to all catecholamines; both adrenaline and noradrenaline give full inhibition of α-MSH release at a 10-fold lower concentration than dopamine.[122] This effect is fully antagonized by sulpiride, a D2 receptor antagonist, while phentolamine, an α-adrenergic receptor antagonist, appears to be a poor antagonist. Moreover, the effect of adrenaline is easily antagonized by the D2 antagonists haloperidol and domperidon. The *Xenopus* catecholamine receptor system clearly shows more analogy to a classical D2 type of receptor than to the D1 type. D1 analogy is unlikely since the dopamine D1 agonist SKF38393 is at least 100 times less potent than apomorphine in inducing inhibition of α-MSH release, and since sulpiride antagonizes this inhibition. Moreover, the dopamine D1 receptor antagonist, SCH23390 has no effect on dopamine-induced α-MSH release inhibition. Similar to classical D2 receptors, apomorphine and LY171555 are potent agonists and their effect can be antagonized by the D2 antagonists sulpiride, domperidon, and haloperidol. In contrast, YM-09151-2 (a D2 antagonist) and N-0437 (a D2 agonist),[122] which have both been reported to be highly effective and selective for mammalian D2 receptors,[125] are devoid of effect in *Xenopus*. The presence of a catecholamine receptor rather than a classical D2 receptor does not explain why the dopamine-induced inhibition of α-MSH release can not be antagonized by haloperidol and domperidon, which antagonize the effects of apomorphine and LY-171555. It also offers no explanation for the fact that phentolamine (an α-receptor antagonist), which had no effect on adrenaline and noradrenaline-induced α-MSH release inhibition, antagonized the effects of apomorphine and LY-171555. Altogether, the *Xenopus* receptors for dopamine appear to be similar to the mammalian D2 receptor, but do not display identical characteristics to their mammalian counterpart. These data suggest that the structure of dopamine receptors in the intermediate lobe of the pituitary may not be perfectly conserved during the course of evolution.

B. Noradrenaline

It is generally accepted that the rat neurointermediate lobe possesses two distinct catecholamine receptors; dopamine D2 receptors which inhibit the release of α-MSH and β-adrenoceptors which stimulate the release of this hormone.[126] While the presence of dopamine receptors negatively coupled to the adenylate cyclase system appears to be a general feature in all vertebrate species except the rabbit (see above), the existence of adrenergic receptors is very controversial. The abundant data accumulated on the effects of adrenomimetic drugs on the intermediate lobe suggest marked species variations in the vertebrate phylum.

1. Anatomical Basis

The organization of the reticulo-infundibular noradrenergic neurons has been studied in detail using the Falck-Hillarp histochemical technique.[91] Noradrenaline perikarya, mainly located in the lower brain stem, send out axons which traverse the medial forebrain bundle and innervate the internal zone of the median eminence. In the rat,[91] some of these neurons reach the pituitary stalk, but do not seem to penetrate the neural lobe since noradrenergic fibers are still observed in the neural lobe after lesion of the median eminence and pituitary stalk. In contrast, the noradrenergic terminals of the neural lobe and of the vascular border zone of the pars intermedia, disappear in sympathectomized rats, indicating that the noradrenergic innervation of the neurointermediate lobe is probably of peripheral, sympathetic origin.[91] The measurement of norepinephrine concentrations in neurointermediate lobe extracts, confirms the decrease in posterior lobe catecholamine levels following bilateral superior cervical ganglionectomy, but also suggests the contribution of axons of brain origin in the norepinephrinergic innervation of the neurointermediate lobe.[127] In teleosts, fluorescent histochemistry of the forebrain reveals the presence of dopamine/noradrenaline perikarya in the *nuclei recessus posterioris* (NRP) and *-lateralis* (NRL) of the paraventricular organ (PVO).[128-131] Lesions of the NRL or *nucleus posterioris paraventricularis* (NPPv) induce degeneration of nerve terminals in the neurointermediate lobe of the goldfish.[83] Radioautographic techniques show the uptake of ^{3}H-noradrenaline into axon terminals at the pars intermedia level.[46] In amphibians, analysis of fluorescent nerve fibers in the pars intermedia of *Rana temporaria* reveals a double innervation containing dopamine and noradrenaline.[58,77] Some of the axon terminals are labeled after injection of ^{3}H-noradrenaline in *Rana esculenta*[132] or ^{3}H-adrenaline in *Xenopus laevis*.[71] In a recent study, Kondo et al., reported that the pars intermedia of *Rana catesbeiana* is devoid of noradrenergic nerve terminals since TH-positive but DBH-negative immunoreactive varicose fibers are found surrounding the melanotropic cells.[133] After treatment with 6-OHDA, Hopkins observed a complete removal of adrenergic nerve endings[71] in the pars intermedia of *Xenopus laevis* and marked depletion of adrenaline and noradrenaline (80%) at the pituitary level. Dopaminergic-containing fibers may originate from PVO and DIN[58] (see above), but the origin of (nor)adrenergic fibers present in the intermediate lobe remains unknown.

2. Control of the Pars Intermedia Secretion by Noradrenaline

In amphibians, both adrenaline and noradrenaline inhibit release of α-MSH in vitro.[24,134-136] In fact, in *Rana pipiens*, adrenaline and noradrenaline induce both inhibition and stimulation of α-MSH secretion depending on the concentration.[24] Bower et al.,[24] suggest that the inhibition of α-MSH release is mediated through α-adrenergic receptors, whereas stimulation acts via β-adrenergic receptors. In agreement with these authors, Tonon et al.,[123] demonstrated that isoproterenol, a β-adrenergic receptors agonist, increases α-MSH release, an effect which is totally abolished by the β-adrenergic receptor antagonist, propranolol (Figure 4). In contrast, MSH cells of *Xenopus laevis* are devoid of β-adrenergic receptors. In this species, isoproterenol, adrenaline, and noradrenaline induce an inhibition of α-MSH release, but the effect of these compounds is totally abolished by sulpiride, a dopamine-D2 receptor antagonist.[136] Thus, in contrast to *Rana*,[24,123] *Xenopus* melanotrophs are not regulated through a β-adrenergic mechanism.

C. GABA

Gamma aminobutyric acid (GABA) is a neuroactive amino acid discovered in brain tissue more than three decades ago.[137-139] The biosynthesis of GABA in specific neurons depends on the presence of an enzyme, called glutamic acid decarboxylase (GAD), which causes the decarboxylation of glutamate. Thus, both the amino acid GABA and the enzyme GAD can be considered to be specific markers of GABAergic neurons.[140] GABA is the most prominent neurotransmitter occurring in the brain: in the central nervous system, 45% of the neurons are GABAergic and these neurons exhibit an ubiquitous distribution in the brain of mammals; in several brain areas GABAergic neurons comprise more than 90% of the total neurons.[141] Based on electrophysiological and pharmacological investigations, GABA is generally considered to be one of the major inhibitory neurotransmitters within the brain.[142] Recent immunohistochemical and pharmacological studies strongly suggest that GABA is

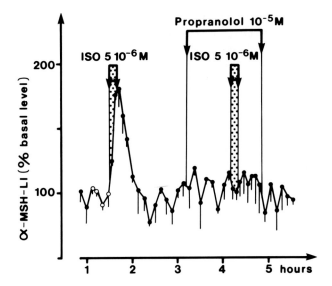

FIGURE 4. Effect of isoproterenol in the absence or in the presence of a β-adrenergic antagonist, propranolol, on the release of α-MSH by perifused frog NIL. The stimulation of α-MSH release induced by the β-adrenergic agonist isoproterenol (5×10^{-6} M), is completely abolished in the presence of propranolol (10^{-5} M). See legend to Figure 2 for additional information. (From Tonon, M.C., et al., *Endocrinology*, 112, 133, 1983. With permission.)

an important neuroendocrine regulator, controlling the secretory activity of both the anterior and neurointermediate lobes of the pituitary.[31,143-146]

1. Anatomical Basis

The distribution of GABAergic neurons has been studied by immunocytochemistry using either antisera raised against GAD[147-151] or against GABA itself.[152-155] Selective uptake of ^{3}H-GABA has also been applied to the autoradiographic localization of GABAergic neurons.[156-160] The advantages and disadvantages of these different methods have been discussed by Mugnaini and Oertel.[141] Detailed mapping of GABAergic neurons has been investigated in the brain of the mouse,[153,160] rat,[153,161] rabbit,[162] and cat.[154] In all species examined, the GABAergic cell bodies and nerve terminals are widely distributed and constitute the most abundant neuronal system of the brain. The highest concentrations of GABAergic cell bodies (density higher than 90% of total neurons) are found in the neocortex, hippocampus, striatum, amygdala, pallidum, and cerebellum (Purkinje cells). In the diencephalon, a high proportion of GABAergic cell bodies are found in the dorso-lateral region of the hypothalamus (zona incerta), the posterior magnocellular nucleus, the preoptic nucleus, and the dorsomedian hypothalamic nucleus. In the basal hypothalamus, the arcuate nucleus contains numerous immunoreactive cell bodies and nerve fibers. A dense plexus exhibiting GAD immunoreactivity reaches the external layer of the median eminence in the mouse, rat, and cat.[151] In the mouse, this network extends from the rostral part of the median eminence towards the pituitary stalk. In all mammalian species examined so far, axon terminals are distributed in the neural lobe and a dense plexus of fibers innervates the intermediate lobe of the pituitary.[31,144] In the goldfish, recent studies at the light- and electron-microscopic levels indicate the presence of GABAergic fibers among the parenchymal cells of the adenohypophysis.[163] In amphibians, GAD activity has been compared in brain and pituitary extracts; the intermediate lobe contains 12% of the GAD activity found in the whole brain while no significant activity can be detected in the adenohypophysis.[164] These results confirm the pioneer work of Oertel et al.,[31] who showed that homogenates of rat neurointermediate lobes contain 10 to 15 times more GAD activity than the anterior lobe. Recently, GABAergic fibers have been directly visualized in the neurointermediate lobe complex of two anuran species, *Xenopus laevis*[165] and *Rana ridibunda*.[164] Concurrently, pharmocological experiments indicated that GABA is likely to be involved in the inhibitory control of melanotropin secretion.

2. Control of the Pars Intermedia Secretion by GABA

Hadley and Bagnara[166] provided the first pharmacological evidence that GABA inhibits MSH secretion. Application of a depolarizing concentration of K^+ was shown to inhibit MSH release by whole rat neurointermediate lobes.[167,168] According to Tilders,[167] high K^+ is responsible for a massive discharge of dopamine by axon terminals innervating pars intermedia cells. However, in two amphibian species, *Xenopus*[169] and *Rana* (unpublished data), K^+-induced inhibition of α-MSH secretion cannot be blocked by dopamine antagonists, suggesting that inhibiting neurotransmitter(s) other than dopamine may be released by nerve endings within the pars intermedia. Indeed, Verburg-van Kemenade has found that high K^+ stimulates release of both ^{3}H-dopamine and ^{3}H-GABA from nerve terminals within the neurointermediate lobe tissue (unpublished observations). The first indication that GABA may be involved in the inhibitory control of the intermediate lobe of amphibians was reported by Davis and Hadley.[170] These authors showed that GABA (as well as dopamine) reduces the frequency of spontaneous action potentials recorded from pars intermedia cells of *Rana berlandieri*. In rat, GABA appears to have a dual effect on α-MSH secretion: a transient stimulation caused by depolarization of melantropic cell membrane followed by a prolonged inhibition.[145,171] Electrophysiological studies indicate the existence of both $GABA_A$ and $GABA_B$ sites on rat and pig pars intermedia cells. GABA and $GABA_A$ receptor agonists increase membrane chloride conductance and thereby depolarize the melantropic cells.[172-174] This depolarization is responsible for a transient release of MSH.[145,171] Conversely, baclofen, a specific $GABA_B$ receptor agonist, produces a decrease in membrane conductance[173] and an inhibition of α-MSH release.[171,174]

Using the perifusion technique, GABA receptors have recently been characterized in two amphibian species. In *Xenopus laevis*, administration of GABA induces a dose-dependent inhibition of α-MSH release from neurointermediate lobes.[165] Moreover, GABA, baclofen (a $GABA_B$ receptor agonist), and the two $GABA_A$ receptor agonists, isoguvacine, and homotaurine, inhibit α-MSH secretion from isolated *Xenopus* melanotrophs.[169] The absence of a stimulatory phase during the onset of GABA administration[165] and the fact that bicuculline, a $GABA_A$ receptor antagonist, has no effect on GABA-induced inhibition of α-MSH secretion in vitro,[175] indicate that, in *Xenopus*, GABA acts mainly through activation of $GABA_B$ receptors. In vivo administration of GABA to *Xenopus laevis* causes a dose-dependent bleaching of the skin,[165] a result consistent with in vitro findings in this species. In *Rana ridibunda*, both GABA and muscimol, a $GABA_A$ receptor agonist, cause a brief stimulation of α-MSH release followed by an inhibition (Figure 5).[164] Picrotoxin, a Cl$^-$ channel blocker, abolishes only the stimulatory effect of GABA, whereas bicuculline totally inhibits the effects of GABA (both stimulatory and inhibitory phases) (Figure 6). Administration of baclofen induces a dose-dependent inhibition of α-MSH secretion. In contrast to GABA and muscimol, baclofen does not cause any stimulatory effect whatever the dose.[164] Thus, in amphibians, as is mammals,[171,173,174] neurointermediate lobe tissue contains both $GABA_A$ and $GABA_B$ receptors.[164,169] Activation of either of these receptors induces inhibition of α-MSH release.[164,169] Nevertheless GABA acts essentially through $GABA_B$ receptors in *Xenopus* melanotrophs, whereas in *Rana*, the effect of GABA is mainly achieved through activation of $GABA_A$ receptors.

The control of α-MSH release shows homologies with the control of prolactin release. In mammals, GABA is a powerful inhibitor of prolactin secretion in vitro.[143,176] The visual-

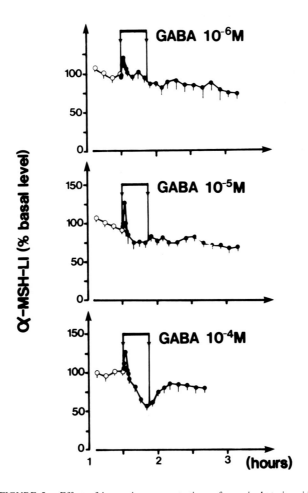

FIGURE 5. Effect of increasing concentrations of γ-aminobutyric acid (GABA) on the release of α-MSH by perifused frog NIL. Administration of GABA (10^{-6} to 10^{-4} M) for 20 min periods, induced a brief stimulation of α-MSH release. At the highest doses of GABA (10^{-5} and 10^{-4} M) this stimulatory phase was followed by a marked inhibition of α-MSH secretion. See legend to Figure 2 for additional information. (From Adjeroud, S., et al., Brain Res. Bull., 17, 1986, in press. With Permission.)

ization of nerve terminals immunoreactive for both GAD[144,177,178] and GABA[155] in the external zone of the rat median eminence and the presence of significant amounts of GABA in the portal blood,[179,180] suggest a neuroendocrine role of GABA in the control of anterior pituitary function. In addition, the existence of GABA-transaminase activity in the anterior pituitary[181] supports the concept that GABA is a neuroendocrine regulator of adenohypophyseal secretions. The receptors involved in the control of rat mammotrophs were initially thought to be $GABA_A$ sites since picrotoxin blocks the inhibitory effect of GABA.[176] More recently, the effect of GABA on prolactin secretion in the rat was examined using a rapid superfusion system.[182] According to Anderson and Mitchell,[182] GABA and muscimol cause a biphasic effect on prolactin secretion. Both stimulatory and inhibitory phases are antagonized by bicuculline, while baclofen is devoid of effect on prolactin release. These authors suggest the existence of two types of $GABA_A$ receptors mediating GABA-induced stimulation and inhibition of prolactin secretion. Since similar results are obtained on α-MSH secretion by perifused frog neurointermediate lobe,[164] we postulate that two types of $GABA_A$ receptors are also present in *Rana ridibunda* melanotrophs.

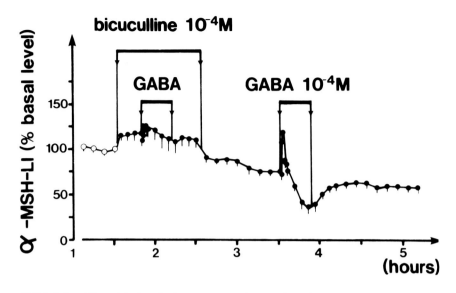

FIGURE 6. Effect of GABA, in the presence or absence of the $GABA_A$ antagonist bicuculline, on the release of α-MSH by perifused frog NIL. Both stimulatory and inhibitory effects of GABA (10^{-4} M) were totally abolished in the presence of bicuculline (10^{-4} M). See legend to Figure 2 for additional information. (From Adjeroud, S., et al., *Brain Res. Bull.*, 17, 1986, in press. With permission.)

D. Serotonin

Several pharmacological studies have suggested the involvement of serotonin in the control of melanotropin secretion in the lizard,[106,183] frog,[170] eel,[184] and rat.[185]

1. Anatomical Basis

The distribution of serotoninergic neurons in the mammalian brain has been the subject of a number of articles (see Steinbusch for review[186-188]). Using either the formaldehyde-induced fluorescence technique[52,189] or the immunohistochemical approach,[186,187,190] the main structures containing serotoninergic perikarya have been identified in mammals: raphe nuclei (pallidus, obscurus, magnus, dorsalis)[52,186,187,191,192] and the nucleus dorsomedialis hypothalami.[187,193] These neurons send ascending projections towards the mesencephalon, diencephalon, and telencephalon.[52,186,194,195] Several descending serotoninergic pathways have been also localized in the spinal cord (dorsal and ventral horn, and intermediate grey).[196,197] In lower vertebrates, the serotoninergic system has been extensively investigated. Most of the results obtained in fish,[198,199] amphibians,[200,201] reptiles,[202] and birds,[203,204] indicate that the organization of serotonin-containing neurons has been highly preserved during evolution. In particular, serotonin-containing cell bodies are found in the superior and inferior raphe nuclei and the mesencephalic tegmentum, in the goldfish, *Caurassius auratus*,[199] the salamander, *Necturus maculosus*,[201] the newt, *Triturus cristatus*,[205] the tadpole of *Xenopus laevis*,[57] and the lizard, *Varanus exanthematicus*.[202] In the diencephalon of cold-blooded vertebrates, one group of serotonin cell bodies is located in the paraventricular organ.[199,202,205] Histochemical studies, using formaldehyde-induced fluorescence, have shown the presence of serotonin-containing fibers in the neural and intermediate lobes of the rat[7] and cat.[205a] The existence of serotoninergic fibers in the rat pituitary has been confirmed by immunohistochemical[32,206,207] and autoradiographic methods.[88] Serotonin-containing fibers innervate the rostral zone of the adenohypophysis; other fibers originating from the neural lobe penetrate into the pars intermedia where they are relatively abundant.[10,32,206,207,207a] These fibers are known to mainly originate from extrinsic neurons located in the brain, both

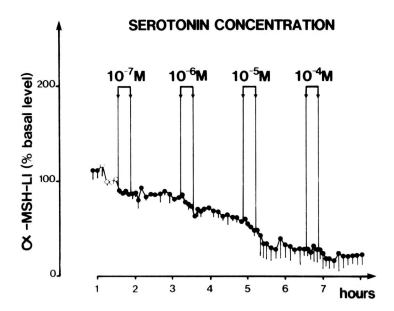

FIGURE 7. Effect of increasing concentrations of serotonin on the release of α-MSH by perifused frog NIL. Whatever the dose administered (10^{-7} to 10^{-5} M), serotonin did not affect α-MSH secretion. See legend to Figure 2 for additional information.

midbrain raphe, and hypothalamic dorsomedial nuclei,[193] as pituitary stalk transection causes a marked decrease in the number of immunoreactive fibers in the pars intermedia.[206] Electrical stimulation of the pituitary stalk causes massive, calcium-dependent release of serotonin, suggesting that this amine may act as a neuromodulator of the intermediate lobe.[208] Thus, a substantial proportion of serotonin in the intermediate lobe probably originates from non-neuronal elements: blood-born elements and mast cells.[209] In the lizard, *Anolis carolinensis*, the intermediate lobe of the pituitary is not innervated,[210,211] but serotoninergic elements innervate the neural lobe and many of them are in contact with the capillaries connecting the pars nervosa to the pars intermedia.[212] Serotonin-containing cells have also been observed among the endocrine cells of the pars intermedia of the amphibian, *Rana catesbeiana*.[133] In the same species, serotonin immunoreactive nerve terminals have been located in the intermediate lobe.[213] In contrast, in the toad, *Xenopus laevis*, serotoninergic fibers are restricted to the neural lobe, and no evidence has been obtained for the presence of serotonin in the intermediate lobe.[136] Altogether, these data would support a role of serotonin in the control of melanotropin secretion.

2. Control of the Pars Intermedia Secretion by Serotonin

In vivo treatment of eels by serotonin or its precursor 5-hydroxytryptophan, causes skin darkening[184] and stimulates MSH cells as revealed by development of the endoplasmic reticulum/Golgi apparatus and the reduction in the number of secretory granules.[214] Administration of parachlorophenylalanine, an inhibitor of tryptophan hydroxylase, depresses the activity of MSH cells in fish[215] and lizards.[216] Other authors have reported that serotonin induces a dose-dependent stimulation of MSH release in the lizard.[183] These results suggest that serotonin may have a physiological role in the regulation of MSH secretion in fish and reptiles. In contract, Verburg-van Kemenade et al.,[136] did not observe any effect of serotonin on α-MSH release by the pars intermedia of *Xenopus laevis* in vitro. Similar results have been obtained in *Rana ridibunda*; for doses ranging from 10^{-7} to 10^{-4} M, serotonin does not modify α-MSH secretion (Figure 7). In the rat, high doses of serotonin have been shown to stimulate ACTH release from dispersed pars intermedia cells[217] or intact neurointermediate lobes.[218] However, Briaud et al.,[219] did not observe any effect of serotonin on perifused rat neurointermediate lobes. In the same species, serotonin is able to stimulate α-MSH release in vitro,[185] but some authors have noted that serotonin fails to alter hormonal secretion.[109,113,220]

These data suggest that serotonin is most likely involved in the control of MSH secretion in fish and lizards, whereas it has only a minor role (if any) in the regulation of the pars intermedia of amphibians and mammals.

E. Other Neurotransmitters

Several classical neurotransmitters, other than dopamine, noradrenaline GABA, and serotonin, have been reported to be implicated in the regulation of melanotropin release. In particular, investigations conducted with two of them, acetylcholine and histamine, suggest that they are involved, in some species, in the control of pars intermedia cells.

1. Anatomical Basis

Measurements of acetylcholine by bioassay and cholinesterase by colorimetric assay, have been performed in the pituitaries of various mammalian species.[221] Both techniques indicate the presence of high concentrations of cholinergic nerve terminals in the region of the pars intermedia-nervosa junction of the rat. Cholinergic innervation of the pars intermedia of the rat has also been reported by other authors.[7,222,223] In the amphibian, *Xenopus laevis*, Hopkins[71] showed that the pars intermedia is devoid of cholinergic fibers. In contrast, electron-microscopic studies have revealed the presence of cholinergic nerve terminals in the pars intermedia of the frog, *Rana pipiens*.[74]

The distribution of histamine in hypothalamic nuclei of the rat has been investigated by an enzymatic-isotopic method.[224] High concentrations of histamine are found in the ventral premammillary nucleus and in the suprachiasmatic and arcuate nuclei. Using antisera against glutamic acid decarboxylase, Takeda et al.,[225] have observed the coexistence of GABA and histamine within neurons of the magnocellular nucleus of the posterior hypothalamus. The median eminence contains the highest concentration of histamine (twice as much as the ventral premammillary nucleus which, among the hypothalamic nuclei, is the richest in amine).[224] However, part of the histamine present in the median eminence may not be of neuronal origin and could possibly be released by mast cells. Saavedra et al.,[90] have quantified histamine in the three lobes of the rat pituitary and have found that the intermediate lobe contains the higher amount of this amine. In fact, the concentration of histamine in the pars intermedia is much higher than those of catecholamines and serotonin.

2. Control of the Pars Intermedia Secretion by TRH

The first demonstration that P-Glu-His-Pro-NH$_2$ stimulates α-MSH secretion, has been established in *Rana ridibunda*.[262,263] Figure 8 illustrates the response of perifused frog neurointermediate lobes to various doses of TRH. Ten-minute pulses of synthetic TRH lesioned or electrically stimulated.[227] However, the muscarinic antagonist, atropin, induces melanophore expension, presenting something of a paradox.[226] In *Xenopus laevis*, acetylcholine has no influence on MSH secretion in vitro,[136] a fact which is consistent with the observation of Hopkins that cholinergic fibers are absent from the intermediate lobe in this species.[71] In the rat, several studies indicate that acetylcholine does not control the activity of pars intermedia cells,[219] whereas one report suggests that exogenous acetylcholine stimulates the release of "biologically active ACTH" from isolated neurointermediate lobes.[218]

Application of histamine to intermediate lobes of *Rana berlandieri* does not alter the frequency of spontaneous membrane potentials,[170] whereas dopamine inhibits the electrical activity of these cells. In the frog, *Rana ridibunda*, histamine does not influence α-MSH secretion by perfused neurointermediate lobes (unpublished observations). Histamine does not significantly influence hormonal secretion of the pars intermedia of the rat either.[113]

III. CONTROL OF THE PARS INTERMEDIA BY REGULATORY PEPTIDES

A. Thyrotropin-Releasing Hormone

The tripeptide, L-pyroglutamyl-L-histidyl-L-prolineamide, has been isolated and characterized from ovine[228] and porcine[229] brains. In mammals, this hypothalamic peptide is primarily a thyrotropin-releasing factor (TRH),[230,231] but it also exhibits prolactin-releasing activity.[232,233] In the brain of submammalian vertebrates, the tripeptide, P-Glu-His-Pro-NH$_2$, has been localized by immunocytochemistry[64] and quantified by radioimmunoassay.[234-237] Extracts from amphibian brains stimulate TSH release from rat pituitary in vitro[234,235] and in vivo.[234] In addition, high concentrations of P-Glu-His-Pro-NH$_2$ have been reported in the plasma of *Rana pipiens*,[238] and the source of this peptide in the systemic blood seems to be the skin where its concentration greatly exceeds those measured in other tissues.[239-243] Although some reports suggest that P-Glu-His-Pro-NH$_2$ stimulates the pituitary-thyroid axis of *Rana ridibunda*,[244] Studies indicate that this peptide does not act as a thyrotropin-releasing factor in lower vertebrates.[235, 245-255] Therefore, wheather the tripeptide, P-Glu-His-Pro-NH$_2$, corresponds to TRH in fish and amphibians, remains questionable. Recent studies indicate that this peptide stimulates prolactin secretion in cold-blooded vertebrates.[256-261] In addition, P-Glu-His-Pro-NH$_2$ is potentially an important regulator of melanotropin secretion in amphibians.[262-265]

1. Anatomical Basis

The distribution of TRH neuronal systems has been extensively studied in the mammalian brain.[266-271] The main group of TRH-containing cell bodies is located in the parvocellular division of the paraventricular nucleus. Other regions of the hypothalamus, including the dorso-medial, periventricular, and perifornical nuclei and the lateral hypothalamus, are also richly endowed with perikarya staining for TRH.[266,268,270] Complete surgical isolation of the hypothalamus, from other brain regions, results in a marked decrease of hypothalamic immunoreactive TRH, indicating that the hypothalamic peptide may be, in part, of extra-hypothalamic origin.[272] Little is known about the distribution of TRH-neuronal systems in lower vertebrates. The complete mapping of TRH neurons in the brain of the frog, *Rana catesbeiana*, has been described by Seki et al;[64] immunoreactive TRH-cell bodies are located in the anterior part of the preoptic nucleus, the dorsal infundibular nucleus, the nucleus of the diagonal band of Broca, and the medial part of the amygdala. The presence of high concentrations of TRH in the hypothalamic region has also been established in various amphibian species by radioimmunoassays.[64,234-236] In the frog, nerve processes are observed in the external layer of the median eminence, in close contact with the capillary looops of the hypophyseal portal vessels,[64] and a dense network of TRH immunoreactive fibers innervates the neurohypophysis.[64,265,273] Immunoreactive processes are also present in the intermediate lobe.[64,273,274] At the electron-microscopic level, TRH-containing terminals are found among the endocrine cells of the intermediate lobe.[274] In some of these nerve endings, coexistence of TRH and mesotocin has been documented.[274] Quantification of TRH by radioimmunoassay indicates that the neurointermediate lobe contains higher concentrations of TRH than the hypothalamic region.[64,265,275] The existence of a direct innervation of the anuran neurointermediate lobe by TRH fibers supports our early finding that TRH is involved in the control of melanotropin secretion in amphibia.[262-265]

2. Control of Pars Intermedia Secretion by TRH

The first demonstration that P-Glu-His-Pro-NH$_2$ stimulates α-MSH secretion, has been established in *Rana ridibunda*.[262,263] Figure 8 illustrates the response of perifused frog

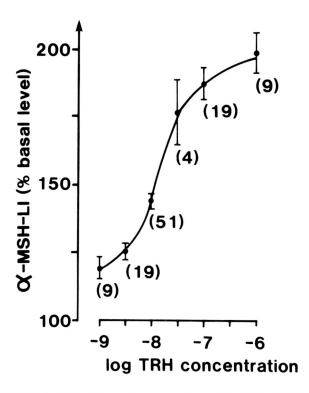

FIGURE 8. Effect of increasing concentrations of thyroliberin (TRH) on the release of α-MSH by perifused frog NIL. Administration of TRH (10^{-9} to 10^{-6} M) for 10 min periods, induced a dose-related stimulation of α-MSH release. Percentage of α-MSH increase was calculated as the net area under the peak (over 30 min) vs. the spontaneous level (mean ± S.E.M.). The number of experiments is indicated between parentheses. (From Tonon, M.C., et al., *Gen. Comp. Endocrinol.*, 52, 173, 1983. With permission.)

neurointermediate lobes to various doses of TRH. Ten-minute pulses of synthetic TRH produce a dose-dependent response for concentrations ranging from 10^{-9} to 10^{-6} M. Half-maximum stimulation occurs at 1.2×10^{-8} M TRH. The melanotropin-releasing activity of TRH has been confirmed in various anuran species including *Rana pipiens*[264] and *Xenopus laevis*.[265] It is interesting to note that, in the toad, TRH stimulates α-MSH release only from neurointermediate lobes of white-background-adapted animals.[265] Among other vertebrates, the stimulatory effect of TRH has only been studied in the goldfish (J. Fryer, personal communication). In the rat, Kraicer[276] has clearly established that TRH has no effect on α-MSH or ACTH secretion by neurointermediate lobes in vitro. Conversely, it has been shown that TRH stimulates ACTH release by pituitary tumor cells.[277] Similarly, TRH has been found to increase ACTH secretion in patients with Cushing's disease or Nelson's syndrome.[278-280] It appears, therefore, that both pituitary melanotrophs of lower vertebrates and ACTH-secreting tumor cells of mammals can express TRH receptors. The α-MSH release inhibiting activity of rat hypothalamic extracts on Frog melanatrophs is markedly strengthened after immunoneutralization of endogenous TRH by TRH antiserum.[263] Structure-activity relationships have been studied using 20 different synthetic TRH analogues.[281] A remarkable correlation exists between the prolactin (or TSH)-releasing potencies of the various analogues on rat anterior pituitary cells and their MSH-releasing activity in amphibians (Figure 9). In particular, the TRH analogue (MK-771[281]) and the recently discovered [thioamide Pro]³-TRH,[282] are equipotent with TRH in their ability to release prolactin or α-

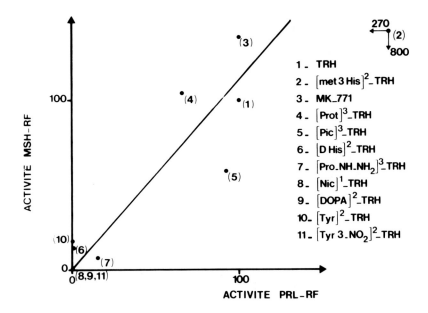

FIGURE 9. Comparison of the biological potency of 11 synthetic analogues of TRH on α-MSH release by perifused frog NIL on the one hand and prolactin (PRL) and/or thyroid-stimulating hormone (TSH) release by rat anterior pituitary cells on the other hand. A significant correlation was observed between MSH-RF activity in amphibians and TSH- or PRL-RF activity in mammals. (From Tonon, M.C., et al., *Ann. Endocrinol.*, 46, 69, 1985. With permission.)

MSH. In addition, the [Met^3-His]^2-TRH analogue is a superagonist on both rat anterior- and frog neurointermediate lobes (Figure 9).

Interactions between TRH and various hormones, neuropeptides, or neurotransmitters on melanotropin secretion, have been examined. A significant uptake of thyroxin (T_4) by frog melanotrophs has been documented.[283] Since thyroid hormones inhibit the stimulatory effect of TRH in rat anterior pituitary cells[284] we have investigated the possible role of T_4 in the regulation of TRH activity on frog melanotrophs.[285] In vivo treatment of frogs with T_4 (0.5 mg/kg twice a day for 9 days) or prolonged infusion of T_4 in vitro, had no effect on TRH-induced α-MSH secretion, indicating that thyroid hormones do not modulate the stimulatory effect of TRH on frog melanotrophs. The presence of high concentrations of immunoreactive somatostatin in the neurointermediate lobe complex of the frog[236,275] and the observation that somatostatin inhibits TRH-induced TSH and prolactin release from rat anterior pituitary cells,[286,287] led us to investigate a possible modulatory action of somatostatin on TRH-induced α-MSH secretion in anphibians. In fact, synthetic somatostatin (10^{-10} to 10^{-6} M) does not affect the spontaneous release of α-MSH secretion and does not modify the stimulatory effect of TRH on α-MSH secretion (Figure 10). Similarly, opioid peptides (β-endorphin, met- and leu-enkephalins) which are present in high concentrations in amphibian pars intermedia-nervosa[275] and which modulate the release of prolactin from anterior pituitary cells,[288-291] do not alter α-MSH secretion in amphibians.[275] Thus, although the stereochemical requirements of TRH for TSH- and prolactin-releasing activities in mammals and α-MSH-releasing activity in amphibians are closely related, the physiological mechanisms involved in the control of the number and/or affinity of TRH receptors are different in mammals and amphibia.

The demonstration that TRH stimulates α-MSH release results from studies which were conducted on whole pars intermedia-nervosa[262,263,265] or in vivo.[264] Since, in these models, the pituitary cells remain tonically inhibited by dopaminergic inputs[123] and since TRH mod-

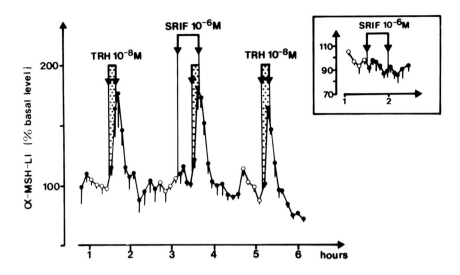

FIGURE 10. Effect of somatostatin (SRIF) on basal and TRH-induced α-MSH release by perifused frog NIL. Administration of SRIF (10^{-6} M) had no effect on basal α-MSH release (inset.) TRH (10^{-8} M), infused in the presence of SRIF (10^{-6} M) induced a stimulation of α-MSH release similar to that obtained in the absence of SRIF. See legend to Figure 2 for additional information. (From Tonon, M.C., et al., *Gen. Comp. Endocrinol.*, 52, 173, 1983. With permission.)

ulates various dopaminergic systems,[292-294] it was essential to determine whether TRH acts directly on frog melanotrophs or whether it exerts its stimulatory effect via presynaptic control of catecholaminergic fibers. In fact, prolonged infusion of the dopaminergic antagonist, haloperidol, or the β-adrenergic antagonist, propranolol, does not reduce the stimulatory effect of TRH.[123] These results, together with the demonstration that TRH stimulates α-MSH release from enzymatically dispersed pars intermedia cells of *Rana ridibunda* and *Xenopus laevis* (unpublished observations) indicate that TRH exerts its stimulatory effect directly on frog melanotrophs.

Several studies have been undertaken to investigate the mechanism of action of TRH at the molecular level. In mammals, it is generally accepted that an elevation of intracellular concentrations of Ca^{2+} serves to couple, at least in part, stimulation to secretion.[295] There is direct evidence indicating that TRH is responsible for an increase in intracellular Ca^{2+} concentrations in GH_3 cells.[296-298] In mammals, TRH causes an increase in intracellular calcium concentration ($[Ca^{2+}]_i$) reflecting a release of Ca^{2+} from intracellular pools and an increase of the influx of extracellular Ca^{2+}. In the frog, TRH looses its ability to induce α-MSH secretion in the absence of extracellular Ca^{2+} or in the presence of Co^{2+}.[299] Conversely, we have observed that administration of Verapamil® or D600, two voltage-sensitive Ca^{2+} channel blockers, does not alter spontaneous α-MSH secretion and has no effect on the response of the glands to TRH (unpublished data). In contrast Nifedipine®, another voltage-sensitive Ca^{2+} channel blocker, induces a slight inhibition of basal α-MSH release, indicating that circulating Ca^{2+} levels may regulate α-MSH release in part by Ca^{2+} influx through voltage-dependent channels. TRH-induced α-MSH release is not affected by Nifedipine® or Dantrolene®, a blocker of calcium release from the intracellular pool[300], (unpublished observations). Since recent studies in GH_3 cells have shown that TRH-induced $[Ca^{2+}]_i$ increase is due to an influx of calcium in part through voltage-dependent and in part through Nifedipine®/Verapamil®-insensitive calcium channels,[296] our results suggest that calcium may be considered as an intracellular messenger in frog melanotrophs and that the effect of TRH may involve stimulation of Ca^{2+} influx through Nifedipine®/Verapamil®-insensitive Ca^{2+} channels. In mammals, it was previously demonstrated that prostoglandins (PGEs) potentiate

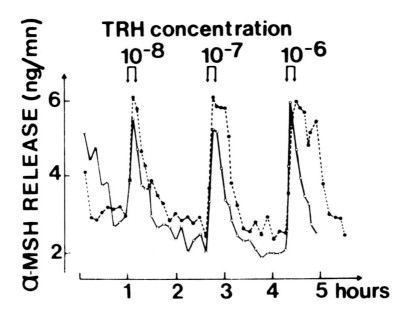

FIGURE 11. Effect of increasing concentrations of TRH in the presence (●- -●) or absence (○-○) of prostaglandin E_1 on the release of α-MSH by perifused frog NIL. PGE_1 (10^{-6} M) was responsible for a prolongation of the effect of TRH (10^{-8} to 10^{-6} M) on α-MSH release. (From Leroux, P., et al., *Prostaglandins*, 21, 599, 1981. With permission.)

the stimulatory effect of TRH on TSH release in vivo[301] and in vitro[302,303] Our data in *Rana ridibunda* show that PGE_1 potentiates the release of α-MSH induced by TRH.[304] It can be mentioned that the perifusion technique makes it possible to determine that PGE_1 is responsible for a prolongation of the effect of TRH, mainly for high concentrations of the neuropeptide (10^{-7} and 10^{-6} M; Figure 11).[304] Thus it appears that PGE_1 enhances the secretory response of mammalian thyrotrophs and amphibian melanotrophs to TRH. A number of studies are consistent with an interaction between dopamine and TRH neuronal systems in the brain.[292-294] Since both dopamine- and TRH-containing nerve terminals are present in the frog neurointermediate lobe[58,64,274] and since both neuronal systems regulate α-MSH secretion[24,120,263,264] a possible interaction between these neuroregulators could be postulated. In fact, administration of these two factors to perifused frog neurointermediate lobes showed that (1) the stimulatory action of TRH is maintained during prolonged administration of dopamine and (2) the effects of TRH and dopamine are additive.[120] Recently, Selingfreund et al.[264] achieved a similar conclusion using the intermediate lobe of *Rana pipiens*; they showed that TRH can stimulate α-MSH release under dopaminergic inhibition. These results are in agreement with those of Matsuskita et al.,[305] who showed that prolactin release induced by TRH is not suppressed by the concomitant infusion of dopamine. Along these lines, Ho et al.,[306] have provided evidence that, in man, TRH and domperidone, a dopaminergic antagonist, stimulate prolactin release through two distinct mechanisms. These results indicate that the intracellular events which are associated with TRH-induced stimulation and dopamine-induced inhibition of α-MSH release in the frog are not linked together. Previous biosynthetic studies have shown that perifused neurointermediate lobes of *Rana ridibunda* release both α-MSH and desacetyl α-MSH (see Jenks et al., Chapter 8, this Volume).[307,308] The nonacetyled form of α-MSH hormone is known to be melanotropically less potent.[309,310] Identification by HPLC of immunoreactive α-MSH released by perifused frog neurointermediate lobes, shows that in basal conditions, 25% of the hormone is in the acetyled form. TRH both stimulates the secretion of melanotropic hormone and favors the formation of the acetyled form.[311] Thus, a preferential release of the most bioactive (monoacetyl) MSH form under TRH-induced stimulation may be of physiological importance in the frog.

B. Neuropeptide Tyrosine

Neuropeptide Y (NPY) is a 36 amino-acid peptide which was originally isolated from porcine brain and sequenced by Tatemoto.[312,313] This peptide, a member of the pancreatic polypeptide (PP) family, shares 70% sequence homology with peptide YY,[314] and 50% homology with PP.[315] Immunohistochemical studies have shown that NPY has a widespread, characteristic distribution in the brain of amphibians,[316,317] rats,[318-321] cats,[322] pigs,[323] monkeys,[324,325] and humans.[326-328] Studies conducted in the cat[322] and in amphibians,[317,329-331] have shown the presence of NPY-containing neurons in the neurointermediate lobe complex. Concurrently, pharmacological studies indicate that ICV administration of NPY impairs LH secretion.[332-334] In addition, there is now clear evidence that this peptide is a powerful inhibitor of melanotropin secretion in amphibians.[329-331]

1. Anatomical Basis

In *Rana ridibunda*, NPY-like immunoreactive perikarya are widely distributed in the central nervous system. The highest concentration of NPY-containing cell bodies is found in the ventral and dorsal infundibular nuclei and in the preoptic nucleus. Numerous perikarya are also present in the posterocentral nucleus of the thalamus, in the anteroventral nucleus of the mesencephalic tegmentum, in the part posterior to the torus semicircularis, in the mesencephalic cerebellar nucleus, and throughout the cerbral cortex.[316] Fibers containing NPY-immunoreactive material are found in all parts of the brain except in the cerebellum, the nucleus isthmi, and the torus semicircularis. An important tract of NPY-containing fibers is located in the infundibulum where a dense medioventral bundle courses caudally towards the median eminence and pituitary stalk. The distribution of NPY neuronal systems in the brain of *Rana catesbeiana*,[317] is identical to that reported in *Rana ridibunda*.[316] The most striking feature is the presence of a dense network of NPY-containing nerve endings among the parenchymal cells of the amphibian pars intermedia which was reported first in *Rana ridibunda*[329,330] and later in *Rana catesbeiana*[317] and *Xenopus laevis*.[331] In contrast, the anterior lobe is totally devoid of NPY-immunoreactive material, and only very rare NPY-positive nerve fibers are noted in the neural lobe of the frog pituitary.[329] The presence of NPY immunoreactivity in noradrenergic and adrenergic cell bodies, has been demonstrated in the rat[319,335] and human brain.[336] Since noradrenaline nerve terminals are found in the amphibian pars intermedia,[58,77] the possible coexistence of NPY and catecholamines in nerve processes of the intermediate lobe remains to be examined. In addition, to the coexistence of NPY and catecholamines, NPY has also been shown to coexist in GABAergic neurons[337] and somatostatinergic neurons[338-340] in the brain of mammals. As GABAergic fibers are present in the intermediate lobe of *Rana*[164] and *Xenopus*,[165] and since somatostatin fibers are found in the distal part of the neural lobe, close to the intermediate lobe,[341,342] coexistence between NPY and GABA and/or somatostatin cannot be excluded. The subcellular localization of NPY has also been studied in fibers innervating the intermediate lobe of the pituitary.[329] Using the highly sensitive immunogold technique, NPY-like material appears to be concentrated in dense core vesicles of about 100 nm in diameter (Figure 12). Using a specific radioimmunoassay technique, serial dilutions of lyophilized acid extracts of frog neurointermediate lobe (as well as the preoptic region, infundibulum, or telencephalon) produce displacement curves which are parallel to synthetic procine NPY.[316,329] Sephadex® G-50 gel chromatography of acid extracts of whole brain or neurointermediate lobes indicate the presence of a major compound which elutes in the same position as synthetic porcine NPY. However, high-performance liquid chromatography (HPLC) analysis of neurointer-

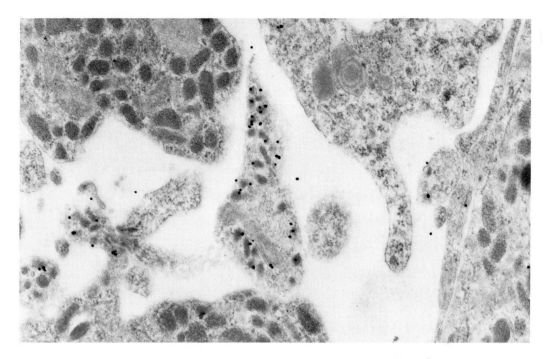

FIGURE 12. Immunoelectron microscopic localization of NPY in the intermediate lobe using the immunogold technique. The immunostaining is restricted to nerve fibers coursing between parenchymal cells. The gold particules are located in dense core vesicles. (Magnification × 30,450).

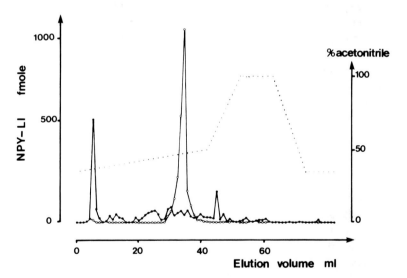

FIGURE 13. Reverse phase HPLC analysis of neuropeptide Y (NPY) in frog NIL extracts. The gradient used is shown on the Figure. HPLC analysis revealed that frog NPY (●–●) has a shorter retention time than the NPY standard (○–○). (From Danger, J.M., et al., Life Sci., 39, 1183, 1986. With permission.)

mediate lobe or brain extracts indicates that frog NPY-like material is less hydrophobic than porcine NPY (Figure 13). To ascertain that NPY-like material detected in the frog brain corresponds to authentic amphibian NPY, the distribution of NPY immunoreactive elements was compared with the distribution of C-PON, the C-terminal-flanking peptide of NPY within the NPY precursor structure.[343] This C-terminal region of the precursors for the PP-

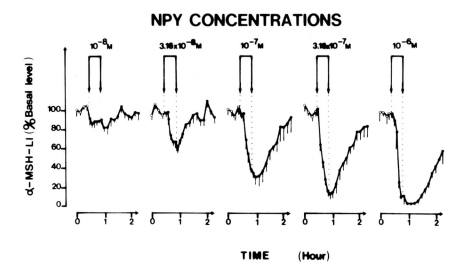

FIGURE 14. Effect of increasing concentrations of NPY on the release of α-MSH by perifused frog NIL. Administration of synthetic porcine NPY (10^{-8} to 10^{-6} M) for 20 min periods, induced a dose-related inhibition of α-MSH release. See legend to Figure 2 for additional information. (From Danger, J.M. et al., *Life Sci.*, 39, 1183, 1986. With permission.)

family of peptides shows considerable structure divergence, and therefore C-PON can be considered a marker peptide for NPY-producing cells. The fact that NPY and C-PON-like immunoreactivities are always associated within the same cell bodies and nerve processes[317,330] indicates that the NPY-like material detected in the brain and pars intermedia of amphibians is authentic NPY and does not result from crossreaction of the NPY antiserum with another peptide of the NPY family.

2. Control of the Pars Intermedia Secretion by NPY

In the perifused frog intermediate lobe system, synthetic porcine NPY inhibits α-MSH secretion in a dose-dependent manner (Figure 14). The lowest effective dose is 10^{-8} M NPY. When the inhibitions of α-MSH secretion elicited by NPY and dopamine are compared, the effect of NPY appears to be more prolonged;[120,329] while the effect of dopamine was concluded in 45 min, it required 180 min to recover a stable baseline after administering NPY at various dose levels. The inhibitory effect of NPY has been observed in three anuran species, *Rana ridibunda*,[329] *Xenopus laevis*,[331] and *Rana pipiens* (Kraicer, personal communication). In *Xenopus laevis*, NPY induces simultaneous inhibition of α-MSH and β-endorphin secretion.[331] In contrast, synthetic NPY does not affect α-MSH secretion in the newt, *Triturus cristatus* (Fasolo, personal communication). A possible effect of NPY on the release of pars intermedia peptides in other vertebrates, including mammals, remains to be investigated.

Structure-activity relationships have been studied in the frog, *Rana ridibunda* using NPY short-chain analogues (synthesized by S. Saint-Pierre and J. C. Martel, University of Sherbrooke, Canada) and various peptides of the NPY family. Deletion of the N-terminus tyrosine residue of procine NPY does not affect the MSH-release-inhibiting potency of the peptide. In contrast, shorter fragments such as NPY,$^{1\text{-}15}$ NPY,$^{16\text{-}36}$ and NPY$^{25\text{-}36}$ were almost inactive. PYY, which exhibits a high degree of sequence homology with NPY, retains 75% of the biological potency of the latter. Avian pancreatic peptide (APP) which shares only 55% sequence homology with NPY, is 5 times less active than NPY on α-MSH secretion.

Since all the studies mentioned above have been conducted on whole neurointermediate lobes which remain tonically inhibited by aminergic inputs,[123] our data raise the question

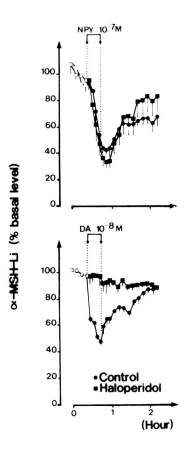

FIGURE 15. Effect of a dopaminergic antagonist, haloperidol on dopamine- or NPY-induced inhibition of α-MSH release by perifused frog NIL. The inhibition of α-MSH release induced by dopamine (10^{-8} M) was totally abolished by constant infusion of haloperidol (10^{-5} M). In contrast, the inhibitory effect of NPY was not significantly altered by haloperidol. See legend to Figure 2 for additional information. (From Danger, J.M., et al., *Life Sci.*, 39, 1183, 1986. With permission.)

of whether NPY acts directly on frog melanotrophs to inhibit α-MSH release or whether NPY is responsible for an indirect action which could be exerted via a dopaminergic or GABAergic mechanism. Figure 15 shows that neither the dopaminergic antagonist, haloperidol, nor the $GABA_A$ receptor antagonist, bicuculline, have any effect on NPY-induced α-MSH inhibition. In *Xenopus*, sulpiride, which antagonizes the inhibitory action of dopamine on perifused neurointermediate lobes,[122] has no effect on NPY-induced inhibition of MSH release.[331] These data give strong support for a direct action of NPY on frog melanotrophs. Immunocytochemical localization of NPY on ultrathin sections of neurointermediate lobes obtained by cryoultramicrotomy[344] made it possible to visualize NPY-binding sites in the cytoplasmic matrix and occasionally in secretory granules of frog melanotrophs. Taken together, these data demonstrate that NPY acts directly on the amphibian pars intermedia to inhibit α-MSH secretion.

Thus, at least two peptidergic systems, namely TRH and NPY, are clearly involved in the control of the amphibian pars intermedia. The intracellular mechanism through which these regulatory peptides control secretion of MSH are still unknown. However, the fact that NPY can totally block the stimulatory effect of TRH (unpublished data) indicates that these two neuropeptides share a common step in their mechanism of action. There is accumulating experimental evidence that calcium mobilization and phosphoinositide breakdown are involved in the action of TRH on mammalian anterior pituitary cells[345-347] and pituitary tumor cell lines.[296,298,348] Whether NPY also acts through the calcium-phospholipid second messenger system remains to be investigated. The pars intermedia, which has been extensively used to investigate dopaminergic D_2 receptors,[111,112,115] may also prove to be a valuable model for studying the mechanism of action of NPY at the molecular level.

C. Corticotropin-Releasing Factor

Corticotropin-releasing factor (CRF) is a 41-amino acid peptide which was initially purified and isolated on the basis of its ability to stimulate the secretion of ACTH and β-endorphin by the pituitary.[349,350] This peptide meets many of the physiological criteria to characterize it as a natural CRF: synthetic CRF induces a dose-dependent release of ACTH and β-endorphin in vivo[351] and in vitro[352,353] without altering the release of other anterior pituitary hormone,[354] and passive immunization of rats with rabbit anti-CRF serum reduces plasma ACTH levels.[351] Two other nonmammalian peptides bearing sequence homologies with CRF, namely urotensin I and sauvagine, have been isolated and characterized from fish urophysis[355] and frog skin,[356] respectively. These peptides exhibit corticotropin-releasing activity in various species: urotensin I and ovine CRF are virtually equipotent in stimulating ACTH release in primary cultures of rat anterior pituitary cells[357] whereas urotensine I is 3 times more potent than CRF to stimulate ACTH secretion from perifused fish pituitary.[358] Evidence have now accumulated to indicate that CRF and CRF-related peptides may participate in the control of intermediate lobe secretion.

1. Anatomical Basis

The availability of synthetic CRF has allowed the generation of antirsera for radioimmunoassay and immunocytochemistry. CRF-containing neurons are widely distributed in the brain, including the telencephalon, mesencephalon, and metencephalon.[359] In the rat, the great majority of CRF-immunoreactive perikarya are located in the parvocellular part of the paraventricular nucleus (PVN)[360-363] as well as in the magnocellular perikarya of the PVN.[364,365] In monkey and humans, CRF neurons are located in both the paraventricular and supraoptic nuclei of the hypothalamus.[366-368] A close topographical relationship is generally found between neurophysin- and CRF-containing neurons,[363,368-370] and the existence of CRF-like material in subpopulations of magnocellular neurons containing vasopressin or oxytocin has been reported. Dense bundles of CRF-containing processes terminate in the external zone of the median eminence[362,363,365,371,372] and the presence of immunoreactive fibers has been observed in the neural lobe of the pituitary.[363,373,374] Along these lines, significant amounts of CRF have been detected in the posterior lobe of the rat.[375] The localization of CRF-neuronal systems has been determined in various submammalian vertebrates including birds,[365,376,377] amphibians,[365,378-380] and fishes.[365,381-384] The distributional map of urotensin-I-like immunoreactivity in the fish brain has been recently provided.[384] In the salmon, CRF-immunoreactive fibers terminate in the distal part of the neurohypophysis and in contact of adenohypophyseal corticotrophs.[385] In amphibians, a large number of CRF-containing fibers are located in the external zone of the median eminence in close contact with the capillaries of the pituitary portal plexus.[380] In *Rana ridibunda*, many fibers are present in the neurohypophysis and a few scattered fibers containing CRF-like material are found in the intermediate lobe of the pituitary,[380] suggesting a possible role of CRF in the control of melanotropin secretion.

2. Control of the Pars Intermedia Secretion by CRF

The first evidence that CRF regulates the release of pars intermedia peptides has been provided by Proux-Ferland et al.[386] in the rat. Intravenous administration of synthetic ovine

CRF leads to a marked increase in plasma concentrations of both ACTH and α-MSH within 5 min. While dexamethasone blocks the stimulatory effect of CRF on anterior lobe corticotrophs, this glucocorticoid agonist has no effect on CRF-induced α-MSH secretion by intermediate lobe melanotrophs. The stimulatory action of CRF has been confirmed using perifused rodent intermediate lobes.[387,388] CRF has a direct effect on pituitary melanotrophs since this peptide stimulates the release of α-MSH from rat intermediate lobe cells in primary culture.[389] CRF has also been found to stimulate α-MSH secretion in the rabbit.[390] In *Xenopus laevis*, CRF, urotensin I, and sauvagine have been shown to stimulate melanotropin secretion from perifused neurointermediate lobes.[391] In contrast, synthetic ovine CRF, which stimulates ACTH secretion by the anterior lobe corticotrophs of *Rana ridibunda*, has no effect on α-MSH release by the neurointermediate lobe of the same species.[392]

D. Neurohypophyseal Peptides

Small capillaries originating from the ventral surface of the neural lobe are the only vascular elements irrigating the intermediate lobe of the pituitary. Therefore, peptides which are contained and likely released by the neural lobe may diffuse more or less actively towards the intermediate lobe and influence the secretory activity of pars intermedia cells. A similar phenomenon may also occur with several biogenic amines which are contained in axon terminals of the neural lobe.

1. Anatomical Basis

Besides the classical neurohypophyseal nonapeptides oxytocin and vasopressin, and the associated neurophysins, the neural lobe of the mammalian pituitary contains several regulatory peptides, such as enkephalins,[393] dynorphin,[393] and CCK.[394] In amphibians, vasoactive intestinal peptide (VIP),[275] somatostatin,[236,275,341,342,395] TRH, CRF, gonadotropin-releasing hormone (LHRH),[60] and dynorphin[396] have been identified by immunocytochemistry and/or quantified by radioimmunoassay in the neural lobe of the pituitary. In addition, mesotocin-containing fibers directly innervate the parenchymal cells of the intermediate lobe in *Rana esculenta*[397] and *Rana ridibunda*.[274,380] A similar observation has been made in leporidae; oxytocinergic neurons have been visualized by immunocytochemistry in the intermediate lobe of the rabbit and the hare.[10] In contrast, in the pars intermedia of the fish, *Lampreta fluviatilis*, immunoreactive neurohypophysial peptidergic fibers are absent.[398] Recently, two novel regulatory peptides have been visualized in the neural lobe of the frog: atrial natriuretic factor (ANF)[399,400] and melanin-concentrating hormone (MCH).[401] Both peptides have also been identified by immunocytochemistry in the neural lobe of mammals,[402,404] and MCH is remarkably abundant in the neural lobe of the fish.[405] These observations suggest that some of the numerous peptides contained in nerve terminals of the pars nervosa may be involved in the regulation of pars intermedia secretion.

2. Control of the Pars Intermedia Secretion by Neurohypophyseal Peptides

The tridecapeptide, Pro-Leu-Gly-NH$_2$, the C-terminal fragment of oxytocin, has been proposed as a melanotropin-release-inhibiting factor (MIF$_1$)[406,407] but it is now generally accepted that this peptide does not influence the secretory activity of intermediate lobe cells.[408-411] In particular, Pro-Leu-Gly-NH$_2$ has no effect on MSH release from perifused neurointermediate lobes of *Rana ridibunda*.[252] We have also studied the effects of various oxytocin fragments (OXT 4-9; OXT 6-9), arginine vasopressin, vasopressin fragments (AVT 4-9; AVT 5-9), and mesotocin. None of these synthetic peptides have any effect on melanotropin secretion in *Rana ridibunda* (unpublished observation). In contrast, AVT stimulates in a dose-dependent manner the release of ACTH by frog anterior pituitary corticotrophs.[392] Somatostatin, which is found in highly concentrated amounts in the neural lobe, is totally devoid of effect on MSH release in the frog.[275] Recently, the synthetic peptide ANF (Arg101-

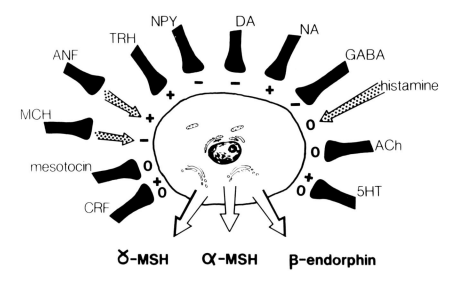

FIGURE 16. Schematic drawing depicting the various aminergic or peptidergic inputs which may participate in the regulation of the melanotropic cell. This scheme is a synoptic view of what has been described in this chapter and represents a synthesis of what has been found in various lower vertebrate species. The dotted arrows are meant to indicate that some neurotransmitters, which have not been found among the fibers innervating the pars intermedia, may originate from the neural lobe or from the systemic blood. The − and + symbols indicate that a given neurotransmitter respectively inhibits or stimulates the release of POMC-derived peptides. The 0 symbols indicate that the neurotransmitter apparently does not affect the secretory activity of melanotrophs.

Tyr126) was found to cause a dose-related stimulation of α-MSH release by perifused frog neurointermediate lobes.[412] In the rat, Howath et al.,[413] showed that ANF stimulates ACTH release from perifused rat anterior pituitary cells. In contrast, ANF was found to attenuate basal and CRF-induced release of pro-opiomelanocortin-derived peptides from rat intermediate lobe cells in primary culture.[414]

Evidence that MCH is involved in the control of melanotropin secretion has been recently obtained in fish.[415,416] In vivo, chronic administration of MCH via an osmotic minipump causes melanin concentration in the skin melanophores of the rainbow trout.[415] Examination of the cytological appearance of pars intermedia cells suggests that MCH inhibits the secretory activity of trout melanotrophs. In addition, measurement of plasma α-MSH levels in trouts administered chronically with MCH, indicates that this peptide prevents the increase in α-MSH secretion normally seen in black-background-adapted animals.[415] Immunoabsorption of endogenous MCH by homologous antiserum significantly enhances α-MSH secretion by eel neurointermediate lobes in vitro.[416] A direct inhibitory effect of MCH on α-MSH release by perifused neurointermediate lobes of *Tilapia* has been recently observed (Balm et al., unpublished observations). Taken together, these results indicate that MCH released by the neural lobe may act on the adjacent intermediate lobe to inhibit α-MSH secretion. Thus, it appears that MCH may exert a coordinate action on the pars intermedia (neuroendocrine action) and on the melanophores (hormonal action) to induce melanin concentration.

IV. CONCLUDING REMARKS

The present review shows that the intermediate lobe of the pituitary can be regulated by multiple peptidergic and aminergic neurotransmitters. An overview of the various factors involved in the control of α-MSH in frogs and toads is schematically shown in Figure 16. The factors regulating melanotropin secretion in amphibians are very similar to those controlling prolactin secretion in mammals; both melanotrophs and mammotrophs are under tonic inhibitory control by the hypothalamus, dopamine and GABA are involved in the inhibitory action of the hypothalamus on prolactin and melanotropin secretion;[120,143,164,169,176] TRH and noradrenaline produce a stimulation of prolactin and melanotropin secretion;[123,232,233,263,417] and finally, in mammals, estradiol stimulates both prolactin- and α-MSH secretion.[418,419] Conversely, other neuropeptides which control prolactin secretion, such as VIP or somatostatin, have no effect on melanotropin secretion by frog intermediate lobe cells.[275] In addition, three neuropeptides are involved in the control of α-MSH secretion in fish and/or amphibians; NPY, ANF, and MCH.[329,412,415,416] Whether these regulatory peptides may also influence prolactin secretion by mammalian pituitary cells remains to be investigated.

ACKNOWLEDGMENTS

This research was supported in part by the Institut National de la Santé et de la Recherche Médicale (grant nos. 58-78-90, 79-1245-4, 82-4019, and 86-4016), the Centre National de la Recherche Scientifique (UA 650), the European Economic Community (grant no. STI-084-J-C), La Direction des Recherches et Etudes Techniques (grant no. 86-1164), the Ministère de l'Industrie et de la Recherche (grant no. 84-H-1335 and 83-C-1026) and by France-Quebec Exchange Programs. We are indebted to Drs. R. Andreatta and K. Scheibli (Ciba-Geigy, Basel), Dr. N. Ling (Salk Institute, San Diego), Dr. D. Coy (Tulane University, New Orleans), Dr. S. Saint Pierre (Sherbrooke University, Quebec) and to Dr. P. Burbarch (Rudolf Magnus Institute, Utrecht) for their generous gifts of synthetic peptides. We thank Ms. R. Bensaadoune and S. Letellier for typing the manuscript.

REFERENCES

1. **Stoeckel, M. E., Schmitt, G., and Porte, A.**, Fine structure and cytochemistry of the intermediate lobe of the mammalian pars intermedia, in *Peptides of the Pars Intermedia*, Pitman Medical, London, 1981, 101.
2. **Pezalla, P. D., Seidah, N. G., Benjanet, S., Crine, P., Lis, M., and Chrétien, M.**, Biosynthesis of beta-endorphin, beta-lipotropin and the putative ACTH-LPH precursor in the frog pars intermedia, *Life Sci.*, 23, 2281, 1978.
3. **Dores, R. M.**, Evidence for a common precursor for a α-MSH and β-endorphin in the intermediate lobe of the pituitary of the reptile, *Anolis carolinensis, Peptides*, 3, 925, 1982.
4. **Martens, G. J. M., Jenks, B. J., and van Overbeeke, A. P.**, Biosynthesis of pairs of peptides related to melanotropin, corticotropin and endorphins in the pars intermedia of the amphibian pituitary gland, *Eur. J. Biochem.*, 122, 1, 1982.
5. **Vaudry, H., Jenks, B. J., and van Overbeeke, A. P.**, Biosynthesis, processing and release of pro-opiomelanocortin related peptides in the intermediate lobe of the pituitary gland of the frog *(Rana ridibunda)*, *Peptides*, 5, 905, 1984.
6. **Bagnara, J. T. and Hadley, M. E.**, Chromatophores and color changes, the comparative physiology of animal pigmentation, Prentice-Hall, Englewood Cliffs, New Jersey, 1973.
7. **Baumgarten, H. G., Björklund, A., Holstein, A. F., and Nobin, A.**, Organization and ultrastructural identification of the catecholamine nerve terminals in the neural lobe and pars intermedia of the rat pituitary, *Z. Zellforsch. Mikrosk. Anat.*, 126, 483, 1972.
8. **Howe, A.**, The mammalian pars intermedia: a review of its structure and function, *J. Endocrinol.*, 59, 385, 1973.
9. **Cameron, E. and Foster, C. L.**, Some light and electron microscopical observations on the pars intermedia of the pituitary gland of the rabbit, *J. Endocrinol.*, 49, 479, 1971.
10. **Schimchowitch, S., Stoeckel, M. E., Klein, M. J., Garaud, J. C., Schmitt, G., and Porte, A.**, Oxytocin-immunoreactive nerve fibers in the pars intermedia of the pituitary in the rabbit and hare, *Cell Tiss. Res.*, 228, 255, 1983.
11. **Etkin, W.**, Hyperactivity of the pars intermedia as a graft in the tadpole, *Anat. Rec.*, Suppl 1, 64, 75, 1935.

12. **Etkin, W. and Sussman, W.**, Hypothalamo-pituitary relations in metamorphosis of *Ambystoma*, *Gen. Comp. Endocrinol.*, 1, 70, 1961.
13. **Etkin, W.**, Hypothalamic inhibition of pars intermedia activity in the frog, *Gen. Comp. Endocrinol.*, Suppl 1, 148, 1962.
14. **Kastin, A. J. and Ross, G. T.**, Melanocyte-stimulating hormone activity in pituitaries of frogs with hypothalamic lesions, *Endocrinology*, 77, 45, 1965.
15. **Ito, T.**, Experimental studies on the hypothalamic control of the pars intermedia activity of the frog, *Rana nigromaculata*, *Neuroendocrinology*, 3, 25, 1968.
16. **Iturriza, F. C.**, Further evidence for the blocking effect of catecholamines on the secretion of melanocyte-stimulating hormone in toads, *Gen. Comp. Endocrinol.*, 12, 417, 1969.
17. **Daniel, P. M. and Prichard, M. M. L.**, The effects of pituitary stalk section in the goat, *Am. J. Pathol.*, 34, 433, 1958.
18. **Hamori, I.**, Tissue reaction and functional changes following hypothalamic and hypophyseal stalk lesions in the intermediate lobe of the rat, *Acta Morphol. Acad. Sci. Hung.*, 9, 155, 1960.
19. **Jégou, S., Tonon, M. C., Leroux, P., Leboulenger, F., Delarue, C., Netchitailo, P., Pelletier, G., Dupont, A., and Vaudry, H.**, Effect of hypophysectomy and pituitary stalk transection on α-melanocyte stimulating hormone like immunoreactivity in the brain of the frog *Rana ridibunda* Pallas, *Brain Res.*, 208, 287, 1981.
20. **Stoppani, A. O. M.**, Pharmacology of color regulation in amphibia and the importance of endocrine glands, *J. Pharmacol. Exp. Therap.*, 76, 118, 1942.
21. **Stoppani, A. O. M.**, Neuroendocrine mechanism of color change in *Bufo arenarum* Hensel, *Endocrinology*, 30, 782, 1942.
22. **Burgers, A. C. J., Leemreis, W., Dominiczak, T., and van Oordt, G. J.**, Inhibition of the secretion of intermedine by D-lysergic acid diethylamide (LSD 25) in the toad *Xenopus laevis*, *Acta Endocrinol.*, 29, 191, 1958.
23. **Iturriza, F. C. and Koch, O. R.**, Effect of the administration of lysergic acid diethylamide (LSD) on the colloid vesicles of the pars intermedia of the toad pituitary, *Endocrinology*, 75, 615, 1964.
24. **Bower, A., Hadley, M. E., and Hruby, V. J.**, Biogenic amines and control of melanophore stimulating hormone release, *Science*, 184, 70, 1974.
25. **Morgan, C. M. and Hadley, M. E.**, Ergot alkaloid inhibition of melanophore stimulating hormone secretion, *Neuroendocrinology*, 21, 10, 1976.
26. **Jenks, B. G.**, Control of MSH synthesis and release in the aquatic toad, *Xenopus laevis*, *Front. Horm. Res.*, 4, 63, 1977.
27. **Tilders, F. J. H. and Smelik, P. G.**, Direct neural control of MSH secretion in mammals: involvement of dopaminergic tuberohypophyseal neurons, in *Frontier in Hormone Research*, Tilders, F. J. H., Swaab, D. F. and van Wimersma Greidanus, T. B., Eds., S. Karger, Basel, 1977, 80.
28. **Schimchowitsch, S., Palacios, J. M., Stoeckel, M. E., Schmitt, G., and Porte, A.**, Absence of inhibitory dopaminergic control of the rabbit pituitary gland intermediate lobe, *Neuroendocrinology*, 42, 71, 1986.
29. **Ramon y Cajal, S.**, Algunas contribuciones al conocimiento de los ganglios del encéfalos, *Ann. Soc. Exp. Hist. Nat.*, 3, 195, 1894.
30. **Wingstrand, K. G.**, Microscopic anatomy, nerve supply and blood supply of the pars intermedia, in *The Pituitary Gland*, Vol. 3, Harris, G. W. and Donovan, B. T., Eds., Butterworths, London, 1966, 1.
31. **Oertel, W. H., Mugnaini, E., Tappaz, M. L., Weise, V. K., Dahl, A. L., Schmechel, D. E., and Kopin, I. J.**, Central GABAergic innervation of neurointermediate pituitary lobe: biochemical and immunocytochemical study in the rat, *Proc. Natl. Acad. Sci. USA*, 79, 675, 1982.
32. **Westlund, K. N. and Childs, G. V.**, Localization of serotonin fibers in the rat adenohypophysis, *Endocrinology*, 111, 1761, 1982.
33. **Bridges, T. E., Fisher, A. W., Gosbee, J. L., Lederis, K., and Santolaya, R. C.**, Acetylcholine and cholinesterases (assays and light- and electron microscopical histochemistry) in different parts of the pituitary of the rat, rabbit and domestic pig, *Z. Zellforsch.*, 136, 1, 1973.
34. **Whitaker, S. and Labella, F. S.**, Cholinesterase in the posterior and intermediate lobes of the pituitary. Species differences as determined by light and electron microscopic histochemistry, *Z. Zellforsch.*, 142, 69, 1973.
35. **Kurosuomi, K., Matsuzawa, T., and Shibasaki, K.**, Electron microscope studies of the fine structures of the pars nervosa and pars intermedia, and their morphological interrelation in the normal rat hypophysis, *Gen. Comp. Endocrinol.*, 1, 433, 1961.
36. **Bargmann, W., Lindner, E., and Andres, K. H.**, Über Synapsen und endokrinen Epithelzellen und die Definition sekretorischer Neurone. Untersuchungen am Zwischenlappen der Katzenhypophyse, *Z. Zellforsch.*, 77, 282, 1967.
37. **Vincent, D. S. and Anand Kumar, T. C.**, Electron microscopic studies on the pars intermedia of the ferret, *Z. Zellforsch.*, 99, 185, 1969.

38. **Falck, B., Hillarp, N. A., Thieme, G., and Torp, A.,** Fluorescence of catecholamines and related compounds condensed with formaldehyde, *J. Histochem. Cytochem.*, 10, 348, 1962.
39. **Nagatsu, I.,** Localization of dopamine-β-hydroxylase in bovine adrenal gland and rat sciatic nerves by the improved enzyme immunocytochemical and enzyme-immunofluorescent methods, *Acta Histochem. Cytochem.*, 7, 147, 1974.
40. **Nagatsu, I., Kondo, Y., Inagali, S., Karasawa, N., Kato, T., and Nagatsu, T.,** Immunofluorescent studies on tyrosine hydroxylase: application for its axoplasmic transport, *Acta Histochem. Cytochem.*, 10, 494, 1977.
41. **Berod, A., Hartman, B. K., Keller, A., Joh, T. H., and Pujol, J. F.,** New double-labeling technique using tyrosine hydroxylase and dopamine β-hydroxylase immunohistochemistry: Evidence for dopaminergic cells laying in the pons of the beef brain, *Brain Res.*, 240, 235, 1982.
42. **Leboulenger, F., Leroux, P., Tonon, M. C., Coy, D. H., Vaudry, H., and Pelletier, G.,** Coexistence of vasoactive intestinal peptide and enkephalins in the adrenal chromaffin granules of the frog, *Neurosci. Lett.*, 37, 221, 1983.
43. **Kawakani-Kondo, Y., Yoshida, M., Karasawa, N., Yamada, K., Takagi, I., Kondo, T., and Nagatsu, I.,** Ontogenetic study on the monoamine and peptide containing cells in the pituitary and hypothalamus of the bullfrog, *Rana catesbeiana*, by immunohistochemistry, *Acta Histochem. Cytochem.*, 17, 387, 1984.
44. **Stoeckel, M. E., Tappaz, M., Hindelang, C., Sewery, C., and Porte, A.,** Opposite effects of monosodium glutamate on the dopaminergic and GABAergic innervations of the median eminence and the intermediate lobe in the mouse, *Neurosci. Lett.*, 56, 249, 1985.
45. **Racké, K. and Muscholl, E.,** Uptake of [^{3}H]dopamine into dopaminergic and noradrenergic neurones of the isolated neurointermediate lobe of the rat hypophysis. Effects of desipramine and nomifensine, *J. Neurochem.*, 41, 1488, 1983.
46. **Follenius, E.,** La localisation fine des terminaisons nerveuses fixant la noradrenaline-^{3}H dans les différents lobes de l'adenohypophyse de l'epinoche (*Gasterosteus aculeatus*), in *Aspects of Neuroendocrinology*, Bargman, W. and Scharrer, B., Eds., New York, 1970, 232.
47. **Cuello, A. C. and Iversen, L. L.,** Localization of tritiated dopamine in the median eminence of the rat hypothalamus by electron microscope autoradiography, *Brain Res.*, 63, 474, 1973.
48. **Descaries, L., Bosler, O., Berthelet, F., and Des Rosiers, M. H.,** Dopaminergic nerve endings visualized by high-resolution autoradiography in adult rat neostriatum, *Nature (London)*, 284, 620, 1980.
49. **Geffard, M., Kah, O., Ontoniente, B., Seguela, P., Le Moal, M., and Delagge, M.,** Antibodies to dopamine: radioimmunological study of specificity in relation with immunocytochemistry, *J. Neurochem.*, 42, 1593, 1984.
50. **Ontoniente, B., Geffard, M., and Calas, A.,** Ultrastructural immunocytochemical study of the dopaminergic innervation of the rat lateral septum with anti-dopamine antibodies, *Neuroscience*, 13, 385, 1984.
51. **Kah, O., Dubourg, P., Ontoniente, B., Geffard, M., and Calas, A.,** The dopaminergic innervation of the goldfish pituitary. An immunocytochemical study at the electron-microscope level using antibodies against dopamine, *Cell Tiss. Res.*, 244, 577, 1986.
52. **Dahlström, A. and Fuxe, K.,** Evidence for the existence of monoamine-containing neurons in the central nervous system. I. Distribution of monoamines in the cell bodies of the brain, *Acta Physiol. Scand.*, 62, suppl 232, 5, 1964.
53. **Lindwall, O. and Björklund, A.,** Organization of catecholamine neurons in the rat central nervous system, in *Handbook of Psychopharmacology*, Vol. 9, Iversen, L. L., Iversen, S. D. and Snyder, S. H., Eds., Plenum Press, New York, 1978, 139.
54. **Björklund, A., Lindwall, O., and Nobin, A.,** Evidence for an incert–hypothalamic dopamine neurone system in the rat, *Brain Res.*, 89, 29, 1975.
55. **Björklund, A. and Skagerberg, G.,** Evidence for a major spinal cord projection from the diencephalic A11 dopamine cell group in the rat using transmitter-specific fluorescent retrograde tracing, *Brain Res.*, 177, 170, 1980.
56. **Fuxe, K. and Hökfelt, T.,** The influence of central catecholamine neurons on the hormone secretion from the anterior and posterior pituitary, in *Neurosecretion*, Stutinsky, F., Ed, Springer-Verlag, Berlin, 1967, 165.
57. **Terlou, M. and Ploemacher, R. E.,** The distribution of monoamines in the tel-, di- and mesencephalon of *Xenopus laevis* tadpoles, with special reference to the hypothalamo-hypophysial system, *Z. Zellforsch.*, 137, 521, 1973.
58. **Prasada Rao, P. D. and Hartwig, H. G.,** Monoaminergic tracts of the diencephalon and innervation of the pars intermedia in *Rana temporaria*, *Cell Tiss. Res.*, 151, 1, 1974.
59. **Goos, H. J.th. and van Halewijn, R.,** Biogenic amines in the hypothalamus of *Xenopus laevis* tadpoles, *Naturwissenschaften*, 8, 393, 1968.
60. **Alpert, L. C., Brawer, J. R., Jackson, I. M. D., and Reichlin, S.,** Localization of LHRH in neurons in frog brain (*Rana pipiens* and *catesbeiana*), *Endocrinology*, 98, 910, 1976.

61. **Doerr-Schott, J. and Dubois, M. P.**, LH-RH-like system in the brain of *Xenopus laevis* Daud. Immunohistochemical identification, *Cell Tiss. Res.*, 172, 477, 1976.
62. **Goos, H. J. T., Ligtenberg, P. J. M., and van Oordt, P. G. W. J.**, Immunofluorescence studies on gonadotropin-releasing hormone (GRH) in the forebrain and the neurohypophysis of the green frog, *Rana esculenta Cell Tiss. Res.*, 168, 325, 1976.
63. **King, J. A. and Milar, R. P.**, Phylogenic and anatomical distribution of somatostatin in vertebrates, *Endocrinology*, 105, 1322, 1979.
64. **Seki, T., Nakai, Y., Shioda, S., Mitsuma, T., Kikuyama, S.**, Distribution of immunoreactive thyrotropin-releasing hormone in the forebrain and hypophysis of the bullfrog, *Rana catesbeiana*, *Cell Tiss. Res.*, 233, 507, 1983.
65. **Danger, J. M., Guy, J., Benyamina, M., Jégou, S., Leboulenger, F., Coté, J., Tonon, M. C., Pelletier, G. and Vaudry, H.**, Localization and identification of neuropeptide Y (NPY)-like immunoreactivity in the frog brain, *Peptides*, 6, 1225, 1985.
66. **Tonon, M. C., Burlet, A., Lauber, M., Cuet, P., Jégou, S., Gouteux, L., Ling, N., and Vaudry, H.**, Immunohistochemical localization and radioimmunoassay of corticotropin-releasing factor in the forebrain and hypophysis of the frog *Rana ridibunda*, *Neuroendocrinology*, 40, 109, 1985.
67. **Oksche, A. and Ueck, M.**, The nervous system, in *Physiology of the Amphibia*, Vol. III, Lofts, B., Ed., Academic Press, New York, 1976, 313.
68. **Iturriza, F. C.**, Electron-microscopic study of the pars intermedia of the pituitary of the toad, *Bufo arenarum*, *Gen. Comp. Endocrinol.*, 4, 492, 1964.
69. **Cohen, A. G.**, Observations on the pars intermedia of *Xenopus laevis*, *Nature (London)*, 215, 55, 1967.
70. **Pehlemann, F. W.**, Ultrastructure and innervation of the pars intermedia of the pituitary of *Xenopus laevis*, *Gen. Comp. Endocrinol.*, 9, 481, 1967.
71. **Hopkins, C. R.**, Localization of adrenergic fibers in the amphibian pars intermedia by electron microscope autotradiography and their selective removal by 6-hydroxydopamine, *Gen. Comp. Endocrinol.*, 16, 112, 1971.
72. **Doerr-Schott, J. and Follenius, E.**, Localisation des fibres aminergiques dans l'hypophyse de *Rana esculenta*. Etude autoradiographique au microscope électronique, *C.R. Acad. Sci. (Paris)*, 269, 737, 1969.
73. **Doerr-Schott, J. and Follenius, E.**, Innervation de l'hypophyse intermédiaire de *Rana esculenta* et identification des fibres aminergiques par autoradiographie au microscope électronique, *Z. Zellforsch.*, 106, 99, 1970.
74. **Nakai, Y. and Gorbman, A.**, Evidence for a double innervated secretory unit in the anuran pars intermedia. II. Electron microscopic studies, *Gen. Comp. Endocrinol.*, 13, 108, 1969.
75. **Ito, T.**, Changes in skin color and fine structure of the intermediate pituitary gland of the frog, *Rana nigromaculata*, after extirpation of the median eminence, *Neuroendocrinology*, 8, 180, 1971.
76. **Smoller, C. G.**, Ultrastructural studies on the developing neurohypophysis of the pacific tree-frog, *Hyla regilla*, *Gen. Comp. Endocrinol.*, 7, 44, 1966.
77. **Enemar, A. and Falck, B.**, On the presence of adrenergic nerves in the pars intermedia of the frog, *Rana temporaria*, *Gen. Comp. Endocrinol.*, 5, 577, 1965.
78. **Enemar, A. Falck, B., and Iturriza, F. C.**, Adrenergic nerves in the pars intermedia of the pituitary in the toad *Bufo arenarum*, *Z. Zellforsch.*, 77, 325, 1967.
79. **Braak, H.**, Biogene Amine im Gehirn vom Frosch *(Rana esculenta)*, *Z. Zellforsch.*, 106, 269, 1970.
80. **Bartels, W.**, Die Ontogenese des aminhaltigen Neuronensystems im Gehirn von *Rana temporaria*, *Z. Zellforsch.*, 116, 94, 1971.
81. **Follenius, E.**, Intégration de la dopamine dans les terminaisons aminergiques de la méta-adénohypophyse de l'épinoche *(Gasterosteus aculeatus)*, *C. R. Acad. Sci. (Paris)*, 273, 1039, 1971.
82. **Kah, O., Dubourg, P., Chambolle, and Calas, A.**, Ultrastructural identification of catecholaminergic fibers in the goldfish pituitary, *Cell Tiss. Res.*, 238, 621, 1984.
83. **Fryer, J. N., Boudreault-Chateauvert, C., and Kirby, R. P.**, Pituitary afferent originating in the paraventricular organ (PVO) of the goldfish hypothalamus, *J. Comp. Neurol.*, 242, 475, 1985.
84. **Kah, O., Chambolle, P., Thibault, J., and Geffard, M.**, Existence of dopaminergic neurons in the preoptic region of the goldfish, *Neurosci. Lett.*, 48, 293, 1984.
85. **Kah, O., Dulka, J. G., Dubourg, P., Thibault, J., and Peter, R. E.**, Neuroanatomical substrate for the inhibition of gonadotropin secretion in goldfish: existence of a dopaminergic preoptico-hypophyseal pathway, *Neuroendocrinology*, in press.
86. **Fuxe, K. and Hökfelt, T.**, Further evidence for the existence of tuberoinfundibular dopamine neurons, *Acta Physiol. Scand.*, 66, 245, 1966.
87. **Björklund, A., Falck, B., Hromek, F., Owman, C., and West, K. A.**, Identification and terminal distribution of the tubero-hypophyseal monoamine fibre systems in the rat by means of stereotaxic and microfluorimetric techniques, *Brain Res.*, 17, 1, 1970.
88. **Calas, A.**, Morphological correlates of chemically specified neuronal interactions in the hypothalamo-hypophyseal area, *Neurochem. Int.*, 7, 927, 1985.

89. **Björklund, A., Falck, B., and Rosengren, E.,** Monoamines in the pituitary gland of the pig, *Life Sci.,* 6, 2103, 1967.
90. **Saavedra, J. M., Palkovits, M., Kizer, J. S., Browstein, M., and Zivin, I. A.,** Distribution of biogenic amines and related enzymes in the rat pituitary gland, *J. Neurochem.,* 25, 257, 1975.
91. **Björklund, A. and Nobin, A.,** Organization of tuberohypophyseal and reticulo-infundibular catecholamine neuron systems in the rat brain, *Brain Res.,* 51, 171, 1973.
92. **Zambrano, D., Nishioka, R. S., and Bern, H. A.,** The innervation of the pituitary gland of teleost fishes, in *Brain Endocrine Interaction. Median Eminence: Structure and Function,* Knigge, K. M., Scott, E. E. and Weindl, A., Eds., S. Karger, Basel, 1972, 50.
93. **Olivereau, M.,** Action de la réserpine chez l'anguille: II. Effet sur la pigmentation et le lobe intermédiaire. Comparaison avec l'effet de l'adaptation sur un fond noir, *Z. Anat. Entwickl. Gesch.,* 137, 30, 1972.
94. **Fremberg, M. and Olivereau, M.,** Melanophore responses and intermediate lobe activity in the eel *Anguilla anguilla* after injection of 6-OH-dopamine, *Acta Zool.,* 54, 231, 1973.
95. **Wilson, J. F. and Dodd, J. M.,** Effects of pharmacological agents on the *in vivo* release of melanophore-stimulating hormone in the dogfish *Scyliorhinus canicula, Gen. Comp. Endocrinol.,* 20, 556, 1973.
96. **Iturriza, F. C.,** Monoamines and control of the pars intermedia of the toad pituitary, *Gen. Comp. Endocrinol.,* 6, 19, 1966.
97. **Goos, H. J. T.,** Hypothalamic control of the pars intermedia in *Xenopus laevis* tadpoles, *Z. Zellforsch.,* 97, 118, 1969.
98. **Abe, K., Butcher, R. W., Nicholson, W. E., Baird, C. E., Liddle, R. A., and Liddle, G. W.,** Adenosine 3',5'-monophosphate (cyclic AMP) as the mediator of the actions of melanocyte stimulating hromone (MSH) and norepinephrine on the frog skin, *Endocrinology,* 84, 362, 1969.
99. **Tilders, F. J. H., van Deft, A. M. L., and Smelik, P. G.,** Reintroduction and evaluation of an accurate, high capacity bioassay for melanocyte-stimulating hormone using the skin of *Anolis carolinensis in vitro, J. Endocrinol.,* 66, 165, 1975.
100. **Longshore, M. and Horowitz, J. M.,** Localization and characterization of adrenergic receptors on frog skin melanophores, *J. Endocrinol. Metab.,* 4, E84, 1981.
101. **Tilders, F. J. H. and Smelik, P. G.,** Effects of hypothalamic lesions and drugs interfering with dopaminergic transmission on pituitary MSH content of rats, *Neuroendocrinology,* 25, 275, 1978.
102. **Penny, R. J., Tilders, F. J. H., and Thody, A. J.,** The effect of hypothalamic lesions on immunoreactive α-melanocyte stimulating hormone secretion in rat, *J. Physiol.,* 292, 59, 1979.
103. **Usategui, R., Oliver, C., Vaudry, H., Lombardi, G., Rozenberg, I., and Mourre, A. M.,** Immunoreactive α-MSH and ACTH levels in rat plasma and pituitary, *Endocrinology,* 98, 189, 1976.
104. **Penny, R. J. and Thody, A. J.,** An improved radioimunoassay for α-melanocyte-stimulating hormone (α-MSH) levels in the rat: serum and pituitary α-MSH after drugs which modify catecholaminergic neurotransmission, *Neuroendocrinology,* 25, 193, 1978.
105. **Levitin, H. P. and Mezzadri-Levitin, M. R.,** Coat color in intact and neurointermediate lobe grafted agouti mice: effect of dopamine agonists and antagonists, *Neuroendocrinology,* 35, 194, 1982.
106. **Levitin, H. P.,** Monoaminergic control of MSH release in the lizard *Anolis carolinensis, Gen. Comp. Endocrinol.,* 41, 279, 1980.
107. **Tilders, F. J. H., van Der Woude, H. A., Swaab, D. F., and Mulder, A. H.,** Identification of MSH release-inhibiting elements in the neurointermediate lobe of the rat, *Brain Res.,* 171, 425, 1979.
108. **Schmitt, G., Stoeckel, M. E., and Koch, B.,** Evidence for a possible dopaminergic control of pituitary alpha-MSH during ontogenesis in mice, *Neuroendocrinology,* 33, 306, 1981.
109. **Jackson, S. and Lowry, P. J.,** Secretion of pro-opiomelanocortin peptides from isolated perifused rat pars intermedia cells, *Neuroendocrinology,* 37, 248, 1983.
110. **Randle, J. C. R., Moor, B. C., and Kraicer, J.,** Dopaminergic mediation of the effect of elevated potassium on the release of pro-opiomelanocortin-derived peptides from the pars intermedia of the rat pituitary, *Neuroendocrinology,* 37, 141, 1983.
111. **Munemura, M., Cote, T. E., Tsuruta, K., Eskay, R. L., and Kebabian, J. W.,** The dopamine receptor in the intermediate lobe of the rat pituitary gland: pharmacological characterization, *Endocrinology,* 107, 1676, 1980.
112. **Meunier, H. and Labrie, F.,** The dopamine receptor in the intermediate lobe of the rat pituitary gland is negatively coupled to adenylate cyclase, *Life Sci.,* 30, 962, 1982.
113. **Stoll, G., Martin, R., and Voigt, K. H.,** Control of peptide release from cells of the intermediate lobe of the rat pituitary, *Cell Tiss. Res.,* 236, 561, 1984.
114. **Stefanini, E., Devoto, P., Marchisio, A. M., Vernaleone, P., and Collu, R.,** [^{3}H]-spiperidol binding to a putative dopamine receptor in rat pituitary gland, *Life Sci.,* 26, 583, 1980.
115. **Frey, E. A., Cote, T. E., Grewe, C. W., and Kebabian, J. W.,** [^{3}H] spiperidol identifies a D-2 dopamine receptor inhibiting adenylate activity in the intermediate lobe of the rat pituitary gland, *Endocrinology,* 110, 1897, 1982.

116. **Sibley, D. R. and Creese, I.,** Dopamine receptor binding in bovine intermediate lobe pituitary membranes, *Endocrinology,* 107, 1405, 1980.
117. **Kebabian, J. W. and Calne, D. B.,** Multiple receptor for dopamine, *Nature,* 277, 93, 1979.
118. **Loh, Y. P., Li, A., Gritsch, H. A., and Eskay, R. L.,** Immunoreactive α-melanotropin and β-endorphin in the toad pars intermedia: dissociation in storage, secretion and subcellular localization, *Life Sci.,* 29, 1599, 1981.
119. **Martens, G. J. M., Jenks, B. G., and van Overbeeke, A. P.,** N-α-acetylation is linked to α-MSH release from pars intermedia of the amphibian pituitary gland, *Nature (London),* 294, 558, 1981.
120. **Adjeroud, S., Tonon, M. C., Gouteux, L., Leneveu, E., Lamacz, M., Cazin, L., and Vaudry, H.,** *In vitro* study of frog *(Rana ridibunda Pallas)* neurointermediate lobe secretion by use of a simplified perifusion system. IV. Interaction between dopamine and thyrotropin-releasing hormone on α-MSH secretion, *Gen. Comp. Endocrinol.,* 64, 428, 1986.
121. **Verburg-van Kemenade, B. M. L., Jenks, B. G., and Driessen, A. G. J.,** GABA and dopamine act directly on melanotrophs of *Xenopus* to inhibit MSH secretion, *Brain Res. Bull.,* 17, 697, 1986.
122. **Verburg-van Kemenade, B. M. L., Tonon, M. C., Jenks, B. G., and Vaudry, H.,** Characteristics of receptors for dopamine in the pars intermedia of the amphibian *Xenopus laevis*, *Neuroendocrinology,* 44, 446, 1986.
123. **Tonon, M. C., Leroux, P., Stoeckel, M. E., Jégou, S., Pelletier, G., and Vaudry, H.,** Catecholaminergic control of α-melanocyte-stimulating hormone (α-MSH) release by frog neurointermediate lobe *in vitro:* evidence for a direct stimulation of α-MSH release by thyrotropin-releasing hormone, *Endocrinology,* 112, 133, 1983.
124. **Hadley, M. E. and Bagnara, J. T.,** Regulation of release and mechanism of action of MSH, *Am. Zool.,* 15, 81, 1975.
125. **Stoof, J. C. and Kebabian, J. W.,** Two dopamine receptors: biochemistry, physiology and pharmacology, *Life Sci.,* 35, 2281, 1984.
126. **Cote, T. E., Eskay, R. L., Frey, E. A., Grewe, C. W., Munemura, M., Stoof, J. C., Tsuruta, K., and Kebabian, J. W.,** Biochemical and physiological studies of the beta-adrenoceptor and D-2 dopamine receptor in the intermediate lobe of the rat pituitary gland: a review, *Neuroendocrinology,* 35, 217, 1982.
127. **Saavedra, J. M.,** Central and peripheral catecholamine innervation of the rat intermediate and posterior pituitary lobes, *Neuroendocrinology,* 40, 281, 1985.
128. **Baumgarten, H. G.,** Biogenic amines in the cyclostome and lower vertebrate brain, *Histochem. Cytochem.,* 4, 1, 1972.
129. **Fremberg, G. T. and Hartwig, H. G.,** Formaldehyde-induced fluorescence in the telencephalon and diencephalon of the eel *(Anguilla anguilla) Cell Tiss. Res.,* 176, 1, 1977.
130. **Kah, O., Chambrolle, P., and Olivereau, M.,** Innervation aminergique hypothalamo-hypophysaire chez *Gambusia* sp. (Téléostéen Poecilidé) étudiée par deux techniques de fluorescence, *C. R. Acad. Sci. (Paris),* 286, 705, 1978.
131. **Terlou, M., Ekengren, B., and Hiemstra, K.,** Localization of monoamines in the forebrain of two salmonid species, with special reference to the hypothalamo-hypophysial system, *Cell Tiss. Res.,* 190, 417, 1978.
132. **Doerr-Schott, J. and Follenius, E.,** Identification et localisation des fibres aminergiques dans l'éminence médiane de la grenouille verte *(Rana esculenta)* par autoradiographie au microscope électronique, *Z. Zellforsch.,* 111, 427, 1970.
133. **Kondo, Y., Nagatsu, I., Yoshida, M., Karasawa, N., and Nagatsu, T.,** Existence of noradrenalin cells and serotonin cells in the pituitary gland of *Rana catesbeiana, Cell Tiss. Res.,* 228, 405, 1983.
134. **Iturriza, F. C. and Kasal-Iturriza, M.,** Noradrenalin inhibition of MSH release in incubates of toad pars intermedia, *Gen. Comp. Endocrinol.,* 3, 108, 1972.
135. **Terlou, M., Goos, H. J. Th., and van Oordt, P. G. W. J.,** Hypothalamic regulation of intermedia activity in amphibians, *Fortschr. Zool.,* 22, 117, 1974.
136. **Verburg-van Kemenade, B. M. L., Jenks, B. G., and van Overbeeke, A. P.,** Regulation of melanotropin release from the pars intermedia of amphibian *Xenopus laevis:* evaluation of the involvement of serotoninergic, cholinergic, or adrenergic receptor mechanisms, *Gen. Comp. Endocrinol.,* 63, 471, 1986.
137. **Awapara, J., Landua, A. J., Fuert, R., and Seale, B.,** Free γ-aminobutyric acid in brain, *J. Biol. Chem.,* 187, 35, 1950.
138. **Roberts, E. and Frankel., S.,** γ-aminobutyric acid in brain: its formation from glutamic acid, *J. Biol. Chem.,* 187, 55, 1950.
139. **Udenfriend, S.,** Identification of γ-aminobutyric acid in brain by the isotope derivate method, *J. Biol. Chem.,* 187, 65, 1950.
140. **Salganikoff, K. and De Robertis, E.,** Subcellular distribution of the enzyme of the glutamic acid, glutamine and gamma-aminobutyric acid cycle in rat brains, *J. Neurochem.,* 12, 287, 1965.

141. **Mugnaini, E. and Oertel, W.**, An atlas of the distribution of GABA-ergic neurons and terminals in the rat CNS as revealed by GAD immunohistochemistry, in *Handbook of Chemical Neuroanatomy*, Vol. 4, GABA and Neuropeptides in the CNS, Björklund, A., and Hökfelt, T., Eds., Elsevier, Amsterdam, 1985, 436.
142. **Krnjékic, K. and Schwartz, S.**, Is γ-aminobutyric acid an inhibitory transmitter?, *Nature (London)*, 211, 1372, 1966.
143. **Schally, A. V., Redding, T. W., Arimura, A., Dupont, A., and Linthicum, G. L.**, Isolation of gamma-aminobutyric acid from pig hypothalami and demonstration of its prolactin release-inhibiting (PIF) activity *in vivo* and *in vitro*, *Endocrinology*, 100, 681, 1977.
144. **Vincent, S. R., Hökfelt, T., and Wu, J. Y.**, GABA neuron systems in hypothalamus and pituitary gland. Immunohistochemical demonstration using antibodies against glutamate decarboxylase, *Neuroendocrinology*, 34, 117, 1982.
145. **Tomiko, S. A., Taraskevich, P. S., and Douglas, W. W.**, GABA acts directly on cells of pituitary pars intermedia to alter hormone output, *Nature (London)*, 301, 706, 1983.
146. **Anderson, R. A. and Mitchell, R.**, Effect of γ-aminobutyric acid receptor agonists on the secretion of growth hormone, luteinizing hormone, adrenocorticotrophic hormone and thyroid stimulating hormone from rat pituitary gland *in vitro*, *J. Endocrinol.*, 108, 1, 1986.
147. **Saito, K., Barber, R., Wu, J. Y., Matsuda, T., Roberts, E., and Vaughn, J. E.**, Immunohistochemical localization of glutamic acid decarboxylase in rat cerebellum, *Proc. Natl. Acad. Sci. U.S.A.*, 71, 269, 1974.
148. **Maitre, M., Blindermann, J. M., Ossola, L., and Mandel, P.**, Comparison of the structure of L-glutamate decarboxylase from human and rat brains, *Biochem. Biophys. Res. Commun.*, 85, 885, 1978.
149. **Perez de la Mora, M., Possani, L. D., Tapia, R., Teran, L., Palacios, R., Fuxe, K. Hökfelt, T., and Ljungdahl, A.**, Demonstration of central γ-aminobutyrate-containing nerve terminals by means of antibodies against glutamate decarboxylase, *Neuroscience*, 6, 875, 1981.
150. **Oertel, W. H., Schmechel, D. E., Mugnaini, E., Tappaz, M. L., and Kopin, I. J.**, Immunocytochemical localization of glutamate decarboxylase in rat cerebellum with a new antiserum, *Neuroscience*, 6, 2715, 1981.
151. **Tappaz, M. L., Wassef, M., Oertel, W. H., Paut, L., and Pujol, J. F.**, Light- and electron-microscopic immunocytochemistry of glutamic acid decarboxylase (GAD) in the basal hypothalamus: morphological evidence for neuroendocrine γ-aminobutyrate (GABA), *Neuroscience*, 9, 271, 1983.
152. **Storm-Mathisen, J., Leknes, A. K., Bore, A. T., Vaaland, J. L., Edminson, P., Hang, F. M. S., and Ottersen, O. P.**, First visualization of glutamate and GABA in neurons by immunocytochemistry.
153. **Ottersen, O. P. and Storm-Mathisen, J.**, Glutmate- and GABA-containing neurons in the mouse and rat brain, as demonstrated with a new immunocytochemical technique, *J. Comp. Neurol.*, 229, 374, 1984.
154. **Somogyi, P., Hodgson, A., Chubb, I. W., Penke, B., and Erdei, A.**, Antisera to γ-aminobutyric acid. II. Immunocytochemical application to the central nervous system, *J. Histochem. Cytochem.*, 33, 240, 1985.
155. **Rabhi, M., Onteniente, B., Kah, O., Geffard, M., and Calas, A.**, Immunocytochemical study of the GABAergic innervation of the mouse pituitary by use of antibodies against gamma-aminobutyric acid (GABA), *Cell Tiss. Res.*, in press.
156. **Follenius, E.**, Intégration sélective du GABA-³H dans la neurohypophyse du poisson téleostéen *Gasterosteus aculeatus*; Etude autoradiographique, *C. R. Acad. Sci. (Paris)*, 275, 1435, 1972.
157. **Chronwall, B. M. and Wolff, J. R.**, Prenatal and postnatal development of GABA-accumulating cells in the occipital neocortex of rat, *J. Comp. Neurol.*, 190, 187, 1980.
158. **Tappaz, M. L., Aguera, M., Belin, M. F., and Pujol, J. F.**, Autoradiography of GABA in the rat hypothalamic median eminence, *Brain Res.*, 186, 379, 1980.
159. **Cuérod, M., Bagnoli, P., Beaudet, A., Rustioni, A., Wiklund, L., and Streit, P.**, Transmitter-specific retrograde labeling of neurons, in *Cytochemical Methods in Neuroanatomy*, Chan-Palay, V., and Palay, S. L., Eds., Alan R. Liss, New York, 1982, 297.
160. **Ottersen, O. P. and Storm-Mathisen, J.**, Neurons containing or accumulating transmitter amino acids, in *Handbook of Chemical Neuroanatomy, Vol. 3: Classical Transmitter and Transmitter Receptor in the CNS, Part II*, Björklund, A., Hökfelt, A., and Kuhar, M. J., Eds., Elsevier, New York, 1984, 141.
161. **Oertel, W. H., Schmechel, D. E., and Mugnaini, E.**, Glutamic acid decarboxylase (GAD): purification, antiserum production, immunocytochemistry, in *Current Methods in Cellular Neurobiology*, Barker, J. L., and McKelvy, J. F., Eds., John Wiley & Sons, New York, 1983, 63.
162. **Penny, G. R., Coneley, M., Diamond, I. T., and Schmechel, D. E.**, The distribution of glutamic acid decarboxylase immunoreactivity in the diencephalon of the opossum and rabbit, *J. Comp. Neurol.*, 228, 38, 1984.
163. **Kah, O., Dubourg, P., Martinoli, M. G., Geffard, M., and Calas, A.**, Morphological evidence for a direct neuroendocrine GABAergic control of the anterior pituitary in teleost, *Experientia*, in press.

164. **Adjeroud, S., Tonon, M. C., Lamacz, M., Leneveu, E., Stoeckel, M. E., Tappaz, M. L., Cazin, L., Danger, J. M., Bernard, C., and Vaudry, H.,** GABAergic control of α-melanocyte stimulating hormone (α-MSH) release by frog neurointermediate lobe in vitro, *Brain Res. Bull.*, 17, 717, 1986.
165. **Verburg-van Kemenade, B. M. L., Tappaz, M., Paut, L., and Jenks, B. G.,** GABAergic regulation of melanocyte-stimulating hormone secretion from the pars intermedia of *Xenopus laevis:* immunocytochemical and physiological evidence, *Endocrinology*, 118, 260, 1986.
166. **Hadley, M. E. and Bagnara, J. T.,** Regulation of release and mechanism of action of MSH, *Am. Zool.*, 15 (suppl.1), 81, 1975.
167. **Tilders, F. J. H., Mulder, A. H., and Smelik, P. G.,** On the presence of a α-MSH-release inhibiting system in the rat neurointermediate lobe, *Neuroendocrinology*, 18, 125, 1975.
168. **Tomiko, S. A., Taraskevitch, P. S., and Douglas, W. W.,** Potassium-induced secretion of melanocyte-stimulating hormone from isolated pars intermedia cells signals participation of voltage-dependent calcium channels in stimulus-secretion coupling, *Neuroscience*, 11, 2259, 1981.
169. **Verburg-van Kemenade, B. M. L., Jenks, B. G., and Driessen, A. G. J.,** GABA and dopamine act directly on melanotropes of *Xenopus* to inhibit MSH secretion, *Brain Res. Bull.*, 17, 697, 1986.
170. **Davis, M. D. and Hadley, M. E.,** Pars intermedia electrical potentials: changes in spike frequency induced by regulatory factors of melanocyte-stimulating hormone (MSH) secretion, *Neuroendocrinology*, 26, 277, 1978.
171. **Demeneix, B. A., Desaulles, E., Feltz, P., and Loeffler, J. P.,** Dual population of $GABA_A$ and $GABA_B$ receptors in rat pars intermedia demonstrated by release of α-MSH caused by barium ions, *Br. J. Pharmacol.*, 82, 183, 1984.
172. **Taraskevich, P. S. and Douglas, W. W.,** GABA directly affects electrophysiological properties of pituitary pars intermedia cells, *Nature (London)*, 299, 733, 1982.
173. **Taraskevich, P. S. and Douglas, W. W.,** Pharmacological and ionic features of γ-aminobutyric acid receptors influencing electrical properties of melanotrophs isolated from the rat pars intermedia, *Neuroscience*, 14, 301, 1985.
174. **Demeinex, B. A., Taleb, O., Loeffler, J.Ph., and Feltz, P.,** $GABA_A$ and $GABA_B$ receptors on porcine pars intermedia cells in primary culture: functional role in modulating peptide release, *Neuroscience*, 17, 1275, 1986.
175. **Verburg-van Kemenade, B. M. L., Jenks, B. G., Lessen, F. J. A., and Vaudry, H.,** Characterization of GABA receptors in the neurointermediate lobe of the amphibian *Xenopus laevis, Endocrinology*, 120, 622, 1987.
176. **Enjalbert, A., Ruberg, M., Fiore, L., Arancibia, S., Priam, M., and Kordon, C.,** Independent, inhibition of prolactin secretion by dopamine and gamma-amino-butyric acid in vitro, *Endocrinology*, 105, 823, 1979.
177. **Tappaz, M., Oertel, W. H., Wassef, M., and Mugnaini, E.,** Central GABAergic neuroendocrine regulations: pharmacological and morphological evidence, in *Chemical Transmission in the Brain, Progress in the Brain Research*, Vol. 55, Buijs, R. M., Pévet, P. and Swaab, D. F., Eds., Elsevier Biochemical Press, Amsterdam, 1982, 77.
178. **Stoeckel, M. E., Tappaz, M., Hindelang, C., Seweryn, C., and Porte, A.,** Opposite effect of monosodium glutamate on the dopaminergic and GABAergic innervation of the median eminence and the intermediate lobe in the mouse, *Neurosci. Lett.*, 36, 249, 1985.
179. **Mulchahey, J. J. and Neill, J. D.,** Gamma aminobutyric acid (GABA) levels in hypophyseal stalk plasma of rats, *Life Sci.*, 31, 453, 1982.
180. **Gudelsky, G. A., Apud, J. A., Masotto, C., Locatelli, V., Cocchi, D., Racagni, G., and Müller, E. E.,** Ethanolamine-O-sulfate enhances γ-aminobutyric acid secretion into hypophyseal portal blood and lowers serum prolactin concentrations, *Neuroendocrinology*, 37, 397, 1983.
181. **Racagni, G., Apud, J. A., Locatelli, V., Cocchi, D., Nistico, G., di Giorgio, R. M., and Müller, E. E.,** GABA of CNS origin in the rat anterior pituitary inhibits prolactin secretion, *Nature (London)*, 281, 575, 1979.
182. **Anderson, R. and Mitchell, R.,** Biphasic effect of $GABA_A$ receptor agonist on prolactin secretion: evidence for two types of $GABA_A$ receptor complex on lactotrophs, *Eur. J. Pharmacol.*, 124, 1, 1986.
183. **Thornton, V. F. and Geschwind, I. I.,** Evidence that serotonin may be a melanophore-stimulating hormone releasing factor in the lizard *Anolis carolinensis, Gen. Comp. Endocrinol.*, 26, 346, 1975.
184. **Olivereau, M. and Olivereau, J.,** Effect of serotonin on prolactin and MSH-secreting cells in the eel. Comparison with the effect of 5-hydroxytryptophan, *Cell Tiss. Res.*, 196, 397, 1979.
185. **Randle, J. C. R., Moor, B. C., and Kraicer, J.,** Differential control of the release of pro-opiomelanocortin-derived peptides from the pars intermedia of the rat pituitary (response to serotonin), *Neuroendocrinology*, 37, 131, 1983.
186. **Steinbusch, H. W. M.,** Distribution of serotonin-immunoreactivity in the central nervous system of the rat-cell bodies and terminals, *Neuroscience*, 6, 557, 1981.

187. **Steinbusch, H. W. M. and Nieuwenhuys, R.**, The raphe nuclei of the rat brain stem: a cytoarchitectomic and immunohistochemical study, in *Chemical Neuroanatomy*, Emson, P. C., Ed., Raven Press, New York, 1983, 131.
188. **Steinbusch, H. W. M.**, Serotonin-immunoreactive neurons and their projections in the CNS, in *Handbook of Chemical Neuroanatomy, Vol. 3: Classical Transmitters and Transmitter Receptors in the CNS, Part II*, Björklund, T., Hökfelt, T. and Kuhar, M. J., Eds., Elsevier, Amsterdam, 1984, 68.
189. **Wiklund, L., Leger, L., and Persson, M.**, Monoamine cell distribution in the cat brain stem. A fluorescence histochemical study with quantification of indolaminergic and locus coeruleus cell groups, *J. Comp. Neurol.*, 203, 613, 1981.
190. **Steinbusch, H. W. M., Verhofstad, A. A. J., and Joosten, H. W. J.**, Localization of serotonin in the central nervous system by immunohistochemistry: description of a specific and sensitive technique and some applications, *Neuroscience*, 3, 811, 1978.
191. **Warembourg, M. and Poulain, P.**, Localization of serotonin in the hypothalamus and the mesencephalon of the guinea-pig. An immunohistochemical study using monoclonal antibodies, *Cell Tiss. Res.*, 240, 711, 1985.
192. **Yamada, H. and Sano, Y.**, Distribution of serotonin nerve cells in the rabbit brain — Immunohistochemistry by the two-step ABC technique using biotin-labeled rabbit serotonin-antibody, *Arch. Histo., Jap.*, 48, 343, 1985.
193. **Mezey, E., Leranth, Cs., Brownstein, M. J., Friedman, E., Krieger, D. T., and Palkovits, M.**, On the origin of the serotoninergic input to the intermediate lobe of the rat pituitary, *Brain Res.*, 294, 231, 1984.
194. **Bobillier, P., Seguin, S., Petitjean, F., Salvert, D., Touret, M., and Jouvet, M.**, The raphe nuclei of the cat brain stem: a topographical atlas of their afferent projections as revealed by autoradiography, *Brain Res.*, 113, 449, 1976.
195. **Parent, A., Descarries, L., and Beaudet, A.**, Organization of ascending serotonin systems in the adult rat brain. A radioautographic study after intraventricular administration of [^{3}H] 5-hydroxytryptamine, *Neuroscience*, 6, 115, 1981.
196. **Bowker, R. M., Steinbusch, H. W. M., and Coulter, J. D.**, Serotoninergic and peptidergic projections to the spinal cord demonstrated by a combined retrograde HRP histochemical and immunocytochemical staining method, *Brain Res.*, 211, 412, 1981.
197. **Björklund, A. and Skagerberg, G.**, Descending monoaminergic projections to the spinal cord, in *Brain Stem Control of the Spinal Mechanisms*, Sjölund, B. and Björklund, A., Eds., Elsevier, Amsterdam, 1982, 55.
198. **Steinbusch, H. W. M. and Nieuwenhuys, R.**, Serotoninergic neuron systems in the brain of the lamprey, *Lampetra fluviatilis*, *Anat. Rec.*, 193, 693, 1979.
199. **Kah, O. and Chambolle, P.**, Serotonin in the brain stem of the goldfish *Carassius auratus*. An immunocytochemical study, *Cell Tiss. Res.*, 234, 319, 1983.
200. **Parent, A.**, Comparative anatomy of the serotoninergic systems. *J. Physiol. (Paris)*, 77, 147, 1981.
201. **Dubé, L. and Parent, A.**, The organization of monoamine-containing neurons in the brain of the salamander, *Necturus maculosus*, *J. Comp. Neurol.*, 211, 21, 1982.
202. **Wolters, J. G., Ten Donkelaar, H. J., Steinbusch, H. W. M., and Verhofstad, A. A. J.**, Distribution of serotonin in the brain stem and spinal cord of the lizard, *Varanus exanthematicus:* an immunohistochemical study, *Neuroscience*, 14, 169, 1985.
203. **Fuxe, K. and Ljunggren, L.**, Cellular localization of monoamines in the upper brain stem of the pigeon, *J. Comp. Neurol.*, 148, 61, 1965.
204. **Dubé, L. and Parent, A.**, The monoamine-containing neurons in avian brain: I. A study of the brain stem of the chicken (*Gallus domesticus*) by means of fluorescence and acetycholinesterase histochemistry, *J. Comp. Neurol.*, 196, 695, 1981.
205. **Fasolo, A., Franzoni, M. F., Gaudino, G., and Steinbusch, H. W. M.**, The organization of serotonin-immunoreactive neuronal systems in the brain of the crested newt, *Triturus cristatus carnifex* Laur, *Cell Tiss. Res.*, 243, 239, 1986.
205a. **Björklund, A. and Flack, B.**, Pituitary monoamines in the cat with special reference to the presence of an unidentified momoamine-like substance in the adenohypophysis, *Z. Zellforsch.*, 93, 254, 1969.
206. **Friedman, E., Krieger, D. T., Léranth, C., Mezey, E., Browstein, M. J., and Palkovits, M.**, Serotonin innervation of the pituitary intermediate lobe decrease after stalk section, *Endocrinology*, 112, 1943, 1983.
207. **Léranth, C., Palkovits, M., and Krieger, D. T.**, Serotonin immunoreactive fibers and terminals in the intermediate lobe of rat pituitary — light and electron microscopic studies, *Neuroscience*, 9, 289, 1983.
207a. **Saland, L. C., Wallace, J. A., and Comunas, F.**, Serotonin-immunoreactive nerve fibers of the rat pituitary: effects of anticatecholamine and antiserotonin drugs on staining patterns, *Brain Res.*, 368, 310, 1986.
208. **Holzbauer, M., Raché, K., and Sharman, D. F.**, Release of endogenous 5-hydroxytryptamine from the neural and the intermediate lobe of the rat pituitary gland evoked by electrical stimulation of the pituitary stalk, *Neuroscience*, 15, 723, 1985.

209. **Palkovits, M., Mezey, E., Chiueh, C. G., Krieger, D. T., Gallaz, K., and Browstein, M. J.,** Serotonin-containing elements of the rat pituitary intermediate lobe, *Neuroendocrinology,* 42, 522, 1986.
210. **Rodriguez, E. M. and La Pointe, J.,** Light and electron microscopic study of the pars intermedia of the lizard, *Klauberina riversiana, Z. Zellforsch. Mikrosk. Anat.,* 104, 1, 1970.
211. **Forbes, M. S.,** Observations on the fine structure of the pars intermedia in the lizard, *Anolis carolinensis, Gen. Comp. Endocrinol.,* 18, 146, 1972.
212. **Meurling, P. and Willstedt, A.,** Vascular connections in the pituitary of *Anolis carolinensis* with special reference to the pars intermedia, *Acta Zool.,* 51, 211, 1970.
213. **Ueda, S., Nojyo, Y., and Sano, Y.,** Immunohistochemical demonstration of serotonin neuron system in the central nervous system of the bullfrog, *Rana catesbeiana, Anat. Embryol.,* 169, 219, 1984.
214. **Olivereau, M. and Chambolle, P.,** Serotoninergic control of MSH secretion in the eel. Ultrastructural study after 5-hydroxytryptophan treatment, *J. Physiol. (Paris),* 75, 109, 1979.
215. **Olivereau, M.,** Serotonin and MSH secretion: effect of parachlorophenylalanine on the pituitary cytology of the eel, *Cell Tiss. Res.,* 191, 83, 1978.
216. **Levitin, H. P.,** Further evidence that serotonin may be a physiological melanocyte-stimulating hormone-releasing factor in the lizard, *Anolis carolinensis, Gen. Comp. Endocrinol.,* 40, 8, 1980.
217. **Kraicer, J. and Morris, A. R.,** *In vitro* release of ACTH from dispersed rat pars intermedia cells. Effect of neurotransmitter substances, *Neuroendocrinology,* 21, 175, 1976.
218. **Fischer, J. L. and Moriaty, C. M.,** Control of bioactive corticotropin release from the neuro-intermediate lobe of the rat pituitary *in vitro, Endocrinology,* 100, 1047, 1977.
219. **Briaud, B., Koch, B., Lutz-Bucher, B., and Mialhe, C.,** *In vitro* regulation of ACTH release from neurointermediate lobe of rat hypophysis. II. Effect of neurotransmitters, *Neuroendocrinology,* 28, 377, 1979.
220. **Baker, B. I.,** Ability of various factors to oppose the stimulatory effect of dibutyryl cyclic AMP on the release of melanocyte-stimulating hormone by the rat pituitary *in vitro, J. Endocrinol.,* 68, 283, 1976.
221. **Bridges, T. E., Fisher, A. W., Gosbee, J. L., Lederis, K., and Santolaya, R. C.,** Acetylcholine and cholinesterases (assays and light and electron microscopical histochemistry) in different parts of the pituitary of rat, rabbit and domestic pig, *Z. Zellforsch.,* 136, 1, 1973.
222. **Cannata, M. A. and Tramezzani, J. H.,** The neural lobe of the neurohypophysis of the rat: several types of nerve endings, *Experientia,* 25, 1281, 1969.
223. **Gallardo, M. R., Cannata, M. A., and Tramezzani, J. M.,** Choline acetyltransferase activity in the neurointermediate lobe of the rat pituitary, *J. Neural. Transm.,* 41, 93, 1977.
224. **Brownstein, M. J., Saawedra, J. M., Palkovits, M., and Axelrod, J.,** Histamine content of hypothalamic nuclei of the rat, *Brain Res.,* 77, 151, 1974.
225. **Takeda, N., Inagaki, S., Shiosaka, S., Taguchi, Y., Oertel, W. H., Tohyama, M., Watanabe, T., and Wada, H.,** Immunohistochemical evidence for the coexistence of histidine decarboxylase-like and glutamate decarboxylase-like immunoreactivities in nerve cells of the magnocellular nucleus of the posterior hypothalamus of rats, *Proc. Natl. Acad. Sci. U.S.A.,* 81, 7647, 1984.
226. **Dierst-Davies, K., Ralph, C. L., and Pechersky, J. L.,** Effects of pharmacological agents on the hypothalamus of *Rana pipiens* in relation to the control of skin melanophores, *Gen. Comp. Endocrinol.,* 6, 409, 1966.
227. **Dierst, K. E. and Ralph, C. L.,** Effect of hypothalamic stimulation on melanophores in frog, *Gen. Comp. Endocrinol.,* 2, 347, 1962.
228. **Burgus, R., Dunn, T., Desiderio, D., and Guillemin, R.,** Structure moléculaire du facteur hypothalamique hypophysiotrope TRF d'origine ovine: mise en évidence par spectrométrie de masse de la séquence PCA His-Pro-NH$_2$, *C.R. Acad. Sci. (Paris),* 269, 1870, 1969.
229. **Nair, R. M. G., Barret, T., Bowers, C. Y., and Schally, A. V.,** Structure of procine thyrotropin-releasing hormone, *Biochemistry,* 9, 1103, 1970.
230. **Vale, W. W., Grant, G., and Guillemin, R.,** Chemistry of the hypothalamic releasing factors: studies on structure-function relationships, in *Frontiers in Neuroendocrinology,* Ganong, W. F. and Martini, L., Eds., Oxford University Press, New York, 1973, 375.
231. **Schally, A. V., Arimura, A., and Kastin, A. J.,** Hypothalamic regulatory hormones, *Science,* 179, 341, 1973.
232. **Bowers, C. Y., Friesen, H., Guyda, H. J., and Folkers, K.,** Prolactin and thyrotropin release in man by synthetic pyroglutamyl-histidyl-prolineamide, *Biochem. Biophys. Res. Commun.,* 45, 1033, 1971.
233. **Rivier, C. and Vale, W. W.,** *In vivo* stimulation of prolactin secretion in the rat by thyrotropin releasing factor, related peptides and hypothalamic extracts, *Endocrinology,* 95, 978, 1974.
234. **Jackson, I. M. D. and Reichlin, S.,** Thyrotropin-releasing hormone (TRH): distribution in hypothalamic and extrahypothalamic brain tissues of mammalian and submammalian chordates, *Endocrinology,* 95, 854, 1974.
235. **Taurog, A., Oliver, C., Eskay, R. L., Porter, J. C., and McKenzie, J. M.,** The role of TRH in the neoteny of the Mexican axolotl (Ambystoma mexicanum), *Gen. Comp. Endocrinol.,* 24, 257, 1974.

284. **De Léan, A., Ferland, L., Drouin, J., Kelly, P. A., and Labrie, F.,** Modulation of pituitary thyrotropin-releasing hormone receptor levels by estrogens and thyroid hormones, *Endocrinology,* 100, 1496, 1977.
285. **Leroux, P., Tonon, M. C., Saulot, P., Jégou, S., and Vaudry, H.,** *In vitro* study of frog (*Rana ridibunda* Pallas) neurointermediate lobe secretion by use of a simplified perifusion system. II. Lack of action of thyroxine on TRH-induced α-MSH secretion, *Gen. Comp. Endocrinol.,* 51, 323, 1983.
286. **Vale, W. W., Rivier, J., Brazeau, P., and Guillemin, R.,** Effects of somatostatins on the secretion of thyrotropin and prolactin, *Endocrinology,* 95, 968, 1974.
287. **Drouin, J., De Léan, A., Rainville, D., Lachance, R., and Labrie, F.,** Characteristics of interaction between thyrotropin-releasing hormone and somatostatin for thyrotropin and prolactin release, *Endocrinology,* 98, 514, 1976.
288. **Lien, E. L., Fenichel, R. L., Garsky, V., Sarantakis, D., and Grant, N. H.,** Enkephalin-stimulated prolactin release, *Life Sci.,* 19, 837, 1976.
289. **Dupont, A., Cuzan, L., Labrie, F., Coy, D. H., and Li, C. H.,** Stimulation of prolactin release in the rat by intraventricular injection of β-endorphin and methionine-enkephalin, *Biochem. Biophys. Res. Commun.,* 75, 76, 1977.
290. **Ferland, L., Fuxe, K., Eneroth, P., Gustafsson, J. A., and Skett, P.,** Effect of methionine-enkephalin on prolactin release and catecholamine levels and turnover in the median eminence, *Eur. J. Pharmacol.,* 43, 89, 1977.
291. **Enjalbert, A., Ruberg, M., Arancibia, S., Priam, M., and Kordon, C.,** Endogenous opiates block dopamine inhibition of prolactin secretion *in vitro, Nature,* 280, 595, 1979.
292. **Shapiro, L. M., Bowes, G. M., and Vaughan, P. F. T.,** Effect of melanostatin and thyroliberin on the biosynthesis and release of dopamine by rat brain striatal P_2 fractions, *Life Sci.,* 27, 2099, 1980.
293. **Heal, D. J., Green, A. R., and Youdim, M. B. H.,** The actions of TRH and its analogues on the mesolimbic dopamine system, in *Thyrotropin-Releasing Hormone,* Griffiths, E. C. and Bennett, G. W., Eds., Raven Press, New York, 1983, 271.
294. **Narumo, S. and Nagawa, Y.,** Modification of dopaminergic transmission by thyrotropin-releasing hormone, in *Molecular Pharmacology of Neurotransmitter Receptors,* Sagawa, T., Ed., Raven Press, New York, 1983, 185.
295. **Rasmussen, H., Kojima, I., Kojima, K., Zawalich, W., and Apfeldorf, W.,** Calcium as intracellular messenger: sensitivity modulation, C-kinase pathway, and sustained cellular response, in *Advances in Cyclic Nucleotide and Protein phosphorylation Research,* Greengard, P. and Robison, G. A., Eds., Raven Press, New York, 1984, 159.
296. **Albert, P. R. and Tashjian, A. H. Jr.,** Relationship of thyrotropin-releasing hormone-induced spike and plateau phases in cytosolic free Ca^{2+} concentrations to hormone secretion, *J. Biol. Chem.,* 259, 15350, 1984.
297. **Drummond, A. H.,** Bidirectional control of cytosolic free calcium by thyrotropin-releasing hormone in pituitary cells, *Nature,* 316, 752, 1985.
298. **Gershengorn, M. C. and Thaw, C.,** Thyrotropin-releasing hormone (TRH) stimulates biphasic elevation of cytoplasmic free calcium in GH_3 cells. Further evidence that TRH mobilizes cellular and extracellular Ca^{2+}, *Endocrinology,* 116, 591, 1985.
299. **Lamacz, M., Leneveu, E., Tonon, M. C., Bernard, C., Gouteux, L., and Vaudry, H.,** Rôle du calcium dans la sécrétion d'α-MSH induite par la TRH chez la grenouille, *Ann. Endocrinol. (Paris),* 47, 54, 1986.
300. **Francis, K. T.,** The effect of dantrolene sodium on the efflux of Ca^{45} from rat heavy sarcoplasmic reticulum, *Res. Commun. Chem. Pathol. Pharmacol.,* 21, 573, 1978.
301. **Brown, M. R. and Hedge, G. A.,** *In vivo* effects of prostaglandins on TRH-induced TSH secretion, *Endocrinology,* 95, 1392, 1974.
302. **Sundberg, D. K., Fawcett, C. P., Illner, P., and McCann, S. M.,** The effect of various prostaglandins and a prostaglandin synthetase inhibitor on rat anterior pituitary cyclic AMP levels and hormone release *in vitro, Proc. Soc. Exp. Biol. Med.,* 148, 54, 1975.
303. **Drouin, J. and Labrie, F.,** Specificity of the stimulatory effect of prostaglandins on hormone release in rat anterior pituitary cells in culture, *Prostaglandins,* 11, 355, 1976.
304. **Leroux, P., Tonon, M. C., Jégou, S., Delarue, D., Leboulenger, F., Netchitailo, P., and Vaudry, H.,** Stimulatory effect of prostaglandin E_1 on thyroliberin-induced α-melanotropin release from perifused neurointermediate lobes of frog pituitary gland, *Prostaglandins,* 21, 599, 1981.
305. **Matsushita, N., Kato, Y., Shimatsu, A., Katakami, H., Yanaihara, N., and Imura, H.,** Effects of VIP, TRH, GABA, and dopamine on prolactin release from superfused rat anterior pituitary cells, *Life Sci.,* 32, 1263, 1983.
306. **Ho, K. Y., Smythe, G. A., and Lazarus, L.,** The interaction of TRH and dopaminergic mechanisms in the regulation of stimulated prolactin release in man, *Clin. Endocrinol.,* 23, 7, 1985.
307. **Vaudry, H., Jenks, B. G., and van Overbeeke, A. P.,** The frog pars intermedia contains only the non-acetylated form of α-MSH: acetylation to generate α-MSH occurs during the release process, *Life Sci.,* 33, 97, 1983.

308. **Vaudry, H., Jenks, B. G., and van Overbeeke, A. P.**, Biosynthesis, processing and release of proopiomelanocortin related peptides in the intermediate lobe of the pituitary gland of the frog *(Rana ridibunda)*, *Peptides*, 5, 905, 1984.
309. **Guttmann, S. T. and Boissonnas, R. A.**, Influence of the structure of the N-terminal extremity of α-MSH on the melanophore stimulating activity of this hormone, *Experientia*, 17, 265, 1961.
310. **Eberlé, A. N.**, Structure and chemistry of the peptide hormone of the intermediate lobe, in *Peptides of the Pars Intermedia*, Ciba Foundation Symposium 81, Pitman Medical, Summit, New Jersey, 1981, 13.
311. **Tonon, M. C., Leroux, P., Jenks, B. G., Gouteux, L., Jégou, S., Guy, J., Pelletier, G., and Vaudry, H.**, Le lobe intermédiaire de l'hypophyse des amphibiens: une glande endocrine à sécrétion multiples et sous contrôle pluri-hormonal, *Ann. Endocrinol.*, 46, 69, 1985.
312. **Tatemoto, K., Carlquist, M., and Mutt, V.**, Neuropeptide Y- a novel brain peptide with structural similarities to peptide YY and pancreatic polypeptide, *Nature*, 296, 659, 1982.
313. **Tatemoto, K.**, Neuropeptide Y: complete amino acid sequence of the brain peptide, *Proc. Natl. Acad. Sci. U.S.A.*, 79, 2514, 1982.
314. **Tatemoto, K. and Mutt, V.**, Isolation of two novel candidate hormones using a chemical method for finding naturally occurring polypeptides, *Nature (London)*, 285, 417, 1980.
315. **Kimmel, J. R., Hayden, L. J., and Pollock, H. G.**, Isolation and characterization of a new pancreatic polypeptide hormone, *J. Biol. Chem.*, 250, 9369, 1975.
316. **Danger, J. M., Guy, J., Benyamina, M., Jégou, S., Leboulenger, F., Côté, J., Tonon, M. C., Pelletier, G., and Vaudry, H.**, Localization and identification of neuropeptide Y (NPY)-like immunoreactivity in the frog brain, *Peptides*, 6, 1225, 1985.
317. **Cailliez, D., Danger, J. M., Andersen, A. C., Polak, J. M., Pelletier, G., Kawamura, K., Kikuyama, S., and Vaudry, H.**, Neuropeptide Y (NPY)-like immunoreactive neurons in the brain and pituitary of the amphibian *Rana catesbeiana*, *Zool. Sci.*, 4, 123, 1987.
318. **Allen, Y. S. and Adrian, T. E.**, neuropeptide Y distribution in the rat brain, *Science*, 221, 877, 1983.
319. **Everitt, B. J., Hökfelt, T., Terenius, L., Tatemoto, K., Mutt, V., and Goldstein, M.**, Differential coexistence of neuropeptide Y (NPY)-like immunoreactivity with catecholamines in the central nervous system of the rat, *Neuroscience*, 11, 443, 1984.
320. **Pelletier, G., Guy, J., Allen, Y. S., and Polak, J. M.**, Electron microscope immunocytochemical localization of neuropeptide Y (NPY) in the brain, *Neuropeptides*, 4, 319, 1984.
321. **Chronwall, B. M., DiMaggio, D. A., Massari, V. J., Pickel, V. M., Ruggiero, D. A., and O'Donohue, T. L.**, The anatomy of neuropeptide-Y-containing neurons in the rat brain, *Neuroscience*, 15, 1159, 1985.
322. **Léger, L., Charnay, Y., Danger, J. M., Vaudry, H., Pelletier, G., Dubois, P. M., and Jouvet, M.**, Mapping of neuropeptide Y-like immunoreactivity in the feline hypothalamus and hypophysis, *J. Comp. Neurol.*, 255, 283, 1987.
323. **Lundberg, J. M., Terenius, L., Hökfelt, T., and Tatemoto, K.**, Comparative immunohistochemical and biochemical analysis of pancreatic polypeptide-like peptides with special reference to presence of neuropeptide Y in central and peripheral neurons, *J. Neurosci.*, 4, 2376, 1984.
324. **Smith, Y., Parent, A., Kerkérian, L., and Pelletier, G.**, Distribution of neuropeptide Y immunoreactivity in the basal forebrain and upper brainstem of the squirrel monkey *(Saimiri sciureus)*, *J. Comp. Neurol.*, 236, 71, 1985.
325. **Hendry, S. H. C., Jones, E. G., and Emson, P. C.**, Morphology, distribution and synaptic relations of somatostatine and neuropeptide Y-immunoreactive neurons in rat and monkey neocortex, *J. Neurosci.*, 4, 2497, 1984.
326. **Adrian, T. E., Allen, J. M., Bloom, S. R., Ghatei, M. A., Rossor, M. N., Roberts, G. W., Crow, T. J., Tatemoto, K., and Polak, J. M.**, Neuropeptide Y distribution in human brain, *Nature*, 306, 584, 1983.
327. **Pelletier, G., Desy, L., Kerkérian, L., and Côté, J.**, Immunocytochemical localization of neuropeptide Y (NPY) in the human hypothalamus, *Cell Tiss. Res.*, 238, 203, 1984.
328. **Dawbarn, D., Hunt, S. P., and Emson, P. C.**, Neuropeptide Y: regional distribution, chromatographic characterization and immunohistochemical demonstration in post-mortem human brain, *Brain Res.*, 296, 168, 1984.
329. **Danger, J. M., Leboulenger, F., Guy, J., Tonon, M. C., Benyamina, M., Martel, J. C., Saint-Pierre, S., Pelletier, G., and Vaudry, H.**, Neuropeptide Y in the intermediate lobe of the frog pituitary acts as an α-MSH-release inhibiting factor, *Life Sci.*, 39, 1183, 1986.
330. **Caillez, D., Danger, J. M., Polak, J. M., Pelletier, G., Andersen, A. C., Leboulenger, F., and Vaudry, H.**, Co-distribution of neuropeptide Y and its C-terminal flanking peptide in the brain and pituitary of the frog, *Rana ridibunda*, *Neurosci. Lett.*, 74, 163, 1987.
331. **Verburg-van Kemenade, B. M. L., Jenks, B. G., Danger, J. M., Vaudry, H., Pelletier, G., and Saint-Pierre, S.**, A NPY-like peptide may function as an MSH-release inhibiting factor in *Xenopus laevis*, *Peptides*, 8, 61, 1987.

332. **Kalra, S. P. and Crowley, W. R.**, Norepinephrine-like effects of neuropeptide Y on LH release in the rat, *Life Sci.*, 35, 1173, 1984.
333. **Kerkérian, L., Guy, J., Lefèvre, G., and Pelletier, G.**, Effect of neuropeptide Y (NPY) on the release of anterior pituitary hormones in the rat, *Peptides*, 6, 1201, 1985.
334. **McDonald, J. K., Lumpkin, M. D., Samson, W. K., and McCann, S. M.**, Neuropepitide Y affects secretion of luteinizing hormone and growth hormone in ovariectomized rats, *Proc. Natl. Acad. Sci. U.S.A.*, 82, 561, 1985.
335. **Hökfelt, T., Lundberg, J. M., Tatemoto, K., Mutt, V., Terenius, L., Polak, J. M., Bloom, S., Sasek, C., Elde, R., and Goldstein, M.**, Neuropeptide Y (NPY) - and FMRFamide like immunoreactivities in catecholamine neurons of the rat medulla oblongata, *Acta Physiol. Scand.*, 117, 315, 1983.
336. **Hökfelt, T., Lundberg, J. M., Lagercrantz, H., Tatemoto, K., Mutt, V., Lindberg, J., Terenius, L., Everitt, B. J., Fuxe, K., Agnati, L., and Goldstein, M.**, Occurrence of neuropeptide Y (NPY)-like immunoreactivity in catecholamine neurons in the human medulla oblongata, *Neurosci. Lett.*, 29, 217, 1983.
337. **Hendry, S. H., Jones, E. G., De Felipe, J., Schmechel, D., Brandon, C., and Emson, P. C.**, Neuropeptide-containing neurons of the central cerebral cortex are also GABAergic, *Proc. Natl. Acad. Sci. U.S.A.*, 81, 6526, 1984.
338. **Vincent, S. R., Skirboll, L., Hökfelt, T., Johansson, O., Lundberg, M., Elde, R. P., Terenius, L., and Kimmel, J.**, Coexistence of somatostatin and avian pancreatic polypeptide (APP)-like immunoreactivity in some forebrain neurons, *Neuroscience*, 7, 439, 1982.
339. **Chronwall, B. M., Chase, T. N., and O'Donohue, T. L.**, Coexistence of Neuropeptide Y and somatostatin in rat and human cortical and rat hypothalamic neurons, *Neurosci. Lett.*, 52, 213, 1984.
340. **Hendry, S. H., Jones, E. G., and Emson, P. C.**, Morphology, distribution, and synaptic relations of somatostatin- and neuropeptide Y-immunoreactive neurons in rat and monkey neocortex, *J. Neurosci.*, 4, 2497, 1984.
341. **Dubois, M. P., Barry, J., and Léonardelli, J.**, Mise en évidence par immunofluorescence et répartition de la somatostatine (SRIF) dans l'éminence médiane des vertébrés (mammifères, oiseaux, amphibiens poissons), *C.R. Acad. Sci. (Paris)*, 279, 1899, 1974.
342. **Vandesande, F. and Dierickx, K.**, Immunocytochemical localization of somatostatin-containing neurons in the brain of *Rana temporaria*, *Cell Tiss. Res.*, 205, 45, 1980.
343. **Minth, C. D., Bloom, S. R., Polak, J. M., and Dixon, J. E.**, Cloning, characterization and DNA sequence of a human cDNA encoding neuropeptide tyrosine, *Proc. Natl. Acad. Sci. USA*, 81, 4577, 1984.
344. **Morel, G., Leneveu, E., Tonon, M. C., Pelletier, G., Vaudry, H., and Dubois, P. M.**, Subcellular localization of thyrotropin-releasing hormone (TRH)- and neuropeptide Y (NPY)-like immunoreactivity in the neurointermediate lobe of the frog pituitary, *Peptides*, 6, 1085, 1985.
345. **Leung, P. C. K., Raymond, V., and Labrie, F.**, Mechanisms of action of TRH: involvement of the phosphatidylinositol (PI) response in the action of TRH in rat anterior pituitary cells, *Life Sci.*, 31, 3037, 1982.
346. **Baird, J. G. and Brown, B. L.**, Thyrotropin-releasing hormone stimulates inositol phosphate production in normal anterior pituitary cells and GH_3 tumor cells in the presence of lithium, *Biosci. Rep.*, 3, 1091, 1983.
347. **Merritt, J. E. and Brown, B. L.**, An investigation of calcium in the control of prolactin secretion: studies with low calcium, methoxyverapamil, cobalt and manganese, *J. Endocrinol.*, 101, 319, 1984.
348. **Rebecchi, M. J. and Gershengorn, M. C.**, Thyroliberin stimulates rapid hydrolysis of phosphatidylinositol 4,5-*bis*-phosphate by a phosphodiesterase in rat mammotropic pituitary cells, *Biochem. J.*, 216, 287, 1983.
349. **Spiess, J., Rivier, J., Rivier, C., and Vale, W.**, Primary structure of corticotropin-releasing factor from ovine hypothalamus, *Proc. Natl. Acad. Sci. U.S.A.*, 78, 6517, 1981.
350. **Vale, W., Spiess, J., Rivier, C., and Rivier, J.**, Characterization of a 41-residue ovine hypothalamus peptide that stimulates secretion of corticotropin and β-endorphin, *Science*, 213, 1394, 1981.
351. **Rivier, C., Brownstein, M., Spiess, J., Rivier, J., and Vale, W.**, *In vivo* corticotropin-releasing factor-induced secretion of adrenocorticotropin, β-endorphin and corticosterone, *Endocrinology*, 110, 272, 1982.
352. **Baird, A., Wehrenberg, W. B., Shibasaki, T., Benoit, R., Chong-Li, Z., Esch, F., and Ling, N.**, Ovine corticotropin-releasing factor stimulates the concomitant secretion of corticotropin, β-lipotropin, β-endorphin and α-melanotropin by the bovine adenohypophysis *in vitro*, *Biochem. Biophys. Res. Commun.*, 108, 859, 1982.
353. **Giguère, V., Labrie, F., Côté, J., Coy, D. H., Sueiras-Diaz, J., and Schally, A. V.**, Stimulation of cyclic AMP accumulation and corticotropin release by synthetic ovine corticotropin-releasing factor in rat anterior pituitary cells: site of glucocorticoids action, *Proc. Natl. Acad. Sci. U.S.A.*, 79, 3466, 1982.
354. **Rivier, C. and Vale, W.**, Effect of corticotropin-releasing factor (CRF) on some endocrine functions, in *Endocrinology*, Labrie, F. and Proulx, L., Eds., Elsevier Science Publishers, Excerpta Medica, Amsterdam, 1984, 959.

355. **Lederis, K., Letter, A., McMaster, D., Moore, G., and Schlesinger, D.,** Complete amino acid sequence of urotensin I, a hypotensive and corticotropin-releasing neuropeptide from *Catostomus, Science,* 218, 162, 1982.
356. **Montecucchi, P. C. and Henschen, A.,** Amino acid composition and sequence analysis of sauvagine, a new active peptide from the skin of *Phyllomedusa sauvagei, Intl. J. Protein Pept. Res.,* 18, 113, 1981.
357. **Lederis, K., Vale, W., Rivier, J., MacCannell, K. L., McMaster, D., Kobayashi, Y., Suess, U., and Laurence, J.,** Urotensin I. A novel CRF like in *Catostomus commersoni, Proc. West. Pharmacol. Sci.,* 25, 223, 1982.
358. **Fryer, J., Lederis, K., and Rivier, J.,** Urotensin I, a CRF-like neuropeptide, stimulates ACTH release from teleost pituitary, *Endocrinology,* 113, 2308, 1983.
359. **Swanson, L. W., Sawchenko, P. E., Rivier, J., and Vale, W.,** Organisation of ovine corticotropin releasing factor immunoreactive cells and fibers in the rat brain: an immunohistochemical study, *Neuroendocrinology,* 36, 165, 1983.
360. **Bugnon, C., Fellmann, D., Gouget, A., and Cardot, J.,** Corticoliberin in rat brain: immunocytochemical identification and localization of a novel neuroglandular system, *Neurosci. Lett.,* 30, 25, 1982.
361. **Merchenthaler, I., Vigh, S., Petrusz, F., and Shally, A. V.,** Immunocytochemical localization of corticotropin releasing factor (CRF) in the rat brain, *Am. J. Anat.,* 165, 384, 1982.
362. **Pelletier, G., Desy, G., Côté, J., Lefèvre, G., Vaudry, H., and Labrie, F.,** Immunoelectron microscopic localization of corticotropin releasing factor in the rat hypothalamus, *Neuroendocrinology,* 35, 402, 1982.
363. **Burlet, A., Tonon, M. C., Tankosic, P., Coy, D. H., and Vaudry, H.,** Comparative immunocytochemical localization of corticotropin-releasing factor (CRF-41) and neurohypophyseal peptides in the brain of the Brattleboro and Long Evans rats, *Neuroendocrinology,* 37, 64, 1983.
364. **Roth, K. A., Weber, E., and Barchas, J. D.,** Immunoreactive corticotropin releasing factor (CRF) and vasopressin are colocalized in a subpopulation of the immunoreactive vasopressin cells in the paraventricular nucleus of the hypothalamus, *Life Sci.,* 31, 1857, 1982.
365. **Fellmann, D., Bugnon, C., Bresson, J. L., Gouget, A., Cardot, J., Clavequin, M. C., and Hadjiyiassemis, M.,** The CRF neuron: immunocytochemical study, *Peptides,* 5, 19, 1984.
366. **Bugnon, C., Fellmann, D., Bresson, J. L., and Clavequin, M. C.,** Immunocytochemical study of the ontogenesis of the CRF-containing neuroglandular system in the human hypothalamus, *C. R. Acad. Sci. (Paris),* 294, 107, 1982.
367. **Pelletier, G., Désy, L., Côté, J., and Vaudry, H.,** Immunocytochemical localization of corticotropin-releasing factor-like immunoreactivity in the human hypothalamus, *Neurosci. Lett.,* 41, 259, 1983.
368. **Paull, W. K., Phelix, C. F., Copeland, M., Palmitter, P., Gibbs, F. P., and Middleton, C.,** Immunohistochemical localization of corticotropin releasing factor (CRF) in the hypothalamus of the squirrel monkey, *Saimiri sciureus, Peptides,* 5, Suppl. 1, 45, 1984.
369. **Tramu, G., Croix, C., and Pillez, A.,** Ability of the CRF immunoreactive neurons of the paraventricular nucleus to produce a vasopressin-like material, *Neuroendocrinology,* 37, 467, 1983.
370. **Sawchenko, P. E., Swanson, L. W., and Vale, W. W.,** Co-expression of corticotropin-releasing factor and vasopressin immunoreactivity in parvocellular neurosecretory neurons of the adrenalectomized rat, *Proc. Natl. Acad. Sci. U.S.A.,* 81, 1883, 1984.
371. **Kawata, M., Hashimoto, K., Takahara, J., and Sano, Y.,** Immunohistochemical demonstration of corticotropin releasing factor containing nerve fibers in the median eminence of the rat and monkey, *Histochemistry,* 76, 15, 1982.
372. **Tramu, G. and Pillez, A.,** Localisation immunohistochimique des terminaisons nerveuses à corticolibérine (CRF) dans l'éminence médiane du cobaye et du rat, *C.R. Acad. Sci. (Paris),* 294, 107, 1982.
373. **Bloom, F. E., Battenberg, E. L. F., Rivier, J., and Vale, W. W.,** Corticotropin releasing factor (CRF): immunoreactive neurons and fibers in rat hypothalamus, *Regul. Pept.,* 4, 43, 1982.
374. **Merchenthaler, I., Vigh, S., Petrusz, P., and Schally, A. V.,** The paraventriculo-infundibular corticotropin releasing factor (CRF) pathway as revealed by immunocytochemistry in long-term hypophysectomized or adrenalectomized rats, *Regul. Pept.,* 5, 295, 1983.
375. **Saavedra, J. M., Rougeot, C., Culman, J., Israel, A., Niwa, M., Tonon, M. C., Vaudry, H., and Dray, F.,** Decreased corticotropin-releasing factor-like immunoreactivity in rat intermediate and posterior pituitary after stalk section, *Neuroendocrinology,* 39, 93, 1984.
376. **Péczely, P. and Antoni, F. A.,** Comparative localization of neurons containing ovine corticotropin releasing factor (CRF)-like and neurophysin like immunoreactivity in the diencephalon of the pigeon *(Columba livia domestica), J. Comp. Neurol.,* 228, 69, 1984.
377. **Bons, N., and Bouillé, C., Vaudry, H., and Guillaume, V.,** Localisation par immunofluorescence des neurones à corticolibérine dans l'encéphale du pigeon, *C.R. Acad. Sci. (Paris),* 300, 49, 1985.
378. **Olivereau, M., Vandesande, F., Boucique, E., Ollevier, F., and Olivereau, J. M.,** Mise en évidence immunocytochimique d'un système peptidergique de type CRF (corticotropin-releasing factor) dans le cerveau des amphibiens. Comparàison avec la répartition du système à somatostatine, *C.R. Acad. Sci. (Paris),* 299, 871, 1984.

379. **Verhaert, P., Marivoet, S., Vandesande, F., and De Loof, A.**, Localization of CRF immunoreactivity in the central nervous system of three vertebrate and one insect species, *Cell Tiss. Res.*, 238, 49, 1984.
380. **Tonon, M. C., Burlet, A., Lauber, M., Cuet, P., Jégou, S., Gouteux, L., and Vaudry, H.**, Immunohistochemical localization and radioimmunoassay of corticotropin-releasing factor in the forebrain and hypophysis of the frog *Rana ridibunda, Neuroendocrinology,* 40, 109, 1985.
381. **Bugnon, C., Cardot, J., Gouget, A., and Fellmann, D.**, Mise en évidence d'un système neuronal peptidergique réactif à un immunsérum anti CRF41, chez les téléostéens dulcicoles et marins, *C.R. Acad. Sci. (Paris),* 296, 711, 1983.
382. **Olivereau, M., Ollevier, F., Vandesande, F., and Verdonck, W.**, Immunocytochemical identification of CRF-like and SRIF-like peptides in the brain and the pituitary of cyprinid fish, *Cell Tiss. Res.*, 237, 379, 1984.
383. **Belenky, M. A., Kuzik, V. V., Chernogovskaya, E. V., and Polenov, A. L.**, The hypothalamo-hypophysial system in acipenseridae. X. Corticoliberin-like immunoreactivity in the hypothalamus and hypophysis of *Acipenser ruthenus, Gen. Comp. Endocrinol.*, 60, 20, 1985.
384. **Yulis, C. R., Lederis, K., Wong, K. L., and Fisher, A. W. F.**, Localization of urotensin-I- and corticotropin-releasing factor-like immunoreactivity in the central nervous system of *Catostomus commersoni, Peptides,* 7, 79, 1986.
385. **Ollevier, F. and Verdonck, W.**, Corticotropin-releasing factor in the pituitary of *Salmo gairdneri, 12th Conf. Europ. Comp. Endocrinol.*, Abstr. 6, 1983.
386. **Proux-Ferland, L., Labrie, F., Dumont, D., Côté, J., Coy, D. H., and Sveiraf, J.**, Corticotropin-releasing factor stimulates secretion of melanocyte-stimulating hormone from the rat pituitary, *Science,* 217, 62, 1982.
387. **Sakly, M., Schmitt, G., and Koch, B.**, CRF enhances release of both α-MSH and ACTH from anterior and intermediate pituitary, *Neuroendocrinol. Lett.*, 4, 289, 1982.
388. **Lutz-Bucher, B. and Koch, B.**, Failure of vasopressin to potentiate the effect of synthetic CRF on ACTH output from intermediate pituitary, *Neuroendocrinol. Lett.*, 5, 111, 1983.
389. **Meunier, H., Lefèvre, G., Dumont, D., and Labrie, F.**, CRF stimulates α-MSH secretion and cyclic AMP accumulation in rat pars intermedia cells, *Life Sci.*, 31, 2129, 1982.
390. **Stoeckel, M. E. and Schimchowitsch, S.**, Peptidergic control of the intermediate lobe of the pituitary gland in the leporidae, *6th Eur. Winter Conf. Brain Res.*, Abstr. 41, 1986.
391. **Verburg-van Kemenade, B. M. L., Jenks, B. G., Tonon, M. C., Vaudry, H., and van Overbeeke, A. P.**, Control of α-MSH release from the pars intermedia of the amphibian *Xenopus laevis;* a microsuperfusion study, *7th Intl. Cong. Endocrinol.*, Quebec, Abstr. 2210, 1984.
392. **Tonon, M. C., Cuet, P., Lamacz, M., Jégou, S., Côté, J., Gouteux, L., Ling, N., Pelletier, G., and Vaudry, H.**, Comparative effects of corticotropin-releasing factor arginine-vasopressin and related neuropeptides on the secretion of ACTH and α-MSH by frog anterior pituitary cells and neurointermediate lobes *in vitro, Gen. Comp. Endocrinol.,* 61, 438, 1986.
393. **Khachaturian, H., Lewis, M. E., Schafer, M. G. H., and Watson, S. J.**, Anatomy of the CNS opiod systems, *Trends Neurosci.*, 7, 111, 1984.
394. **Vanderhaeghen, J. J., Goldman, S., Lotstra, F., van Reeth, O., Deschepper, C., Rossier, J., and Dchiffmanon, S.**, Co-existence of cholecystokinin- or gastrin-like peptides with other peptides in the hypophysis and the hypothalamus, in *Ann. Acad. Sci., Neuronal Cholecystokinin,* Vol. 448, Vanderhaegen, J. J. and Crawley, J. N., Eds., New York, 1985, 334.
395. **Yui, R.**, Immunohistochemical studies on peptide neurons in the hypothalamus of the bullfrog *Rana catesbeiana, Gen. Comp. Endocrinol.*, 49, 195, 1983.
396. **Cone, R. I.**, Dynorphin immunoreactivity in the toad neurointermediate lobe, *Life Sci.*, 31, 1801, 1982.
397. **Vandesande, F. and Dierickx, K.**, Immunocytochemical demonstration of separate vasotocinergic neurons in the amphibian hypothalamus magnocellular neurosecretory system, *Cell Tiss. Res.*, 175, 289, 1976.
398. **Goossens, N., Dierickx, K., and Vandesande, F.**, Immunocytochemical demonstration of the hypothalamo-hypophysial vasotocinergic system of *Lampetra fluviatilis, Cell Tiss. Res.*, 117, 317, 1977.
399. **Netchitailo, P., Feuilloley, M., Pelletier, G., Cantin, M., Leboulenger, F., Andersen, A. C., and Vaudry, H.**, Localization of atrial factor (ANF)-immunoreactive material in the hypothalamo-pituitary complex of the frog, *Neurosci. Lett.*, 72, 141, 1986.
400. **Netchitailo, P., Feuillolet, M., Pelletier, G., Leboulenger, F., Cantin, M., Gutkowska, J., and Vaudry, H.**, Atrial natriuretic factor-like immunoreactivity in the central nervous system of the frog, *Neuroscience,* 22, 341, 1986.
401. **Andersen, A. C., Pelletier, G., Eberle, A. N., Leroux, P., Jégou, S., and Vaudry, H.**, Localization of melanin-concentrating hormone-like immunoreactivity in the brain and pituitary of the frog *Rana ridibunda, Peptides,* 7, 941, 1986.
402. **Skofitsch, G., Jacobowitz, D. M., and Zamir, N.**, Immunohistochemical localization of a melanin-concentrating hormone-like peptide in the rat brain, *Brain Res. Bull.*, 15, 635, 1985.

403. **Kawata, M., Nakao, K., Morii, V., Kiso, Y., Yamashita, H., Imura, H., and Sano, Y.**, Atrial natriuretic polypeptide: topographical distribution in the rat brain by radioimmunoassay and immunohistochemistry, *Neuroscience*, 16, 521, 1985.
404. **Skofitsch, G., Jacobowitz, D. M., Eskay, R. L., and Zamir, N.**, Distribution of atrial natriuretic factor-like immunoreactive neurons in the rat brain, *Neuroscience*, 16, 917, 1985.
405. **Naito, N., Nakai, H., Kawauchi, H., and Hayashi, Y.**, Immunocytochemical identification of melanin-concentrating hormone in the brain and pituitary gland of the teleost fishes *Oncorhyncus keta* and *Salmo gairdneri, Cell Tiss., Res.* 242, 41, 1985.
406. **Celis, M. E., Taleisnik, S., and Walter, R.**, Regulation of formation and proposed structure of the factor inhibiting the release of melanocyte-stimulating hormone, *Proc. Natl. Acad. Sci. U.S.A.*, 68, 1428, 1971.
407. **Nair, R. M. G., Kastin, A. J., and Schally, A. V.**, Isolation and structure of hypothalamic MSH release inhibiting hormone, *Biochem. Biophys. Res. Commun.*, 43, 1376, 1971.
408. **Bower, S. A., Hadley, M. E., and Hruby, V. J.**, Comparative MSH release-inhibiting activities of tocinoic acid (the ring chain of oxytocin), *Biochem. Biophys. Res. Commun.*, 45, 1185, 1971.
409. **Thody, A. J., Shuster, S., Plummer, N. A., Bogie, W., Leigh, R. L., Goolamali, S. K., and Smith, A. G.**, The lack of effect of MSH release-inhibiting factor (MIF) on the secretion of α-MSH in normal men, *J. Clin. Endocrinol. Metab.*, 38, 491, 1974.
410. **Thornton, V. F. and Geschwind, I. I.**, The characteristics of melanocyte stimulating hormone release from incubated pituitaries of the liazard, *Anolis carolinensis, Gen. Comp. Endocrinol.*, 26, 336, 1975.
411. **Donnadieu, M., Laurent, M. F., Luton, J. P., Bricaire, H., Girard, F., and Binoux, M.**, Synthetic MIF has no effect on β-MSH and ACTH hypersecretion in Nelson's syndrome, *J. Clin. Endocrinol. Metab.*, 42, 1145, 1976.
412. **Lamacz, M., Netchitailo, P., Tonon, M. C., Feuilloley, M., Cantin, M., Pelletier, G., and Vaudry, H.**, Atrial natriuretic (ANF) stimulates the release of α-MSH from frog neurointermediate lobes *in vitro*, interaction with dopamine, GABA and neuropeptide Y, *Life Sci.*, 40, 1853, 1987.
413. **Horvath, J., Ertl, T., and Schally, A. V.**, Effect of atrial natriuretic peptide on gonadotropin release in superfused rat pituitary cells, *Proc. Natl. Acad. Sci. U.S.A.*, 83, 3444, 1986.
414. **Shibasaki, T., Naruse, M., Yamauchi, N., Masuda, A., Imaki, T., Naruse, K., Demura, H., Ling, N., Inagami, T., and Shizume, K.**, Rat atrial natriuretic factor suppresses proopiomelanocortin-derived peptides secretion from both anterior and intermediate lobe cells and growth hormone release from anterior lobe cells of rat pituitary *in vitro, Biochem. Biophys. Res. Commun.*, 135, 1035, 1986.
415. **Baker, B. I., Bird, D. J., and Buckingham, J. C.**, Effects of chronic administration of melanin-concentrating hormone on corticotropin, melanotropin, and pigmentation in the trout, *Gen. Comp. Endocrinol.*, 63, 62, 1986.
416. **Barber, L. D., Baker, B. I., Penny, J. C., and Eberle, A. N.**, Melanin-concentrating hormone (MCH) inhibits the release of α-MSH from teleost pituitary glands, in press.
417. **Denef, C. and Baes, M.**, β-adrenergic stimulation of prolactin release from superfused pituitary cell aggregates, *Endocrinology*, 356, 1982.
418. **Raymond, V., Beaulieu, M., Labrie, F., and Boissier, J. R.**, Potent antidopaminergic activity of estradiol at the pituitary level on prolactin release, *Science*, 200, 1173, 1978.
419. **Thody, A. J., Wilson, C. A., Lucas, P. D., and Fischer, C.**, Variations in plasma concentrations of α-melanocyte-stimulating hormone during the oestrous cycle of the rat and after administration of ovarian steroids, *J. Endocrinol.*, 88, 73, 1981.

Chapter 10

MELANOTROPIN ENZYMOLOGY

Ana Maria de L. Castrucci, Mac E. Hadley, and Victor J. Hruby

TABLE OF CONTENTS

I.	Introduction	172
II.	Relative Stability of Melanotropic Peptides in Sera	172
III.	Relative Stability of Melanotropic Peptides Incubated in Purified Proteolytic Enzymes	176
IV.	Relative Stability of Melanotropic Peptides in Brain Homogenates	178
V.	Discussion and Summary	178
Acknowledgments		181
References		181

I. INTRODUCTION

α-Melanotropin (α-melanocyte stimulating hormone, α-MSH; hereafter referred to as MSH) plays an essential role in the control of melanin pigmentation of the skin of many animals. Although not present as a product of the human pituitary gland, this melanotropic peptide may also subserve other extrapigmentary roles, possibly as a neurohormone within the brain.[1] Certain analogues of MSH may prove useful clinically in humans. For example, there is hope that MSH or a related analogue may be used medicinally as an antipyretic agent (see Lipton, Chapter 7, Volume II).[2] Therefore, an understanding of the enzymology of MSH and related analogues may lead to the design of clinically useful melanotropins that exhibit resistance to inactivation by digestive or serum enzymes.

The half-life of iodinated MSH has been shown by radioimmunoassay to be 120 min in the frog, *Xenopus laevis*[3], 7.1 min in the rat,[4] and 4.8 min in humans.[5] The half-time disappearance of ^{3}H-labeled MSH, which is structurally more similar to MSH than iodinated MSH, was reported to be 20.8 min[6] and 25 min[5] in humans. These radioimmunoassays do not, however, determine whether the melanotropin monitored is still bioactive.

Several studies using MSH in the presence of proteolytic enzymes have suggested that the hormone was resistant to exopeptidases, probably due to N-terminal acetylation and C-terminal amidation. Cleavage of the peptide is apparently initiated by endopeptidases such as chymotrypsin and trypsin. During cleavage of MSH by rat serum and brain homogenates, large amounts of phenylalanine and arginine are generated.[7] Trypsin shows predominant specificity for peptide bonds of basic residues such as lysine, histidine, and arginine, whereas the actions of chymotrypsin are specific for uncharged, generally aromatic amino acids such as tyrosine, phenylalanine, and tryptophan.[8] Figure 1 shows the primary structure of MSH and the probable sites of protease attack. An initial cleavage of MSH by endopeptidases would provide N-terminal and C-terminal fragments which would then be further susceptible to cleavage by exopeptidases. An initial cleavage between either His^6-Phe^7 or Phe^7-Arg^8 would be sufficient to inactivate the melanotropin since the MSH 1-6 and 1-7 fragments are inactive, and the 7-13 fragment is only minimally active at very high concentrations.[9]

II. RELATIVE STABILITY OF MELANOTROPIC PEPTIDES IN SERA

Chemical messengers such as hormones, whatever their method of delivery to their target cells, must often be metabolized rather rapidly if a precise control is to be established. Peptide hormones released into the blood are susceptible to inactivation by proteolytic enzymes present in the serum or on cells. Several studies have determined the relative stability of melanotropins incubated for varying periods of time in sera.[10,11] Both MSH and β-melanotropin (β-MSH) were inactivated when incubated in either frog (Figure 2) or rat serum (Figure 3). MSH was more susceptible to inactivation than was β-MSH.[12] This observation was somewhat unexpected since, unlike MSH, the N- and C-terminals, respectively, of β-MSH are not acetylated and amidated. Both melanotropins were inactivated more rapidly in frog serum (Figure 2) than in rat serum (Figure 3). Again, this was somewhat surprising since incubations in frog serum were performed at room temperature (20°C), whereas the incubations in rat serum were done at 37°C.

A number of melanotropins were synthesized with the hope that they might be radioiodinated for use in receptor studies.[13] These analogues were inactivated by several methods of iodination employed.[14,15] Substitution of norleucine for methionine resulted in peptides that were resistant to inactivation by the conditions of radioiodination.[14,15] All Nle^4-substituted analogues of MSH were more active than their natural homologues containing methionine.[16] These Nle^4-substituted analogues did not differ, however, from the Met^4-analogues relative to their stability in sera (Figure 4).

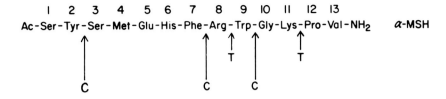

FIGURE 1. Primary structure of α-MSH showing the possible sites of cleavage by α-chymotrypsin (C) and trypsin (T).

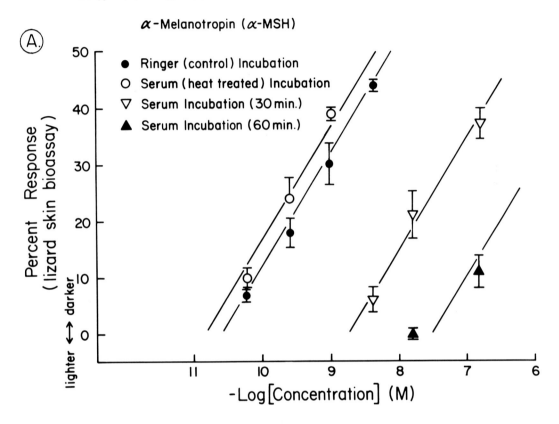

FIGURE 2. In vitro demonstration of the relative resistance of (A) α-MSH and (B) β-MSH to inactivation by 50% frog serum. Each value represents the mean, ± S. E., darkening response of *Anolis carolinensis* skins (N = 12) to the melanotropins after incubation in serum for the times indicated.

For reasons discussed elsewhere (see Hruby et al.; Chapter 7, Volume III) a large number of MSH analogues was synthesized in which the D-phenylalanine diastereomer was substituted for L-Phe at position 7, in addition to Nle⁴ at position 4 of the peptides.[17,18] All these analogues proved to be extremely resistant (possibly nonbiodegradable) to inactivation by sera (Figure 4). These [Nle^4,D-Phe^7]-substituted analogues were much more potent than the Met^4- or Nle^4-substituted analogues (Table 1). They also exhibited prolonged melanotropic activity as will be discussed further (see below).

Again, for reasons that are discussed elsewhere (see Cody et al.; Chapter 7, Volume III), a large number of cyclic melanotropins were synthesized.[19-21] In certain melanocyte bioassays these [Cys^4,Cys^{10}]-substituted analogues were superpotent. In other bioassays they were

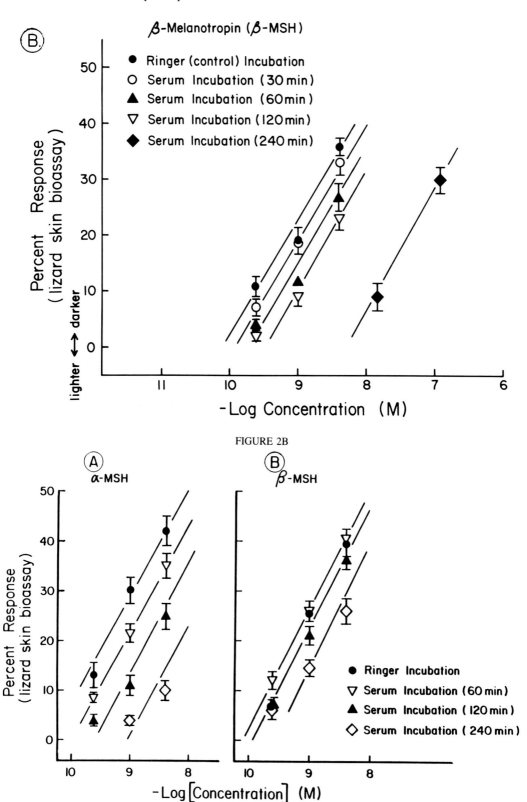

FIGURE 3. In vitro demonstration of the relative resistance of (A) α-MSH and (B) β-MSH to inactivation by 50% rat serum. Each value represents the mean, ± S. E., darkening response of *Anolis carolinesis* skins (N = 12) to the melanotropins after incubation in serum for the times indicated.

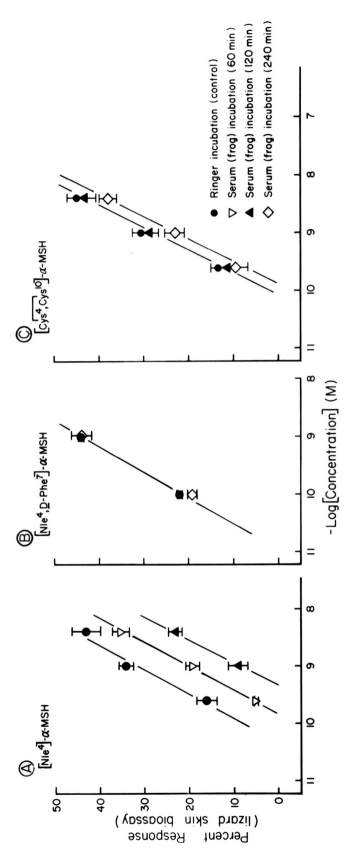

FIGURE 4. In vitro demonstration of the relative resistance of (A) [Nle^4]-α-MSH, (B) [Nle^4,D-Phe^7]-α-MSH and (C) [Cys^4,Cys^{10}]-α-MSH to inactivation by 50% frog serum. Each value represents the mean, ± S. E., darkening response of *Anolis carolinensis* skins (N = 6) to the melanotropins after incubation in serum for the times noted.

Table 1
RELATIVE POTENCIES, RESIDUAL ACTIVITIES, AND STABILITY OF MELANOTROPIC PEPTIDES

Melanotropin	Potency		Residual Activity[a]		Stability[b]	
	Frog	Lizard	Frog	Lizard	Chymotrypsin	Trypsin
α-MSH	1.0	1.0	−	−	−	−
β-MSH	0.8	1.0	−	−	−	−
[Nle^4]-α-MSH	4.0	1.5	−	−	−	−
[Nle$_4$,D-Phe^7]-α-MSH	60.0	5.0	+	+	+	+
Ac-[Nle^4]-α-MSH$_{4-10}$NH$_2$	0.002	0.06	−	−	−	−
[Cys^4,Cys^{10}]-α-MSH	10.0	2.0	−	−	−	−
Ac-obCys^4,Cys^{10}]-α-MSH$_{4-13}$NH$_2$	30.0	0.6	−	−	−	−
Ac-[Cys^4,D-Phe^7,Cys^{10}]-α-MSH$_{4-13}$NH$_2$	6.0	6.0	+	+	+	+
Ac-[Cys^4,Cys^{10}]-α-MSH$_{4-11}$NH$_2$	0.16	0.07	−	−	−	−
Ac-[Cys^4,D-Phe^7,Cys^{10}]-α-MSH$_{4-11}$NH$_2$	2.5	3.0	−	−	+	+
Ac-[Cys^4,Cys^{10}]-α-MSH$_{4-10}$NH$_2$	0.06	0.003	−	−	−	+
Ac-[Cys^4,D-Phe^7,Cys^{10}]-α-MSH$_{4-10}$NH$_2$	0.75	0.5	−	−	+	+

[a] (−) The analogue has no residual activity; (+) the analogue shows prolonged activity.
[b] (−) The melanotropin is rapidly degraded in the presence of the proteolytic enzyme (+); the peptide is resistant to enzyme degradation.

approximately equipotent to the parent hormone, MSH. Although these cyclic melanotropins exhibited superpotency as did the [Nle^4,D-Phe^7]-substituted analogues, they did not exhibit prolonged melanotropic activities as did the D-Phe^7-containing analogues (Table 1). Nevertheless, like these latter peptides, some cyclic melanotropins were extremely resistant to inactivation by serum enzymes (Figures 4 and 5).

III. RELATIVE STABILITY OF MELANOTROPIC PEPTIDES INCUBATED IN PURIFIED PROTEOLYTIC ENZYMES

The foregoing results demonstrated the remarkable resistance of certain melanotropin analogues to inactivation by serum enzymes. To more clearly quantitate the stability of these melanotropins, the residual biological activities of the peptides following incubation with two purified proteolytic enzymes, trypsin and α-chymotrypsin, were determined. Again, the natural melanotropins, MSH and β-MSH, were rapidly inactivated; after 30 min of incubation with either enzyme little or no melanotropic activity could be detected (Table 1). The [Nle^4, D-Phe^7]-substituted analogues, in contrast, were very resistant to inactivation by α-chymotrypsin. [Nle^4,D-Phe^7]-α-MSH, for example, still retained full melanotropic activity even after 60 min incubation with the enzyme (Table 1). Although the shorter [Nle^4,D-Phe^7]-substituted analogues lost some of their activity after 60 min incubation with α-chymotrypsin, none of these analogues were inactivated to any degree in the presence of trypsin (Table 1).

The [Cys^4,Cys^{10}]-substituted cyclic melanotropins, although more stable than the natural melanotropins, did become inactivated by α-chymotrypsin (Table 1). Trypsin, however, hardly affected the activity of the cyclic melanotropins even after 60 min of incubation with the enzyme (Figure 6). [Cys^4,D-Phe^7,Cys^{10}]-substituted analogues were very resistant to inactivation by α-chymotrypsin and totally so to trypsin (Figure 6, Table 1).

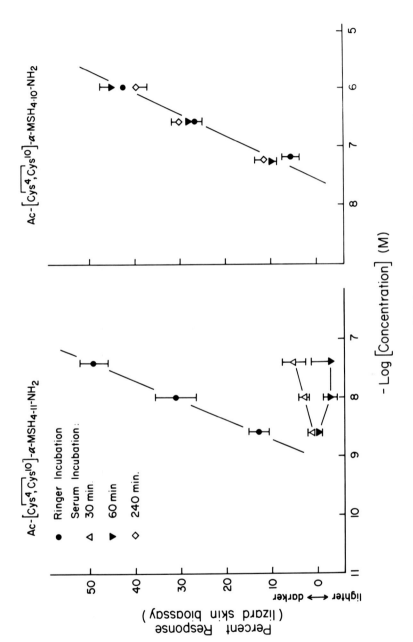

FIGURE 5. In vitro demonstration of the relative stability of (A) Ac-[Cys⁴,Cys¹⁰]-α-MSH$_{4\text{-}10}$-NH$_2$ and (B) Ac-[Cys⁴,Cys¹⁰]-α-MSH$_{4\text{-}11}$-NH$_2$ in 50% frog serum. Each value is the mean, ± S. E., darkening response of *Anolis carolinensis* skins (N = 12) to the melanotropins after incubation in serum for the times indicated.

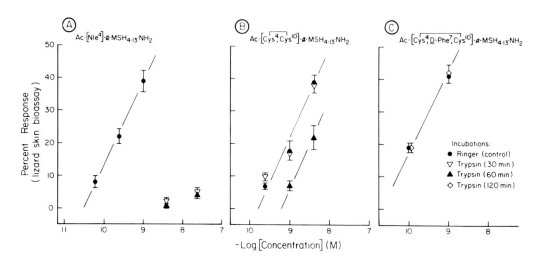

FIGURE 6. In vitro demonstration of the relative resistance of (A) Ac-[Nle⁴]-α-MSH$_{4-13}$-NH$_2$, (B) Ac-[Cys⁴,Cys¹⁰]-α-MSH$_{4-13}$-NH$_2$ and (C) Ac-[Cys⁴,D-Phe⁷,Cys¹⁰]-α-MSH$_{4-13}$-NH$_2$ in trypsin (100 μ/mℓ). Each value is the mean, ± S. E., darkening response of *Anolis carolinensis* skins (N = 6) to the melanotropins after incubation in the enzyme for the times indicated.

IV. RELATIVE STABILITY OF MELANOTROPIC PEPTIDES IN BRAIN HOMOGENATES

Immunocytochemical studies have shown that MSH and/or related peptides are localized in neurons within the mammalian brain, and MSH has been implicated in processes related to memory, learning, and attention in mammals, including humans.[1]

If melanotropins are to be used clinically to treat deficits in certain cognitive abilities, then it is important to determine the stability of MSH and related analogues in brain extracts. Several studies have determined the half-life and cleavage products of MSH and related peptides during incubation in rat and mouse brain homogenates.[7,22] The residual biological activities of MSH and other melanotropins were quantatively determined by Akiyama et al. (Table 2).[22] As for serum enzymes, MSH and [Nle⁴]-substituted analogues were rapidly inactivated by brain homogenates. Cyclic [Cys⁴,Cys¹⁰]-substituted analogues were also inactivated by the brain extracts (Table 2). Incorporation of D-phenylalanine into either linear or cyclic analogues, resulted in total resistance of the peptides to inactivation by brain enzymes (Figure 7, Table 2).

V. DISCUSSION AND SUMMARY

Peptide hormones have relatively short half-lives as would be expected if they are to provide a precise control of target tissues. Unfortunately, therefore, the use of natural peptides is a costly means of replacement therapy for those individuals requiring hormone supplementation. As discussed above, certain melanotropins were synthesized that possess several unique and clinically important attributes. Some of these melanotropins are (1) superpotent (up to 10,000 times more active (minimum effective dose) than MSH (2) prolonged-acting, both in vitro and in vivo, that is, their melanotropic actions are still manifested after they are "removed" (3) superactive and prolonged-acting even for very short fragment analogues and (4) resistant to inactivation by proteolytic enzymes. This last attribute is most important if these analogues are to prove to be useful in clinical medicine.

An early discovery was that heat-alkali treatment of posterior pituitary glands (neurointermediate lobes) resulted in potentiation and prolongation of the melanotropic activities of

Table 2
RELATIVE STABILITIES OF MELANOTROPINS IN RAT BRAIN HOMOGENATES

Melanotropins	Stability[a]			
	30	60	120	240
α-MSH	20	1	0.0	—[b]
Ac-α-MSH$_{4-10}$NH$_2$	10	10	—	—
[Nle^4]-α-MSH	10	1	0.0	—
Ac-[Nle^4]-α-MSH$_{4-11}$NH$_2$	1	0.2	—	—
Ac-[Nle^4]-α-MSH$_{4-10}$NH$_2$	5	3	0.0	—
[Nle^4,D-Phe^7]-α-MSH$_{4-11}$NH$_2$	—	100	20	10
Ac-[Nle^4,D-Phe^7]-α-MSH$_{4-11}$NH$_2$	100	100	100	100
Ac-[Nle^4,D-Phe^7]-α-MSH$_{4-10}$NH$_2$	100	100	100	50
[Cys^4,Cys^{10}]-α-MSH	—	2	0.2	—
Ac-[Cys^4,Cys^{10}]-α-MSH$_{4-10}$NH$_2$	20	10	—	—
Ac-[Cys^4,D-Phe^7,Cys^{10}]-α-MSH$_{4-10}$NH$_2$	100	100	100	100

[a] Percent of remaining biological activity after incubation in rat brain homogenates for the times (min) noted
[b] Not determined.

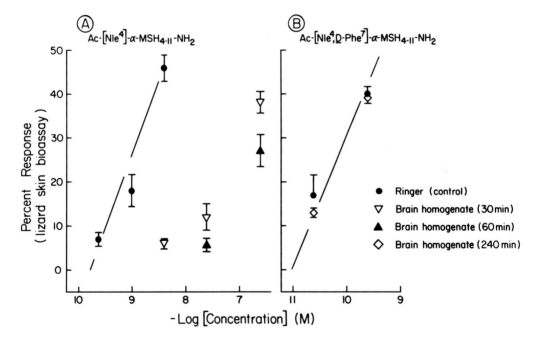

FIGURE 7. In vitro demonstration of the relative resistance of (A) Ac-[Nle^4]-α-MSH$_{4-11}$-NH$_2$ and (B) Ac-[Nle^4, D-Phe^7]-α-MSH$_{4-11}$-NH$_2$ to incubation in 10% rat brain homogenate. Each value is the mean, ± S. E., darkening response of *Anolis carolinensis* skins (N = 12) to the melanotropins after incubation in brain homogenate for the times indicated.

the extracts.[23] Synthetic MSH similarly treated, also exhibited these characteristics of increased potency and residual (prolonged) activity.[24] Based upon the extent of racemization of the individual amino acids comprising MSH, and other relevant facts and conformational considerations, D-phenylalanine was substituted for its L-enantiomer at position-7 of the hormone. In addition, norleucine was also substituted for methionine at position-4 since it had already been determined that this replacement increased the melanotropic activity of the peptide as well as protected the peptide from inactivation by oxidative and reductive influences. The resulting [Nle^4,D-Phe^7]-containing melanotropins were more potent than their L-Phe^7-containing melanotropins and proved to possess the unique attributes enumerated above.[17] Subsequently, it was determined that all [D-Phe^7]-containing melanotropins were more potent than their L-Phe^7-containing homologues.[18] In fact, Ac-[Nle^4,D-Phe^7]-α-MSH$_{4-10}$-NH$_2$ appears to be the most potent of these analogues, in several assay systems.[18] From the point of view of economics of peptide synthesis, this is a consideration of considerable importance, since shorter peptides are easier to produce.

It was hypothesized that the resulting superpotency could be related to the fact that the [D-Phe^7]-containing analogues may possess a *beta* or other reverse turn conformation, the resulting constraint helping the peptides maintain a conformation which may be more compatible with the melanotropin receptor (see Hruby et al. Chapter 5, Volume III). Based upon this premise, MSH and related fragment analogues were conformationally constrained by incorporation of a [$\overline{\text{Cys}^4,\text{Cys}^{10}}$]-disulfide bond. These cyclic melanotropins also exhibited superpotency in several bioassays, but unlike the [D-Phe^7]-containing analogues did not exhibit residual melanotropic activity.[19,20]

The phenomenon of prolongation may relate to the amphiphilicity that the conformationally constrained peptides may possess, that is, the hydrophilic residues may project to one side of the molecule and the lipophilic residues (e.g., histidine, tryptophan, phenylalaine) may project to the other side. This amphipatic property may allow these peptides to more strongly interact with melanotropin receptors or to enter and remain sequestered into the lipid layer of the melanocyte plasma membrane where they are then available to activate the regulatory proteins involved in receptor signal transduction and adenylate cyclase activation (see Sawyer et al., Chapter 5, Volume III).

Since both the [Nle^4,D-Phe^7]-substituted and cyclic analogues were very resistant to inactivation by serum or brain enzymes,[10,22] it would appear that these constrained conformations possessed by these analogues prevent access by proteolytic enzymes to cleavage sites within the peptides which are present in the native hormones. The observation that α-chymotrypsin is much more active than trypsin in inactivating the melanotropins, suggests that the initial cleavage site is between the aromatic amino acid and its adjacent C-terminal-related amino acid. The fact that D-phenylalanine renders the molecule resistant to inactivation, suggests that the initial cleavage site is probably between the Phe^7-Arg^8 sequence. The observation that Ac-[$\overline{\text{Cys}^4,\text{Cys}^{10}}$]α-MSH$_{4-10}$-NH$_2$ is resistant to inactivation by sera, whereas the Ac-[$\overline{\text{Cys}^4,\text{Cys}^{10}}$]α-MSH$_{4-11}$-NH$_2$ analogue is rapidly inactivated (Figure 5), suggests that other factors within the primary structure of a melanotropin will also affect its susceptibility to inactivation by proteolytic enzymes.

When the unique properties of superpotency and prolonged melanotropic activity exhibited by these peptide hormone analogues are discussed, it might be argued that these qualities relate to the fact that these peptides' are resistant, to inactivation by proteolytic enzymes. This is not true, at least relative to the melanotropic properties manifested in vitro. The important characteristics of the peptides may relate to conformational aspects of the molecules that make them more active at melanotropin receptors. For example, all [Cys^4,D-Phe^7,Cys^{10}]-substituted melanotropins are, without exception, resistant to inactivation by proteolytic enzymes. Nevertheless, whereas Ac-[Cys^4,D-Phe^7,Cys^{10}]-α-MSH$_{4-13}$-NH$_2$ exhibits prolonged activity, the 4-11 and 4-10 fragment analogues do not (Table 1).[25] Some very weak melanotropins have also been synthesized that still exhibit ultraprolonged biological activity.[16,18] For example, whereas Ac-α-MSH$_{5-11}$-NH$_2$ is a very weak agonist compared to Ac-[D-Phe^7]-α-MSH$_{5-11}$-NH$_2$, the former is prolonged-acting whereas the D-Phe^7 enantiomer does not exhibit residual activity. Also, the very weak fragment analogue, Ac-α-MSH$_{8-13}$-NH$_2$ (which lacks Phe), is prolonged-acting. These results clearly point out that the phenomena of superpotency and prolonged activity are separate characteristics which do not relate to whether the peptide is or is not resistant to inactivation by proteolytic enzymes. Nevertheless, those melanotropins that are enzyme-resistant will generally have a longer half-life in vivo and this will make them more potent when used in animals and humans.

It has now been demonstrated that [Nle^4,D-Phe^7]-substituted melanotropin analogues can be delivered systemically by topical application (transdermal delivery).[26] These melanotropins were, in fact, quite active when delivered at relatively low doses (see Chaturvedi, Chapter 9, Volume III). When administered by these routes melanogenesis was initiated in hair follicle melanocytes throughout the skin. It is believed that the efficacy of these melanotropins in activating integumental melanocytes is related most likely to the fact that these peptides are essentially totally resistant to inactivation by serum enzymes. These hormone analogues may prove clinically useful in the treatment of vitiligo and other hypopigmentary disorders. Several melanotropin-drug conjugates have been synthesized that are also resistant to inactivation by proteolytic agents, and these peptides may prove useful in site-specific delivery of drugs or diagnostic ligands.[27,28] The recent report that [Nle^4,D-Phe^7]-α-MSH is a very potent antipyretic agent (see Lipton, Chapter 7, Volume II), raises the hope that other medically important uses of these unique melanotropins may become a reality.

ACKNOWLEDGMENTS

These studies were supported in part by grants from the Public Health Service (MH-27257, MH-30626, AM-17420) and from the National Science Foundation (PCM-883300, PCM-811220, PCM-770731, PCM-8412-84, PCM-810078), U.S.A., and from the Conselho Nacional de Desenvolvimento Cientifico e Tecnólogico (410066/86), Brasil.

REFERENCES

1. **O'Donohue, T. K. and Dorsa, D. M.**, The opiomelanotropinergic neuronal and endocrine systems, *Peptides*, 3, 353, 1982.
2. **Glun, J. R. and Lipton, J. M.**, Hypothermic and antipyretic effects of centrally administered ACTH (1-24) and α-melanotropin, *Peptides*, 2, 177, 1981.
3. **Goos, H. J. Th., DeGraan, P., and Van Oordt, P. G. W. J.**, Half-life of melanocyte-stimulating hormone and hormone content of the pars intermedia in relation to background adaptation time in *Xenopus laevis*, *J. Endocrinol.*, 72, 26P, 1977.
4. **Wilson, J. F. and Harry, F. M.**, Release, distribution and half-life of α-melanotropin in the rat, *J. Endocrinol.*, 86, 61, 1980.
5. **Redding, T. W., Kastin, A. J., Nikolics, K., Schally, A. V., and Coy, D. H.**, Disappearance and excretion of labeled α-MSH in man, *Pharmacol. Biochem. Behav.*, 9, 207, 1978.
6. **Ashton, H., Millman, J. E., Telford, R., Thompson, J. W., Davies, T. F., Hall, R., Shuster, S., Thody, A. J., Coy, D. H., and Kastin, A. J.**, Psychopharmacological and endocrinological effects of melanocyte-stimulating hormones in normal man, *Psychopharmacology*, 55, 165, 1977.
7. **Marks, N., Stern, F., and Kastin, A. J.**, Biodegradation of α-MSH and derived peptides by rat brain extracts, and by rat and human serum, *Brain. Res. Bull.*, 1, 591, 1976.
8. **Foster, R. L.**, *The Nature of Enzymology*, Croom Helm Ltd., London, 1980.
9. **Hruby, V. J., Wilkes, B. C., Cody, W. L., Sawyer, T. K., and Hadley, M. E.**, Melanotropins: structural, conformational and biological considerations in the development of superpotent and superprolonged analogs, in *Peptide and Protein Reviews*, Vol. 3, Hearn, M. T. W., Ed., Marcel Dekker, New York, 1984, 1.
10. **Castrucci, A. M. L., Hadley, M. E., Sawyer, T. K., and Hruby, V. J.**, Enzymological studies of melanotropins, *Comp. Biochem. Physiol.*, 78B, 519, 1984.
11. **Castrucci, A. M. L., Hadley, M. E., Yorulmazoglu, E. I., Wilkes, B. C., Sawyer, T. K., and Hruby, V. J.**, Synthesis and studies of superpotent melanotropins resistant to enzyme degradation, in *Biological, Molecular and Clinical Aspects of Pigmentation*, Bagnara, J., Klaus, S. N., Paul, E., and Schartl, M., Eds., University Tokyo Press, Tokyo, 1985, 145.

12. **Castrucci, A. M. L., Hadley, M. E., and Hruby, V. J.**, Melanotropin bioassays: *in vitro* and *in vivo* comparisons, *Gen. Comp. Endocrinol.*, 55, 104, 1984.
13. **Heward, C. B., Kreutzfeld, K. L., Hadley, M. E., Larsen, B., Sawyer, T. K., and Hruby, V. J.**, Preparation of radiolabeled melanotropin suitable for use as a tracer in a radioreceptor assay: [^{125}I-Tyr^2, Nle^4]-α-MSH, in *Phenotypic Expression in Pigment Cells*, M. Seiji, Ed., University of Tokyo Press, Tokyo, 1981, 339.
14. **Heward, C. B., Yang, Y. C. S., Sawyer, T. K., Bregman, M. D., Fuller, B. B., Hruby, V. J., and Hadley, M. E.**, Iodination-associated inactivation of α-melanocyte-stimulating hormone, *Biochem. Biophys. Res. Commun.*, 88, 266, 1979.
15. **Heward, C. B., Yang, Y. C. S., Ormberg, J. F., Hadley, M. E., and Hruby, V. J.**, Effects of chloramine T and iodination on the biological activity of α-melanotropin, *Hoppe-Seyler's Z. Physiol. Chem.*, 360, 1851, 1979.
16. **Wilkes, B. C., Sawyer, T. K., Hruby, V. J., and Hadley, M. E.**, Differentiation of the structural features of melanotropins important for biological potency and prolonged activity *in vitro*, *Intl. J. Pept. Protein Res.*, 22, 313, 1983.
17. **Sawyer, T. K., Sanfilippo, P. J., Hruby, V. J., Engel, M. H., Heward, C. B., Burnett, J. B., and Hadley, M. E.**, [Nle^4, D-phe^7]- α-melanocyte-stimulating hormone: a highly potent α-melanotropin with ultralong biological activity, *Proc. Natl. Acad. Sci. U.S.A.*, 77, 5754, 1980.
18. **Sawyer, T. K., Hruby, V. J., Wilkes, B. C., Draelos, M. T., Hadley, M. E., and Bergsneider, M.**, Comparative biological activities of highly potent active-site analogues of α-melanotropin, *J. Med. Chem.*, 25, 1022, 1982.
19. **Sawyer, T. K., Hruby, V. J., Darman, P. S., and Hadley, M. E.**, [Half-Cys^4, half-Cys^{10}]-α-melanocyte stimulating hormone: a cyclic melanotropin-exhibiting superagonist biological activity, *Proc. Natl. Acad. Sci. U.S.A.*, 79, 1751, 1982.
20. **Knittel, J. J., Sawyer, T. K., Hruby, V. J., and Hadley, M. E.**, Structure-activity studies of highly potent cyclic [Cys^4, Cys^{10}]-melanotropin analogues, *J. Med. Chem.*, 28, 125, 1983.
21. **Cody, W. L., Wilkes, B. C., Muska, B. J., Hruby, V. J., Castrucci, A. M. L., and Hadley, M. E.**, Cyclic melanotropins, Part V: Importance of the C-terminal tripeptide, (Lys-Pro-Val), *J. Med. Chem.*, 27, 1186, 1984.
22. **Akiyama, K., Yamamura, H. I., Wilkes, B. C., Cody, W. L., Hruby, V. J., Castrucci, A. M. L., and Hadley, M. E.**, Relative stability of α-melanotropin and related analogues to rat brain homogenates, *Peptides*, 5, 1191, 1984.
23. **Smith, P. E. and Graeser, J. B.**, A differential response of the melanophore stimulant and oxytocic autocoid of the posterior pituitary, *Anat. Rec.*, 27, 187, 1924.
24. **Bool, A. M., Gray, G. H., II, Hadley, M. E., Heward, C. B., Hruby, V. J., Sawyer, T. K., and Yang, Y. C. S.**, Racemization effects of melanocyte-stimulating hormones and related peptides, *J. Endocrinol.*, 88, 57, 1981.
25. **Cody, W. L., Mahoney, M., Knittel, J. J., Hruby, V. J., Castrucci, A. M. L., and Hadley, M. E.**, Cyclic melanotropins Part IX: 7-D-phenylalanine analogues of the active-site sequence, *J. Med. Chem.*, 28, 583, 1985.
26. **Dawson, B. V., Hadley, M. E., Don, S., and Hruby, V. J.**, *In vitro* animal and human cadaver skin models for percutaneous delivery of melanotropin analogues. *Abstr. XIIIth Intl. Pigment Cell Conf.*, 32, 1986.
27. **Chaturvedi, D. N., Knittel, J. J., Hruby, V. J., Castrucci, A. M. L., and Hadley, M. E.**, Highly potent peptide hormone analogues: synthesis and biological actions of biotin-labeled melanotropins, *J. Med. Chem.*, 27, 1406, 1984.
28. **Chaturvedi, D. N., Hruby, V. J., Castrucci, A. M. L., and Hadley, M. E.**, Synthesis and biological actions of a fluorescein-labeled superpotent melanotropin, *J. Pharm. Sci.*, 74, 237, 1985.

Chapter 11

PERIPHERAL AND CENTRAL PHARMACOKINETICS OF THE MELANOTROPINS

John Fawcett Wilson

TABLE OF CONTENTS

I.	Introduction	184
II.	Quantitation of Melanotropin Concentrations	184
III.	Dosage	184
IV.	Formulations	185
V.	Peripheral Pharmacokinetics	186
	A. Absorption and Bioavailability	186
	B. Distribution	187
	C. Tissue and Plasma Protein Binding	189
	D. Metabolism and Clearance	194
VI.	Specialized Barriers and Secretions	197
VII.	Central Pharmacokinetics	201
VIII.	Pharmacodynamics	203
References		204

I. INTRODUCTION

The first chemical synthesis of a melanotropin, that of α-melanotropin,[1] produced material unstable on storage. The stability problem was overcome by the development of new chemical techniques for peptide synthesis which resulted in approaches for the preparation of both α- and β-melanotropins.[2] A wide range of melanotropin analogues have subsequently been produced by these and other conventional synthesis routes and by solid-phase methods.[3] The commercial availability of these highly purified, synthetic peptides has enabled studies involving the administration of exogenous melanotropins to be undertaken in many fields of biology and medicine, and several extrapigmentary effects have been demonstrated.[4] As documented by subsequent volumes of this series, important roles for the melanotropins have been proposed on the basis of this and other evidence in several aspects of the adaptive physiology of vertebrates. Evaluation of these proposals requires knowledge of natural or experimentally induced changes in hormone secretion and those produced by peptide administration. The present chapter will therefore review the current knowledge concerning the pharmacokinetics of melanotropins and will consider the relationship between measured tissue concentrations of peptide and biological response.

II. QUANTITATION OF MELANOTROPIN CONCENTRATIONS

Pharmacokinetic analyses of drug concentration data are ultimately dependent upon accurate measures of the analyte concentration in tissue samples. In the case of melanotropic peptides which are subject to rapid metabolism in vivo, separate measurement of the parent compound and its metabolites requires a method that includes a powerful chromatographic separation procedure. Such techniques have been applied in investigations into the metabolic breakdown of melanotropins as described below but, because of their technical difficulty, they have not been widely adopted for studies in other areas. The three methods of quantitation in more general use are radiological, immunological, and biological assays. In the former, the radioactive counts resulting from the administration of labeled peptide can be followed in trichloroacetic acid extracts without regard for the nature of the molecule to which the label remains attached. This is recognized as the least satisfactory of the three techniques in view of peptide metabolism, and is commonly improved by partial fractionation of labeled products by immunoprecipitation,[5,6] by a simple chromatographic step, or by adsorption to a silicate.[8] Estimates of half-lives made simultaneously by the three methods for corticotropin 1-39,[5,9-11] corticotropin 1-24,[12] and α-melanotropin,[13] show that radiological estimates exceed those by bio- or immunoassay, while bioassay consistently returns lower half-lives compared to immunoassays. Differences of the latter type have been widely reported in the immunoassay literature and can be traced to variations in the spectra of crossreactivity shown by antipeptide sera and biological receptors of target organs.[14] Although a good correlation can be demonstrated between biological and immunological activities for melanotropin assays,[15] they clearly differ, and it is therefore important to distinguish between data based on the different techniques. It is incorrect, however, to seek to establish a firm preference for one technique when one has to compare, for example, the use of a melanocyte bioassay vs. an immunoassay in studies of the behavioral activity of natural or even novel synthetic peptide sequences.

III. DOSAGE

While a comprehensive review of the in vivo dosages required to produce the wide range of responses described for melanotropins is beyond the scope of the present chapter, some selected examples concentrating on α-melanotropin are appropriate to indicate the range involved. Readers are referred to later volumes for detailed information regarding specific responses.

Considering first an example of the classical pigmentary response by melanocytes in lower vertebrates; melanin granule dispersion can be produced in *Xenopus laevis* by dorsal lymph sac injection of 0.1 to 1 nmol/kg of α-melanotropin.[16] Dose-related increases in sebum secretion were caused by daily subcutaneous administration of 18 to 600 nmol/kg of α-melanotropin in the rat.[17] Dose-related natriuresis in water-loaded rats was induced by intraperitoneal injection of 2 to 100 nmol/kg α-melanotropin, whereas responses to β-melanotropin required higher doses, i.e., above 77 nmol/kg.[18] The methodological details used in a large number of studies of behavioral actions of melanotropins have been reviewed by Beckwith and Sandman.[19] The most commonly used dose of α-melanotropin in rats was some 20 to 40 nmol/kg, depending on the precise animal body weights. The dose of α-melanotropin in man was 86 nmol/kg if a body weight of 70 kg is assumed.[20,21] Values for other peptides differ somewhat, being higher for shorter-chain analogues of the natural peptides, but considerably lower for D-amino-acid-substituted analogues of high potency. For example, 4 - 10 corticotropin showed activity on the conditioned avoidance- and reward-based learning behavior of rats at 97 to 242 nmol/kg,[22,23] and it enhanced attention in man at 223 nmol/kg.[24] In contrast, 4-Met(O_2),8-D-Lys,9-Phe corticotropin 4-9 was effective at a 1000-fold lower dose of 9 to 27 pmol/kg in the rat.[25] These latter behaviorally effective doses refer to peripheral routes of administration. Intraventricular administration of peptides will mimic the effects at much lower doses.[26] Certain additional responses such as a stretching and yawning syndrome[27] and dose-related hypothermia[28] occur only via this route of administration and they require dose levels of 6 nmol/kg and 0.25 to 1.0 nmol/kg, respectively.

The doses of α-melanotropin effective in diverse areas of physiology thus span some 3 log units in the lower nmol/kg range. The apparently higher sensitivity of the lower vertebrate melanocyte response compared to mammalian responses and the differences in relative potencies of natural and synthetic analogues on behavior, will be shown in later sections, to be the result, at least in part, of differences in the susceptibility to enzymatic breakdown of the peptides in the different systems, pharmacokinetic differences being a key factor in influencing effective dose ranges.

IV. FORMULATIONS

The two commonly used vehicles for the administration of melanotropins are nonpyrogenic 0.9% NaCl solution or saline acidified by the addition of 0.01 M acetic acid.[19] The latter acidification follows the early acetic-acid-based extraction methods for the isolation of melanotropins and the adoption of dilute acetic or hydrochloric acid for short-term storage of melanotropin solutions.

The rapid clearance of administered peptides has prompted the use of several pharmaceutical formulations of melanotropins that were developed for the prolongation of the action of corticotropin.[29] The zinc phosphate complex[30] used commercially in Synacthen Depot® (Ciba) has been the usual choice in studies of melanotropins,[19] but requires to be freshly prepared for each injection. A method of stabilizing the zinc complex with histidylhistidine has been demonstrated for 1-Gly corticotropin 1-18 amide which may be applicable to other short-chain peptides.[31] In the behavioral field, however, responses to the zinc phosphate vehicle itself precludes its use.[32]

One final alternative means of formulating melanotropins for administration has been described in a paper on the preparation of suppositories of 1-Gly corticotropin 1-18 amide for rectal administration.[33] Comparisons of hydrophobic and hydrophilic bases with and without surfactants were made and a prolongation of the corticosteroidogenic response demonstrated.

V. PERIPHERAL PHARMACOKINETICS

The material reviewed in this section originates from studies in which pharmacokinetic data have sometimes been produced merely as a side-effect of the investigation. The methods and extent of the data analysis are therefore so variable that further processing of published values has been necessary. At the simplest level, units of measurement have been standardized and values such as mean body weights incorporated when appropriate. A body weight of 70 kg has been assumed for human data where necessary. Much more fundamental data analysis has, however, been undertaken. In several cases, the data available were in the form of published peptide concentration time graphs. This has been digitized and analysed in two ways. First, data were fitted to a one- or two-compartment open model[34] by nonlinear least-squares regression using the reciprocal of the peptide concentration squared as weights.[35] Rate and concentration constants were determined for each compartment and the rate constants are presented here as half-lives. Secondly, the area under the concentration-time curve (AUC) was calculated by the trapezoid rule with extrapolation to infinite time.[34] Use was then made of the following relationships in the estimation of clearance and volumes of distribution from intravenous data:

$$\text{Clearance}_{(area)} = \frac{\text{Peptide dose administered}}{\text{AUC}_{(iv)}}$$

$$\text{Volume of distribution}_{(\beta)} = \frac{\text{Clearance}}{\beta}$$

where β is the terminal-rate constant from the compartmental model fit. When intravenous infusion experiments provided only measures of steady-state peptide concentrations (CSS), clearance was calculated as:

$$\text{Clearance} = \frac{\text{Infusion rate}}{\text{CSS}}$$

Finally, bioavailabilities for various routes of administration were calculated as the ratio of the AUC for the route under consideration to the AUC estimate for intravenous data.

A. Absorption and Bioavailability

It is usual to assume that orally administered peptides are degraded to amino acids by gastrointestinal enzymes. Large doses of corticotropin will, however, cause steroidogenesis in adult[36] or suckling[37] rats. The dose required to produce a response equivalent to that produced intravenously was 3000 times the intravenous dose.[36] It was further demonstrated that breakdown of corticotropin 1-24 did not occur to any great extent in the stomach, but was due to intestinal attack by rapid endopeptidase cleavage of the 7-Phe,8-Arg bond combined with partial cleavage at several other sites. The potential for the development of useful synthetic molecules that are orally active by virtue of resistance to such enzyme attack, has been demonstrated by the 10.5% oral bioavailability achieved by 4-Met(O_2),8-D-Lys,9-Phe corticotropin 4-9.[38]

A less disadvantageous site for absorption is the rectum where peptide digestion is much less than in the intestine.[33] An equipotent dose of 1-Gly corticotropin 1-18 amide, a molecule without D-amino acid modifications to resist enzyme attack, was some 100 times the intravenous dose, and significant improvements in bioavailability could be achieved by the use of suppositories containing an anionic surfactant.

A third alternative to injected parenteral administration for which data is available is the intranasal route in man. Considerable intersubject variability in absorption was observed, with maximum bioavailabilities estimated from the AUCs of published graphical immunoassay data being 6 to 9% for both corticotropin 1-24 and the enzyme-resistant 1-D-Ser,17–Lys,18– Lys corticotropin 1-18 amide.[12] Comparable, variable, low absorption with a maximum bioavailability of 7.6% has also been reported for immunoreactive corticotropin 4-10.[39] However, in studies of 1-Ala,17-Lys corticotropin 1-17 4-amino-n-butyl-amide, reproducible low serum corticotropin levels were obtained.[40]

While the above routes may have advantages of convenience in certain clinical applications, their low and sometimes unpredictable bioavailabilities would seem to make them a poor substitute for the usual intravenous, subcutaneous, intraperitoneal, and intramuscular parenteral administration routes. Studies of the parenteral routes have, however, shown that systemic absorption from these sites, though rapid, is not necessarily high (Table 1). Peak concentrations in both rat and man occurred within some 2 to 20 min of the injection of aqueous solutions. The use of zinc phosphate depot formulations delayed absorption only marginally when given subcutaneously,[41] but gave a sustained release over several hours from intramuscular sites.[12,31,43] The fraction of the administered dose reaching the systemic circulation was surprisingly low by the parenteral routes (Table 1). The considerable loss of peptide was attributed to metabolic breakdown at the injection site or in passage to the systemic circulation, the clearance rates being higher in rat than man. As described below, skin and muscle are major sites for melanotropin inactivation, even more so than the liver. The better absorption from the intraperitoneal site in the rat is consistent with this idea, passage through the liver being unusually less damaging than through peripheral tissues. The bioavailabilities calculated for the enzyme-resistant D-substituted analogues were higher, as would indeed be predicted if metabolism was an important factor in determining absorption from injection sites. Thus, levels of the nonenzymatically resistant forms of melanotropins reaching the systemic circulation are greatly attenuated when given by the common parenteral routes, and only direct intravenous injection or infusion will circumvent this physiological barrier.

B. Distribution

Following intravenous injection of melanotropins, observed blood concentrations of peptide decline rapidly in a manner that can be described mathematically by a sum of exponential terms. In studies where blood sampling and assay sensitivity have allowed observations to be extended past some 10 min after administration, the observed pattern of the fall in blood levels was often biphasic (Figure 1 and Table 2). Exceptions were seen in human studies where three phases were distinguished using iodine-labeled α-melanotropin;[48] in a study of N-terminal 1-76 pro-opiomelanocortin[63] and in a study with corticotropin 1-24 and an enzymatically resistant corticotropin analogue[12] where the early phase was not covered by the samples, but two later phases were seen. Evidence of such multiple later-clearance phases has also been produced with intraperitoneal and subcutaneous doses of α-melanotropin and longer sampling times in the rat when an immunological half-life of 19 to 26 min was measured[41] which exceeds the 7.1 min reported using the same assay for intravenous administration.[45]

The first most-rapid phase (or phases) with a half-life of approximately 2 min (Table 2) is the result of distribution of peptide throughout the body. Melanotropins, with molecular weights of a few thousand, will readily pass through capillary walls and distribute into the extracellular fluid. Further passive intracellular penetration would not be expected from their physiochemical properties. The stability of the distribution half-lives measured across a range of species and peptides serves to emphasize the dependence of the distribution mechanisms upon physical processes such as diffusion. The suggestion by McMartin and Peters[52] that the rapid phase observed at the end of infusions into rats was due to clearance rather than distributive processes, was incorrect. Although steady-state had been reached and clearance had indeed balanced the infusion rate, in a two-compartment model where a substantial proportion of clearance is from the peripheral compartment, distribution between compartments does not reach equilibrium during the infusion. The half-life values calculated for the

Table 1
ABSORPTION RATE AND BIOAVAILABILITY OF PRO-OPIOMELANOCORTIN PEPTIDES FROM PERIPHERAL INJECTION SITES

Peptide	Site	Species	Assay method	Half-life for absorption (min)	Bioavailability (%)	Ref.
α-Melanotropin	Intraperitoneal	Rat	Immunological	7.3	7.0	41
α-Melanotropin	Subcutaneous			5.6	2.8	41
α-Melanotropin ZnPO$_4$				17.7	2.9	41
4-Met(O$_2$),8-D-Lys,9-Phe corticotropin 4-9			Chromatographic	2.3	>89	38
1-D-Ser,17-Lys,18-Lys corticotropin 1-18 amide		Man	Immunological	7.3	31	12
1-D-Ser,17-Lys,18-Lys corticotropin 1-18 amide			Biological	—	46	12
Corticotropin 1-24 ZnPO$_4$	Intramuscular	Rat	Immunological	—	31	12
β-Endorphin 6-17	Subcutaneous		Chromatographic	1—2	31	42
β-Endorphin 6-17	Intramuscular			0.5—1	8.5	42

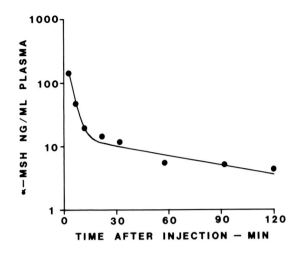

FIGURE 1. Variation in concentrations of α-melanotropin (α-MSH) in plasma following intravenous injection of 1.2 nmol synthetic α-MSH into an eel, *Anguilla anguilla*, of 450 g body weight. (Data from Gilham and Baker.[47])

rapid phase after the end of the infusion are, therefore, correctly included in Table 2 under the heading, distribution half-life.

The outcome of the distribution process can be assessed by the parameter Vd_β which is a measure of the apparent volume of distribution that relates the peptide concentration in plasma to the total amount of peptide in the body at all times during the β- or elimination phase of clearance. Values of Vd_β, which is sometimes called $Vd_{(area)}$ from its method of calculation, are given in Table 2, and they show considerable variation. Some of this variability can be ascribed to differences in sampling periods between studies as if the slow later-clearance phases, noted above, are used to provide the estimate of the terminal-rate constant (β) in the calculations, larger volumes will be produced. For example, if the longer elimination half-life of 23 min[41] for α-melanotropin in the rat is used to calculate Vd_β, an increase from 93 to 301 mℓ/kg results. Notwithstanding these problems, interspecies and interpeptide differences in distribution volumes are apparent from comparisons between studies using similar methodologies (Table 2) and from data from single studies with a range of peptides.[12,21,44,52] The calculated volumes vary between values below that of extracellular fluid to some well in excess of total body water. The latter are the more usual and are indicative of the existence of important processes in extravascular tissues for the reversible sequestration of melanotropins.

C. Tissue and Plasma Protein Binding

The most general studies providing information regarding both the reversible and non-reversible accumulation of melanotropins by specific tissues are those which have followed the fate of radiolabeled material. An autoradiographic study 5 min after injection of iodine-labeled α-melanotropin into rats and mice demonstrated specific uptake by the pineal and pituitary glands and by the kidney.[65] However, at the same time, large accumulations of radioactivity had also occurred in the thyroid gland and stomach. Time-course studies with iodine-labeled corticotropin 1-24[58,66] also showed early specific accumulation of radioactivity by the kidney and to a lesser extent by the liver and adrenals, but demonstrated that the thyroid and stomach uptake was a later-developing phenomenon probably due to uptake of simple iodine following peptide breakdown.

High rapid uptake of radioactivity by the kidney with lower, but significant concentrations accumulating in the liver has also been demonstrated using tritium-labeled corticotropin 4-

Table 2
PHARMACOKINETIC PARAMETERS FOR PRO-OPIOMELANOCORTIN PEPTIDES FOLLOWING INTRAVENOUS ADMINISTRATION

Peptide	Injection type	Species	Assay method	Distribution half-life (min)	Elimination half-life (min)	Volume of distribution $Vd_{(\beta)}$ (mℓ/kg)	Mean clearance rate (mℓ/min/kg)	Ref.
α-melanotropin	Bolus	Rat	Biological	1.6	20.8	526.4	18.5	13
	Infusion	Man	Immunological		11.6	1214.3	68.3	21
	Bolus	Rabbit		2.9	7.1	93.0	9.1	44
		Rat		1.1	13.3	428.9	22.3	45
				2.6				46
				2.4				13
		Anguilla		2.2	61.1	158.4	1.8	47
		Man	Radiological-^{125}I	1, 4.8	180.0			48
			Radiological-^{3}H	1.0	25.0			48
		Rat	Radiological-^{131}I	7.4				13
			Radiological-^{3}H	1.1—4.8				49
		Xenopus	Radiological-^{125}I		120.0			50
Desacetyl-α-melanotropin		Rabbit	Immunological	1.9	7.3	3922.0	340.6	44
Diacetyl-α-melanotropin				2.1	16.0	1076.7	49.5	44
4-Met(O$_2$),8-D-Lys,9-Phe corticotropin 4-9		Rat	Chromatographic	1.6	18.2	518.4	19.7	38
Corticotropin 4-10		Man	Immunological	0.4	3.5	2598.8	519.8	39
	Infusion						689.9	39
1-D-Ser,17-Lys,18-Lys corticotropin 1-18 amide	Bolus		Biological		11.4, 165.8	1251.9	5.2	12
			Immunological		18.9, 99.0	222.7	1.6	12
1-Aib,17,18,19-tri-Lys Corticotropin 1-19 amide	Infusion	Rat	Biological	1.0	16.5	1032.7	43.5	52
	Bolus			4.8	42.5	877.8	14.3	51
Corticotropin 1-24	Infusion				14.1	20860.0	1026.3	52
	Bolus	Man	Immunological	1.6	4.7, 64.7	2344.0	25.1	12
					10.0	387.8	26.9	53

Peptide	Mode	Method	Species						Ref.
	Infusion		Fetal sheep	3.3			121.0		54
	Bolus		Dog				227.0		55
			Man				6.6		56
			Rat				39.0		57
		Chromatographic Radiological-^{125}I		1.2	7.0	46820.0	322.0		7
				1.0	6.6				58
									8
β-melanotropin	Infusion	Immunological	Man	2.7	17.2	83.8	27.7		8
	Bolus		Rabbit	2.8	11.6	474.3	3.4		21
β-lipotropin	Bolus		Man	5.7	54.8	399.4	28.5		59
				6.0	45.1	552.4	4.5		60
			Rat	4.2			8.1		61
N-terminal 1-76 pro-opiomelanocortin				1.0	6.7, 74.0	917.0	6.6		62
							8.6		63
Pig pituitary extract		Biological	Dog	10.9	159.6	1674.0	7.3		64
Dog pituitary extract				12.8	129.5	716.4	3.8		64

9,[38] 1-18,[67] 1-24,[67] and 1-39[11] analogues. The processes involved appeared nonsaturable since the proportion of the dose accumulated was independent of the dose administered.[11] Comparable concentrations to those found in the liver were seen in the spleen and lungs, but when expressed as a percentage of the dose administered, these latter two tissues, together with the heart, testes, and brain, all accumulated less than 1% of the dose.[11,67] Quantitatively large accumulations of radioactivity were demonstrated in muscle, skin, fat, and intestines.[11,67] Although absolute concentrations were 5 to 17% of the highest found in the kidney, the relatively large total mass of these tissues in the body meant that, by 1 min after intravenous injection, they had taken up approximately 50% of the injected dose. Muscle accounted for some 19% of the radioactivity, skin 17%, and the intestines 13%, whereas the kidney and liver held 9 and 8%, respectively, and 34% was still in the plasma.[67] Since chromatography of plasma showed circulating peptides to be largely intact at this time,[68,69] the initial uptake by organs was of intact material.

What the above percentage radioactivity values do not convey is the quite different degree of degradation and the anatomical location of the extravascular peptide in the tissues. Detailed chromatographic,[67-69] bioassay,[51] and autoradiographic[68,70] studies of the fate of exogenous corticotropic peptides have differentiated at least three patterns of peptide disposition. First, the radioactivity in muscle and skin was already extensively degraded by 1 min after administration, but a small proportion of intact peptide persisted in these tissues at 1 hr, seemingly bound in a way that protected it from enzyme attack. The similarity in the spectrum of metabolites found in muscle and skin to those appearing in plasma, suggested that these tissues were in reversible equilibrium with plasma, and it has been proposed that they are responsible for supplying the peptide that maintains the later slow phases of decline in plasma peptide concentrations.[67,71] In contrast to the rapid degradation seen in skin and muscle; kidney, liver and, to a lesser degree, the intestines contained a large proportion of intact peptide 1 min after injection. These tissues differed in that the kidney continued to accumulate radioactivity between 1 and 10 min after injection, while the liver and intestines showed a decline with time more akin to the pattern of total radioactivity in skin and muscle.[67,68] This latter similarity in time-course may suggest that peptides in skin, muscle, liver, and intestines are all in reversible equilibrium with plasma, and that binding to these tissues is the major determinant of the large apparent volumes of distribution of melanotropins. The kidney plays no part in the redistribution processes. As will be described in the following section, uptake is by pinocytosis into lysosomes, so in the long-term it is an irreversible clearance process in which no intact material is returned to the circulation.

A tissue not covered by the above discussion but one that has been the subject of individual attention, is blood itself. Equilibrium dialysis of corticotropin 1-23 using bioassay has shown saturable binding to human albumin, human γ-globulin, and lyophilized whole human plasma with binding falling from 80 to 90% down to 45 to 60% over a range of peptide concentrations in the low μM range.[72] A second study of ^{14}C-labeled α-melanotropin binding to defibrinated, dialyzed, and lyophilized bovine serum proteins also by equilibrium dialysis, demonstrated two classes of binding sites.[73] A saturable high-affinity low-capacity site was seen with concentrations between 0.18 and 0.3 nmol/100 mg protein and a low-affinity high-capacity site from 0.6 nmol/100 mg protein, upwards. The latter site appeared to be nonsaturable, binding a constant 10% of the α-melanotropin in the system. Similar binding was observed with analogues modified at the 4 and 10 positions, suggesting that the groups responsible for binding were located elsewhere.

Our own studies[74] of plasma protein binding to fresh heparinized plasma by ^{125}I-labeled α-melanotropin[75] have been made by equilibrium dialysis against 0.15 M pH 7.4 phosphate/NaCl buffer at 37°C in a Dianorm® dialyzer (MSE Scientific Instruments) using spectrapor-2 membrane (molecular weight cut-off 12,000 to 14,000). Enzyme activity was inhibited by the addition of 1,000 Kiu/mℓ Trasylol® (Bayer). Dialysis proceeded for 5 hr

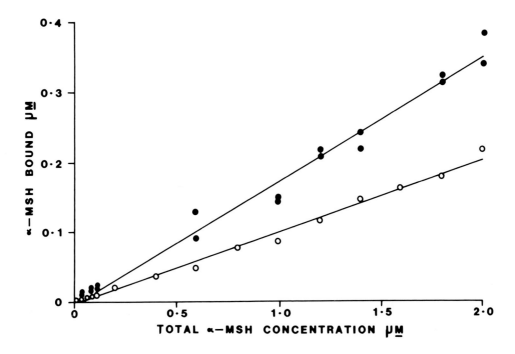

FIGURE 2. Variations in binding of α-melanotropin (α-MSH) at different concentrations to heparinized rat (○) and human (●) plasma. (Data from Smith and Wilson.[74])

with mixing at 12 rpm, a time that preliminary buffer-to-buffer dialysis had shown satisfactorily to reach equilibrium. Radioactive counts of unwashed membranes at the end of the experiments were on average 8% of the total counts added, indicating low nonspecific adsorption. Examples of binding data are displayed in Figure 2. A nonsaturable linear binding was observed between 10 pM and 2 μM concentrations. The mean percentage binding was 10.6 (n = 2) and 17.9 (n = 2) in rat and human plasma, respectively, with the total plasma protein concentrations being 65.4 and 79 g/ℓ, respectively. The observations were thus similar to those for the low-affinity site described by Medzihradszky,[73] but although our experiments spanned the concentration range covered by Medzihradszky, no high-affinity saturable site was evident. This inconsistency may indicate that the attachment of an iodine atom at the 2-Tyr position disrupts the high-affinity interaction. A study of the attachment site on bovine serum albumin by photoaffinity labeling with a corticotropin analogue located the albumin site on the amino-terminal 1-183 segment.[76] Comparisons between additional analogues suggested that the 4-Met residue of corticotropin was not important to binding, but that the 9-Trp-containing segment was involved. This is not at odds with the previously suggested involvement of 2-Tyr as binding most likely involves hydrophobic interactions with the cyclic residues in the peptide.

The influence of these reported plasma protein binding reactions on the pharmacokinetics of peptides in vivo is equivocal. To be important, binding has to be of the order of 90% or more. Such levels may occur when circulating concentrations are in the nM range, but there is no evidence of any resulting restriction of melanotropins to the vascular compartment in the observed volumes of distribution.

It is interesting to speculate finally on the relationship between the peptide binding to tissues described above and binding of melanotropins at biologically active receptors. Studies with blood-borne ^{125}I-corticotropin 1-24 have identified specific binding sites on axon terminals in the rat median eminence that were blocked competitively with both α- and β-melanotropin.[77,78] Similar studies with ^{125}I-β-melanotropin demonstrated specific binding to cultured melanocytes from a mouse melanoma cell line in the postsynthetic G2 phase of the cell cycle.[79] Binding was to cell-surface receptors and a cyclic AMP-mediated stimulation of tyrosinase activity was seen. The association constant for binding was estimated as 3×10^8 mℓ/mol and there were in the order of 10^4 receptors per cell. If one takes a value of cell volume to be 250 fl or 4×10^9 cell/mℓ, then this number of receptors implies a maximum binding of melanotropin to melanocytes of 0.066 nmol/g. This value is of the same order as that observed 1 min after injection of 30 nmol corticotropin 1-24 when, for example, skin contained 0.095 nmol/g.[67] At physiological concentrations, therefore, binding to high-affinity-specific receptors could play an important, but as yet undocumented role in the pharmacokinetics of melanotropins, while following administration of supraphysiological doses, nonsaturable, nonspecific binding to the large tissue masses in the body will dominate.

D. Metabolism and Clearance

As far as the melanotropins are concerned, clearance and metabolism are almost synonymous. In human studies, less than 2% of the radioactivity from an intravenous dose of tritiated corticotropin 1-39 was excreted in urine,[11] while between 2 and 8% of the dose of a pituitary preparation of melanotropins appeared in urine in a biologically active form.[80] Urinary levels of endogenous immunoreactive β-melanotropin were similarly low compared to circulating concentrations in man.[81] Results with the 76 amino acid N-terminal fragment of pro-opiomelanocortin containing the γ-melanotropin sequence were less clear cut.[82] Although extensive degradation had occurred, immunoreactive concentrations of intact and high molecular weight fragments in urine were 15 to 40% of the plasma concentrations in patients with elevated hormone levels. The second or multiple later phases of the decline in blood concentrations of melanotropins following intravenous injection (Figure 1 and Table 2), are thus mainly the result of metabolic clearance.

The log-linear nature of the decline seen during the elimination phase(s) suggests that the metabolic processes are not saturated. This has usually been confirmed in infusion studies where increased infusion rates lead to linear increases in circulating peptide levels as with corticotropin 1-18, 1-24, and 1-39 analogues in rats,[52] with corticotropin 1-24[57] and 1-39[83] in dogs, and with corticotropin 1-39 in man.[84] In a study with corticotropin 1-39 in rats, however, a three-fold increase in metabolic clearance has been reported at infusion rates below approximately 10 pmol/min/kg.[85] Saturation of one component of the metabolic clearance processes is thus indicated for corticotropin, but its relevance in melanotropin clearance is unknown.

Two parameters relating to metabolic clearance of a range of melanotropins and related peptides are displayed in Table 2. The elimination half-life is based on the hybrid rate constant β, and so will reflect both peptide elimination and distribution between compartments, specifically the fraction of the dose in the metabolizing compartment. It follows that conclusions about differences in metabolism cannot be based solely upon changes in half-life. The second parameter is clearance, which is the fraction of the apparent volume of distribution (Vd_β) cleared by the body in unit time. Inspection of the data in Table 2 leads to several conclusions. As noted in Section II, clearance of biological and chromatographically measured melanotropins is greater than for immunoreactive peptides, which in turn is greater than for radioactive material, reflecting the differences in specificity between methods. More interestingly, modification of molecules at their N-terminus by mono- (α-melanotropin) and di-acetylation or by D-amino acid substitution, causes a marked reduction in clearance. These structural changes can be seen to be associated with both lengthened elimination half-lives and decreases in the apparent volumes of distribution. This latter phenomenon is particularly well displayed by the comparison between desacetyl-α-melanotropin and α-melanotropin in the rabbit,[44] and in data from the infusion studies by McMartin and Peters in rats[52] where terminal-elimination curves of different peptides were nearly parallel, but large differences were seen in distribution volumes. Changes in clearance can be attributed to some degree to a reduction in metabolic rate produced by the protective action of the N-terminal changes against amino-peptidase attack which, as described below, is involved in melanotropin catabolism. The altered volume of distribution will, however, also play a vital role. Since more peptide remains in the plasma compartment, less will be exposed to the catabolic activities of the peripheral tissues which are the principle site of melanotropin clearance. The mechanism behind the alteration in volume of distribution is unknown. Rudman and colleagues[44] suggested that an increase in plasma protein binding might be responsible, but equally, some reduction in peripheral accumulation processes could account for the change. Whatever the cause, it is clear that the pattern of tissue binding has a substantial influence on the clearance of melanotropins, and hence is largely responsible for determining the levels circulating in plasma, and that metabolism is more an accessory, partly governing the rate of terminal decline.

Knowledge concerning the anatomical location of the tissues responsible for the metabolic clearance of melanotropins has been derived from several types of study. First, evidence regarding the in vitro stability of melanotropins in whole blood and plasma has demonstrated enzymatic breakdown in this tissue.[9,86,87] Factors which activated the clotting mechanism increased breakdown so that only data obtained with fresh, heparinized material is meaningful for in vivo comparisons. Involvement of cellular components was evident from the more rapid breakdown seen in whole blood compared with plasma.[9,88] Comparisons between a range of analogues have shown faster breakdown with shorter-chain peptides. For example, the immunological half-life of β-melanotropin in human plasma was approximately 24 hr, while that of β-lipotropin was much longer.[88] Very approximate biological half-lives for corticotropin 1-26, 1-19, and 1-18 were 120, 50, and 30 min, respectively.[86] Further comparisons of structurally modified analogues have thrown light on the nature of the enzymes responsible for the attack. The addition of a C-terminal amide to corticotropin 1-18 produced a more resistant structure.[86] N-terminal acetylation and diacetylation of desacetyl-α-melanotropin increased immunological half-lives in rabbit plasma from 28 to 92 and 146 minutes,[44] while an N-terminal 4-methionine sulphoxide substitution increased the chemical half-life of D-substituted corticotropin 4-9 from 25 to 74 min in rat plasma.[25] Thus, both carboxy- and amino-peptidase activity appeared to play a role in the inactivation by blood. Studies in which the identity of metabolites has been investigated, demonstrated, however, that the primary site of enzyme action was located within the peptide chain. When amino acid residues released from α-melanotropin, corticotropin 4-10, and 4-Met(O_2),8-D-Lys,9-Phe corticotropin 4-9 were determined, it was shown that rat and human serum released the internal amino acids preferentially, though rat plasma was also active in cleaving the N-terminal methionine from the unprotected corticotropin 4-10 structure.[89] The susceptible site for endopeptidase attack was thought to be the 7-Phe,8-Arg bond in α-melanotropin and corticotropin 4-10 but the 6-His,7-Phe bond in 4-Met(O_2),8-D-Lys,9-Phe corticotropin 4-9. The latter site has independent support in work demonstrating the 7-9 tripeptide as the main metabolite.[25] This pattern of primary endopeptidase cleavage combined with less-active exopeptidase attack is identical to that described in the intestine.[36]

Although the inactivation of melanotropins by blood is well documented, it plays only a small role in clearance in vivo. Comparison of the half-life in plasma to in vivo values amply demonstrates the importance of peripheral tissues in clearance. For example, when measured by the same immunoassay, the in vitro half-life of α-melanotropin in rat plasma was 39 min[75] as compared to 7.1 in vivo[45] or 92 vs. 11.6 min in the rabbit.[44] Studies of metabolism by peripheral tissues have used a range of methodological approaches. There are many reports of peptide digestion in tissue homogenates,[90,91] but these are of little help in the assessment of the relative importance of tissues in vivo because they involve enzymes to which extracellular peptides are never necessarily exposed. Studies with tissue slices and isolated cells suffer similar problems, but to a lesser degree. For example, inactivation of corticotropin 1-24 by adrenal-cell suspensions was the result of enzymes present in extracellular fluid, but they were thought unlikely to play a role in catabolism under normal physiological conditions.[92] Corticotropin 1-39 has been shown to be inactivated by adipose tissue slices from rat, but not from rabbit or guinea pig.[93] Again the physiological significance of such data is unknown.

Two types of in vivo experiment that have produced more informative data relevant to the present topic are those which have measured the arteriovenous difference in peptide concentration across various organs, and those which have eliminated certain vascular beds from the circulation by occlusion and monitored changes in clearance parameters. The first technique demonstrated no involvement for the lungs in clearance in rats as they neither bound nor degraded corticotropin 1-24 in a single passage.[71] The liver was shown to be involved in the clearance of pituitary extracts of melanotropins in dogs by the occurrence of a portal vein-hepatic vein gradient and by the retardation in the decline in blood concentrations following intravenous administration in hepatectomized animals.[64] A role for the liver has not always been observed. Ligature of the blood supply to the liver and viscera in rats caused only a 2.8% increase in the elimination half-life of α-melanotropin.[94] A significant renal involvement in clearance has been a more consistent finding. Following occlusion of the renal arteries in rats, steady-state plasma concentrations of infused corticotropin 1-39 and of a 1-18 analogue were increased, but that of corticotropin 1-24 fell marginally.[52] The increases were primarily mediated through a change in peptide distribution rather than altered elimination. Renal ligature increased the elimination half-life of α-melanotropin in rats by 10.7%, though the change was not significant.[94] Human data from patients with renal failure supports a role for the kidney with certain peptides. Clearance of β-lipotropin was decreased in uremia, but clearance of corticotropin 1-39 was within the normal range.[95] Conflicting data have been reported for skin, muscle, and fat. Ligation of the blood supply to the hind quarters in the rat had no significant effect on the shape of infusion curves for corticotropin 1-18, 1-24, and 1-39 analogues, though the steady-state plasma concentrations all increased by at least 50%.[52] In contrast, the half-life for elimination of α-melanotropin was increased by 63.4% following similar surgical procedures.[94] An explanation for the apparent inconsistencies in these data is not immediately available. It should be noted, however, that the observed changes can be due to influences on both binding and metabolism of peptide by the tissues, and tissue binding was definitely implicated in one of the above studies. A clear picture of the metabolic activity of a tissue may, therefore, be being obscured by interpeptide differences in tissue binding, the importance of which has already been described.

Unquestionably, the best data on the metabolism of melanotropins in peripheral tissues have come from the chromatographic and autoradiographic investigations into the fate of corticotropin analogues in tissue extracts or sections following in vivo peptide administration.[67-70] These studies have been introduced at length in the previous section in relation to the tissue binding of melanotropins. They have demonstrated that skin and muscle have the greatest capacity to catabolize corticotropic peptides. One minute after an intravenous injection, these tissues, which between them had accumulated approximately 50% of the dose administered, had metabolized more than 99% of their content of corticotropin 1-24 and 82% of 1-D-Ser,17-Lys,18-Lys corticotropin 1-18 amide with its blocked termini. Detailed chromatographic analysis of the metabolic products showed rapid release of amino acids from the N-terminus of corticotropin 1-24, indicating an important role for aminopeptidases in skin and muscle in the breakdown of peptides with unprotected N-termini. This contrasts with the breakdown in plasma where aminopeptidases were less important. Not surprisingly, release of N-terminal amino acids from the blocked 1-18 analogue by all tissues was slow. Identification of the major peptide metabolites was performed for muscle - tissue samples at 1 min post-injection. This confirmed the importance of sequential aminopeptidase attack, but demonstrated equally important endopeptidase cleavage sites in the basic region of the molecule centered on the 15-Lys,16-Lys and 20-Val,21-Lys bonds. The cleavage products pointed to secondary carboxypeptidase attack at these sites. A third site of primary cleavage was at the 8-Arg,9-Trp bond. This latter internal site is likely to be the single primary site of attack in α-melanotropin with its blocked termini. Internal endopeptidase cleavage was also shown to be the main degradation pathway in the in vivo metabolism of 4-Met(O_2),8-D-Lys,9-Phe corticotropin 4-9.[38] The 8-D-Lys substitution clearly prevented attack at the position corresponding to the Arg-Trp bond, the major cleavage sites being at 6-His,7-Phe and 7-Phe,8-D-Lys.[38] The tissue site for these changes is, however, unknown.

In decreasing order of importance in catabolism to muscle and skin, the intestines are the next most significant site for corticotropic peptide clearance. One minute after injection, they had accumulated some 13% of the peptide dose, and at this time had degraded over 90% of the peptides with either blocked or unblocked termini. This similarity in handling peptides with dissimilar termini suggests a dominant role for endopeptidases. Release of N-terminal amino acids from corticotropin 1-24, however, showed significant aminopeptidase activity was also present in the ileum and colon. Lung tissue metabolized a similar proportion of its corticotropin 1-24 content, but the low uptake makes this organ of little consequence to overall clearance by the body.

The remaining two organs of note were the liver and kidneys. Between them they had accumulated 17% of the injected dose of corticotropic peptides at 1 min post-injection and of this material, approximately 30% was in the intact state. The percentage was similar for both blocked and unblocked analogues. The mechanism of clearance by these organs may, however, differ radically. The pattern of release of amino acids in the kidney displayed a much greater initial release of amino acids from the termini with the central core remaining intact. Separate autoradiographic studies of the kidney[68,70] have shown that parent peptide and larger circulating fragments are first filtered and then rapidly taken up into endocytotic vesicles in the proximal tubule cells. Later, transfer to lysosomes takes place where complete catabolism seems likely to occur. While the kidney may, therefore, play no role in the redistribution of peptides, it functions as one of several tissues which are metabolic clearance sites for melanotropins.

Finally, it is worth emphasizing that the relative importance of the organs elucidated for the corticotropic series of peptides cannot be extrapolated to the β-melanotropin/β-lipotropin group where the human data points to a relatively greater metabolic role for the kidneys.[95]

VI. SPECIALIZED BARRIERS AND SECRETIONS

Within the body not all regions of the capillary vasculature are equally permeable. The capillaries in most areas of the brain are specialized, having tight junctions between the endothelial cells and a layer of closely applied astrocyte processes. This structure presents a physical barrier to molecules of molecular weight above 500, which includes the major melanotropins, and is the basis of what is termed the blood-brain barrier (BBB). Similar physical barriers exist between blood and cerebrospinal fluid (CSF) and in the placenta, though the precise anatomical details vary in each case.[96] Cell barriers also exist between plasma and various body secretions such as saliva and milk. In all these situations, passage of solute molecules from blood into the protected compartment or fluid behaves as if they were traversing a lipid membrane, and hence a key factor in determining permeability is the lipid solubility of the molecule. In addition to this physical barrier, the BBB and placenta have enzymatic mechanisms to destroy certain metabolites in their passage through the barrier. Interesting exceptions occur to this general scheme. The BBB is, for example, lacking in certain sites, the circumventricular organs, which include the area postrema and the median eminence, and facilitated diffusion by carrier-mediated mechanisms exist to transport specific classes of compounds across the barrier.[97] Transport of solutes into bile is largely by active secretion via a capacity-limited system for polar compounds with molecular weight above 300.[96] The effectiveness of the placental barrier is known to reduce towards term and damage or inflammation can increase the permeability of all barriers.

If one ignores these exceptions initially, then according to the theories of passive diffusion through cell membranes, the permeability of a molecule is directly proportional to its partition coefficient between the membrane and plasma water, and inversely proportional to the square of its molecular weight.[96] The in vivo partition coefficients are satisfactorily reflected by in vitro measures between solvents such as n-octanol and water. A few experimentally determined coefficients have been published for pro-opiomelanocortin peptides. Octanol/water values for four enkephalin and endorphin analogues ranged from 0.0017 to 0.066,[98] 2-D-Ala,5-D-Leu enkephalin was higher at 1.1,[99] while β-endorphin, TRH, oxytocin, and arginine vasopressin were <0.1, <0.005, 0.05, and <0.0015, respectively.[99] The partition coefficient for α-melanotropin was 0.0027 when measured in an n-octanol/0.1 M, pH 7.5 phosphate buffer system at 23°C; quantitation was by immunoassay.[16] With the exception of the Leu-enkephalin analogue, values of this order lead to low permeability values, especially with the unfavorable molecular weights of the larger peptides, and by analogy with data from the therapeutic drug field, are likely to result in poor penetration of lipid membranes by simple diffusion.

Observations on placental transfer of corticotropin 1-39 have been in accord with theory. No significant placental transfer could be detected under normal conditions in the rat,[100] sheep,[6] monkey,[101] or man.[102] Levels of biologically active 1-Aib,17,18,19 tri-Lys corticotropin 1-19 amide were also undetectable in fetal extracts after intravenous administration to pregnant rats.[51] Corticotropin-like activity has been reported in rat milk,[37] which because of its acidic pH is known to favor the accumulation of basic compounds.

Transport of polar molecules into bile is, as noted above, by active secretion. No appreciable amount of an intravenous dose of pituitary-extracted melanotropins was found in the bile of dogs,[64] but significant biologically active concentrations of 1-Aib,17,18,19 tri-Lys corticotropin 1-19 amide were detected in rats.[51] Concentrations were always below plasma concentrations and, based on observed bile-flow rates in rats, the total clearance would be 0.1 to 0.2% of the injected dose. Enterohepatic circulation seems unlikely in view of the high rate of peptide degradation in the intestine.[36]

There has been keen interest in studies concerning the penetration of melanotropins into the central nervous system (CNS), since entry into brain tissue is one obvious mechanism whereby peripherally administered melanotropins could exert their important behavioral effects (see Chapter 5, Volume II).[103] The two distinct barriers in the brain are the BBB between blood and the overwhelming bulk of brain tissue, and the blood-CSF barrier in the choroid plexus which has only 0.02% the surface area of the BBB.[99] Three groups of techniques for investigating the permeability of the barriers can be distinguished.[104] First, are essentially semiquantitative methods such as autoradiography or immunoassay which can show that a certain amount of material has arrived in the brain or CSF following peripheral administration. Second, are quantitative methods used during the initial passage of a bolus dose through the cerebral vasculature which measure the fraction of the solute passing unidirectionally through the barrier. The best-known technique of this class is the brain uptake index (BUI) method of Oldendorf. Third, are longer term kinetic studies determining a rate constant for transfer across the barrier.

Several studies of the first type have demonstrated the limited penetration of melanotropins into CSF. Autoradiography 5 min after carotid administration to rats of ^{125}I-α-melanotropin showed radioactive material in meninges and the choroid plexus which suggested penetration into CSF.[105] In human studies, no detectable immunoreactive corticotropin (<7 pmol/ℓ) was present in CSF after a 48 hr infusion of corticotropin 1-24 which resulted in high plasma concentrations of 539 pmol/ℓ.[53] CSF concentrations of corticotropin were also "essentially unchanged" by bolus injection of 256 nmol of peptide, though a small rise appears in the graphically presented CSF data at 2.5 min after the injection.[53] Further studies by the same authors with ^{125}I-labeled corticotropin 1-39 infusions in cats, resulted in a steady-state CSF/plasma ratio of glass-extractable radioactivity of approximately 0.01. More accurate quantitation of permeability has come from bolus intravenous injection studies in small mammals. Injection of α-melanotropin in rats was followed by a significant rise in immunoreactive CSF concentrations with a peak approximately 10 min after the injection.[106] Calculations using the peak concentration and an estimate of the apparent central volume of distribution from intracerebroventricular (icv) injections,[107] put the amount crossing into CSF at below 0.01% of the dose injected. Studies of β-melanotropin in rabbits provided similar results. An early peak in CSF concentrations was seen at 2 min when the CSF/plasma ratio was 0.028.[59] If the central volume of distribution of β-melanotropin is in proportion to body weight to that in the rat of α-melanotropin, then the peak was equivalent to 0.04% of the injected dose having entered the CNS. Subcutaneous injection of 4-Met(O_2),8-D-Lys,9-Phe corticotropin 4-9 in rats gave a different time-course in CSF peptide levels.[38] Concentrations of this enzymatically resistant analogue rose gradually up to 2 hr after the injection, but, given the 89% subcutaneous bioavailability (Table 1), the CSF content was in line with the other estimates at <0.03% of the injected dose. The only fully quantitative estimate of the permeability of the blood-CSF barrier has been for α-melanotropin in the rat.[46] The study was of the third type producing an estimated rate-constant for peptide entry into CSF of 0.00087 min^{-1}. CSF concentrations were measured in a ventricular perfusate and the technique verified by calibration against inulin, a substance of known low permeability. The rate-constant for α-melanotropin entry was not significantly above that for inulin and it was concluded that α-melanotropin entered CSF by ultrafiltration at the choroid plexus. The data for entry of melanotropins into CSF are thus in agreement with entry by passive processes, the rate being limited by the adverse physicochemical properties of peptides.

The interpretation of studies concerning the permeability of the BBB to melanotropins has been controversial. Autoradiography and tissue counting with ^{125}I-α-melanotropin showed a low uptake of radioactivity into the brains of rat and mouse.[65] More detailed autoradiography demonstrated radioactivity in tissue surrounding blood vessels and a diffuse labeling throughout the brain 5 min after intracarotid injections in rats.[105] The distribution of tritiated α-melanotropin in rat brain was studied at 15 sec and 30 min following intracarotid injection of 0.12 and 1.2 nmol doses, respectively.[49] The percentage of the dose found in the brain was low. In intact animals, it was 0.54% at 15 sec and 0.19% at 30 min. Both uptake values were lower than for control injections of tritiated tyrosine at the same time-points, the tyrosine percentages being 1.49 and 0.37%, respectively. The latter comparison is important in relation to the suggested carrier-mediated brain uptake of peptides discussed below, since tyrosine is known to be transported into the brain by a saturable carrier-mediated system with a Michaelis-Menten constant (Km) of 120 μM.[97] Chromatographic evidence showed that only some of the radioactivity in the brain was associated with material with the rf of intact α-melanotropin. A proportion of intact tritiated peptide has also been characterized in brain tissue following administration of 4-Met(O_2),8-D-Lys,9-Phe corticotropin 4-9 to rats by various routes.[38] Uptake of radioactivity by the brain was also low, being 0.1 to 0.13% of the 50 nmol dose for various brain regions 5 min after intravenous administration. Only slightly lower concentrations followed subcutaneous and, interestingly, oral administration. The proportion of intact peptide fell rapidly with time, being some 16% of the radioactivity 5 min after injection. The concentration of intact peptide in the brain was thus estimated to be 0.018% of the dose at 5 min after injection. An earlier study had shown a 0.07% uptake of a 0.04 nmol dose into whole brain after intravenous administration of the same peptide in rats.[108]

The first claim that melanotropins might penetrate the BBB at a rate higher than the levels predicted by their low lipid solubility, came in a study by Greenberg and colleagues.[109] Employing a modified version of the Oldendorf technique with a dose level of 0.583 nmol, they quantified the BUI of α-melanotropin to be 9.6% in rats, which is more than three times the usual background level for poorly soluble analytes. In terms of the percentage of the injected dose, the brain content at 15 sec after the injection was 1.39%, assuming as shown above, the brain to be 0.77% of body weight.[110] This value was also higher than in the studies cited above. The same group also reported a high BUI for Met-enkephalin,[111] but unfortunately in neither case was calibration with a low-uptake solute such as inulin performed. Enkephalin studies have also been reported by Oldendorf's group using Leu-enkephalin concentrations between 22 and 550 nM and with Met-enkephalin from 21 to 270 nM.[112] Extraction values were not measurably above background levels. Some uncertainty, therefore, exists about the validity of the BUI technique as applied to peptides, and it is true that the method is of limited accuracy for BUI values below 10%.[104]

It is, however, possible to reconcile the differences between the various BUI studies by consideration of the doses used. The higher BUI values were obtained with lower doses, and by postulating a saturable transport system with a Km of 0.35 nM it is possible to envisage a system that allows an extraction of 3.5% by simple diffusion, the level that would be obtained with saturating doses of peptide, and an extraction rising to the higher values within the nonsaturated range.[113] The Km value proposed is noticeably much lower than for any of the known BBB transport systems,[97] and as yet, there is no direct evidence for such a carrier-mediated system for peptides. Indeed, the percentage-uptake values cited above from the semiquantitative studies show little evidence to support the proposed dose-related uptake. An attempted demonstration of the possible transport into or binding by isolated brain capillaries using Leu-enkephalin at doses down to 2 nM, failed to show the existence of a high-affinity transport system, but demonstrated instead the presence of a high-capacity aminopeptidase in brain capillaries.[114] Such enzymatic components of the BBB are well known for metabolites such as monoamines, but their possible existence for melanotropins has not been clarified. Were a carrier-mediated system saturating within the pM to nM range to exist, then it would have little impact on the transport kinetics of exogenous melanotropins at the experimental dosages normally used (see Section III, above), but it would be significant at physiological plasma concentrations.

The only studies specifically of melanotropins using a kinetic– as opposed to the BUI– method, are unpublished observations of rats pretreated with monosodium glutamate to deplete endogenous central melanotropin concentrations.[16] The rate-constant for penetration of immunoreactive α-melanotropin into the brain was 0.0081 min^{-1} using a high 30 nmol dose. Although higher than the control measure for inulin of 0.0009 min^{-1}, the value was low and equivalent to a BUI extraction of 1.44%, assuming a rate of blood flow of 0.00933 cm^3 s^{-1} g^{-1}.[113]

To summarize, the current consensus regarding the permeability of the BBB to melanotropins is that passive diffusion limited by molecular weight and lipid solubility dominate at nM concentrations, but an as yet unconfirmed saturable transport system may exist with a Km of 0.35 nM.

A third and separate route of entry of melanotropins into the brain avoiding the two barriers, is through the regions formed by the circumventricular organs. In a paper concerned principally with possible pituitary to brain transport of peptides, tritiated 4-Met(O_2),8-D-Lys,9-Phe corticotropin 4-9 was administered to rats into the anterior pituitary gland, into the surrounding sella cavity, and intravenously.[108] When the radioactivity was determined in various brain regions, the hypothalamus contained the highest brain content 30 min after intrapituitary injection, but not following injection by the other routes. Neuronal transport did not seem to be involved in the hypothalamic uptake since pituitary stalk section did not abolish the high uptake at 8 days, though it was disturbed in the short-term 24 hr after section. Retrograde vascular flow to the median eminence[115] was thought to be the most likely route of transport through the poorly developed BBB of the region. By analogy, locally high rates of peptide transport from blood to brain are to be expected in all circumventricular organs which will contribute to the overall brain and CSF levels observed. More physiologically relevant, however, is the possibility that certain of the central actions of melanotropins are mediated by peptide acting on receptors located within the circumventricular organs such as the median eminence,[77,78] or the area postrema,[116] and that penetration of the BBB is not required.

A further alternative mechanism whereby melanotropins could exert central effects without crossing the BBB, requires consideration as it has been shown that melanotropins alter the permeability of the blood-CSF barrier and BBB to other compounds. Intracisternal injections in rabbits of α-melanotropin, β-melanotropin, corticotropin 1-24, and 1-39 caused a selective increase in entry of labeled albumin, inulin, mannitol, and sucrose into CSF, but not of glucose, α-aminoisobutyric acid or valine.[117] High-dose intravenous injections of α-melanotropin in rats were without effect on the brain uptake of albumin, but increased that of the small ^{99m}Tc-labeled pertechnetate anion.[118,119] In an homologous system to the blood-CSF barrier, intravenous doses of α-melanotropin, β-melanotropin, corticotropin 1-10, N-acetyl corticotropin 1-10, and corticotropin 1-39 caused an increased aqueous flare response in the rabbit eye due to an increase in the entry of protein through the blood-aqueous humor barrier.[120] Changes in permeability have not always been in the one direction. The 4-Met(O_2),8-D-Lys,9-Phe corticotropin 4-9 analogue decreased permeability of the BBB to antipyrine and morphine when given intravenously to rats.[121,122] The effect was shown to be due to a change in permeability and not blood flow, and although small in absolute terms, the responses were sufficient to require doubling of the analgesic dose of morphine. One is left with the final interesting possibility, therefore, that melanotropins could regulate their own entry into brain tissue through alterations in passive transport mechanisms.

VII. CENTRAL PHARMACOKINETICS

Upon entry into the CNS either from the periphery or by direct administration, melanotropins distribute throughout the CSF and extracellular fluid as if it were a single compartment;[106] changes in the concentration of 4-Met(O_2),8-D-Lys,9-Phe corticotropin 4-9 in the brain and CSF running in parallel with time.[38] There would appear to be no generalized barrier to movement across the ependymal lining of the ventricles. Distribution through the CSF following icv injection proceeds rapidly in a caudal direction, presumably assisted by the overall flow in CSF. In the rat, α-melanotropin injected at the lateral ventricle could be sampled in high concentrations at the cisterna magna within 2 min.[106] Labeled Met-enkephalin and β-endorphin traveled down the length of the spinal cord in mice, concentrations peaking in sacral regions approximately 10 min after icv injection into a lateral ventricle.[123] Fitting the observed CSF concentrations in rats following icv injection of α-melanotropin[38] to a one-compartment open model gives a Vd_β value of 4.03 mℓ/kg. This value, being greater than the likely combined volume of brain extracellular fluid and CSF, indicates a degree of tissue binding within the CNS.

Direct evidence of tissue binding has come from several sources. A passive uptake of α-melanotropin by synaptosomes has been demonstrated, which was presumed to be a nonspecific process since uptake showed linear increases with increasing peptide concentration over the 10^{-10} to 10^{-5} molar range.[109] Regional differences in melanotropin binding to the brain in the rat have been studied after both peripheral and icv administration. An increase in biologically active melanotropins of 30 to 150% was measured in the cerebral cortex, cerebellum, pons, and medulla 1 hr after intraperitoneal injection of either β-melanotropin or corticotropin 1-39, while no increase was observed in hypothalamic and thalamic tissue.[124] An autoradiographic localization of ^{125}I-α-melanotropin was reported in cells of the striatum and reticular nucleus of the thalamus 5 min after intracarotid administration.[105] Higher tissue concentrations of radioactivity were found in the occipital cortex, cerebellum, and pons-medulla 15 sec after intracarotid injection of tritiated α-melanotropin.[49] A similar pattern was observed with intravenous injection of 4-Met(O_2),8-D-Lys,9-Phe corticotropin 4-9, the highest concentrations being found in the olfactory bulb, cortex, and cerebellum.[108] In a separate study with the 4-9 analogue using intracarotid injection, high concentrations were found in the olfactory bulb and hind brain.[125] There is reasonable doubt, however, that these regional distributions following peripheral administration, can present a clear picture of binding within the CNS. The pattern may well be influenced by the differential penetration of peptides into the brain. Results from studies with icv injection are therefore to be preferred if the objective is to observe the relative binding between various brain regions. Studies in rats with the tritiated corticotropin 4-9 analogue demonstrated a different pattern of uptake via this route as compared with peripheral administration.[125] Binding in gross brain regions 30 min after injection was high in the olfactory bulbs and diencephalon, but little was found in the cortex and cerebellum. Investigations by microdissection of individual brain nuclei showed a heterogeneity in binding. A general relationship between the radioactive uptake by a nucleus and its distance from the cerebroventricular system was evident, which demonstrated the influence of differential penetration on binding. One group of nuclei in the septal region stood out from this relationship. The dorsal and fimbrial septal nuclei took up a greater amount of label than their anatomical position would predict. Further uptake studies in the presence of unlabeled peptides demonstrated septal uptake to be the result of a specific saturable process.[126] Autoradiography showed the radioactivity to be located predominantly over the neuropile, but some cellular accumulation was observed in a population of small, elongate cells.[127] Papers describing the specific saturable binding of corticotropin 1-24 in the median eminence have been mentioned previously.[76,77]

Clearance of melanotropins from the CNS proceeds more slowly than from the periphery. The half-life for the disappearance of immunoreactive α-melanotropin from CSF was 33 min in the rat,[106] while that of the enzymatically resistant 4-Met(O_2),8-D-Lys,9-Phe corticotropin 4-9 was almost identical, 51.3% of the total radioactivity being present in CSF 30 min after icv injection.[125] The similarity between the values for these two peptides suggests a minor role for metabolism in central clearance processes. Following icv injection, α-melanotropin,[106] corticotropin 4-9[125] and corticotropin 1-39[53] appear rapidly in plasma. Comparison of the AUC in plasma following icv injection[106,125] with AUC's measured after intravenous injection,[38,45,46] show that over 90% of the injected dose passes into plasma with the maximum plasma content being 2 to 8% of the icv dose. AUC values were computed from the published graphs. The scale of the transfer is in direct contrast to the penetration in the opposite direction and results from the passive bulk flow of CSF through the arachnoid villi.[128] There is, therefore, a minimal role for metabolism in clearance of melanotropins from the CNS. This same conclusion was reached for β-endorphin in monkeys where the AUC of immunoreactive peptide in CSF was identical to that of the inert marker, inulin.[129]

As with peripheral tissues, there are a number of reports describing degradation of melanotropins by brain-tissue extracts or subcellular components.[90,91] In vivo studies are few. After the injection of tritiated α-melanotropin via the carotid artery, chromatographically intact material was present in brain-tissue extracts, but the amount was not specified.[49] Increases in bioactive material were also measured after intraperitoneal administration.[124] The only quantitative data are for 4-Met(O_2),8-D-Lys,9-Phe corticotropin 4-9. Apparently, rapid degradation of peptide was observed in brain tissue after intravenous administration since the percentage of intact peptide in tissue extracts fell from 15.7 to 11.3 to 6.2% at 1, 15, and 30 min after injection, respectively.[38] However, the percentages of intact peptide

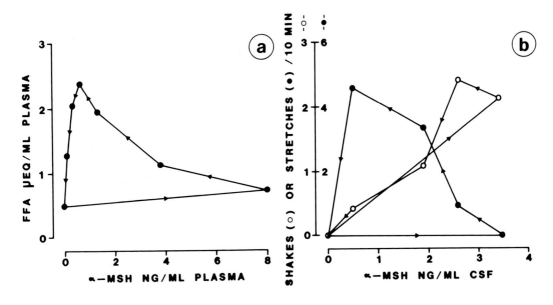

FIGURE 3. Hysteresis loops for (a) variations in free-fatty-acid (FFA) concentrations in plasma of rabbits plotted against changes in plasma concentrations of α-melanotropin (α-MSH); and (b) variation in numbers of body shakes (○) and stretches (●) in rats plotted against changes in CSF concentrations of α-MSH. The points are connected in time order, arrows pointing to later times. (FFA data from Rudman et al.,[44] behavioral parameters from Yamada and Furukawa,[130] and CSF peptide measurements from de Rotte et al.[106])

in the brain after subcutaneous administration, were even lower. It seems, therefore, that a greater brain uptake of smaller peripherally generated peptide fragments may be occurring to distort the ratios of intact to total radioactivity observed. The same problem could influence icv data since material cleared to the periphery will be metabolized and the products will re-enter the CNS. Nevertheless, the proportion of intact peptide in brain tissue after icv administration was 26.5% at 2 hr and 27.3% at 4 hr after the injection, which was consistent with the high degree of metabolic stability predicted by the kinetic analysis.

VIII. PHARMACODYNAMICS

It is apparent from the preceding sections that the melanotropic peptides are short-lived molecules in vivo both in the periphery and the CNS. The time-course of the responses that they produce are not necessarily so ephemeral. To conclude this chapter, therefore, some examples of the pharmacokinetic-pharmacodynamic relationship will be given to illustrate some of the interesting differences between the biological responses to melanotropins.

The data of Rudman and colleagues describing the time course of the in vivo lipolytic response to α-melanotropin in the rabbit demonstrate the normal situation.[44] The changes in free-fatty-acid concentrations in plasma followed circulating peptide concentrations, with only a short delay attributable to the time needed for penetration to the receptor sites on the adipocytes and the lag in generation of free-fatty-acid release. The peak response following intravenous bolus injection was approximately 30 min after the injection. When displayed as a response against peptide concentration in plasma plots with the points joined in sequence, this data therefore displays a marked anticlockwise hysteresis (Figure 3a). A similar time relationship can be shown for certain of the central actions of melanotropins. Combining the CSF concentration data following icv injection of α-melanotropin[106] with response data for body shaking and stretching responses produced by such treatment in the rat,[130] gives the hysteresis plots displayed in Figure 3b. The body shaking response shows minimal hysteresis, the response closely following CSF peptide concentrations. Stretching responses are delayed in comparison with a peak response 90 to 100 min after icv injection. An anticlockwise hysteresis suggestive of a different etiology was therefore present. In all the above examples, however, some direct relationship between peptide concentration and response seems likely. Measurable effects occurred during the period when plasma or CSF concentrations were elevated. This was not the case with other centrally mediated responses to melanotropins. Dose-related effects of corticotropin 4-10 in man were measured at times starting 45 min after injection of the peptide when circulating concentrations were undetectable.[39] In comparisons of the duration of action of corticotropin 4-10 and 4-Met(O_2),8-D-Lys,9-Phe corticotropin 4-9 on active avoidance behavior in rats, the 4-10 analogue delayed extinction of pole-jumping avoidance responses 2 and 4 hr after subcutaneous administration, while the enzyme-resistant 4-9 analogue was still effective at 24 hr, though not at 48 hr after administration.[131] The duration of action was thus related to the stability of the molecules, but responses were evident at times when the concentrations of intact peptides in either plasma or brain tissue would have decayed substantially.[38,39] The apparent dissociation between peptide concentration and response could easily be explained by some indirect mechanism of action whereby initial high-peptide concentrations stimulate an intermediate longer-lasting metabolic response. The observed difference in time-course between the 4-10 and 4-9 analogues, however, points to a direct peptide effect and hence to speculation of the existence of localized concentrations of intact peptide protected by binding in key central sites.

REFERENCES

1. **Guttmann, St. and Boissonnas, R. A.**, Synthèse de l'α-mélanotropine (α-MSH) de porc, *Helv. Chim. Acta*, 42, 1257, 1959.
2. **Schwyzer, R.**, Chemische Synthese der Melanotropine und des corticotropen Hormone der Hypophyse, *Naturwissenschaften*, 53, 189, 1966.
3. **Yajima, H. and Yoshiaki, K.**, Structure-function relationships of the melanotropins, *Pharmacol. Ther. B*, 1, 529, 1975.
4. **Thody, A. J.**, *The MSH Peptides*, Academic Press, London, 1980.
5. **Murphy, S. S., Donald, R. A., and Nabarro, J. D. N.**, The half-life of porcine corticotrophin in pigs, *Acta Endocrinol.*, 61, 525, 1969.
6. **Jones, C. T., Luther, E., Ritchie, J. W. K., and Worthington, D.**, The clearance of ACTH from the plasma of adult and fetal sheep, *Endocrinology*, 96, 231, 1975.
7. **Wolf, R. L., Mendlowitz, M., Soffer, L. J., Roboz, J., and Gitlow, S. E.**, Metabolism of corticotropin in man, *Proc. Soc. Exp. Biol. Med.*, 119, 244, 1965.
8. **Kaneko, M., Kaneko, K., Shinsako, J., and Dallman, M. F.**, Adrenal sensitivity to adrenocorticotropin varies diurnally, *Endocrinology*, 109, 70, 1981.
9. **Besser, G. M., Orth, D. N., Nicholson, W. E., Byyny, R. L., Abe, K., and Woodham, J. P.**, Dissociation of the disappearance of bioactive and radioimmunoreactive ACTH from plasma in man, *J. Clin. Endocrinol. Metab.*, 32, 595, 1971.
10. **Matsuyama, H., Ruhmann-Wennhold, A., Johnson, L. R., and Nelson, D. H.**, Disappearance rates of exogenous and endogenous ACTH from rat plasma measured by bioassay and radioimmunoassay, *Metabolism*, 21, 30, 1972.
11. **Nicholson, W. E., Liddle, R. A., Puett, D., and Liddle, G. W.**, Adrenocorticotropic hormone biotransformation, clearance, and catabolism, *Endocrinology*, 103, 1344, 1978.
12. **Jeffcoate, W. J., Phenekos, C., Ratcliffe, J. G., Williams, S., Rees, L., and Besser, G. M.**, Comparison of the pharmacokinetics in man of two synthetic ACTH analogues: α^{1-24} and substituted α^{1-18} ACTH, *Clin. Endocrinol.*, 7, 1, 1977.
13. **Gardner, R. C., Odell, W. D., and Ross, G. T.**, Comparison of t1/2 of α-MSH by three methods, *Clin. Res.*, 14, 280, 1966.

14. **Imura, H., Sparks, L. L., Grodsky, G. M., and Forsham, P. H.,** Immunologic studies of adrenocorticotropic hormone (ACTH): dissociation of biologic and immunologic activites, *J. Clin. Endocrinol. Metab.,* 25, 1361, 1965.
15. **Wilson, J. F. and Morgan, M. A.,** α-Melanotropin-like substances in the pituitary and plasma of *Xenopus laevis* in relation to colour change responses, *Gen. Comp. Endocrinol.,* 38, 172, 1979.
16. **Wilson, J. F.,** unpublished observations.
17. **Thody, A. J. and Shuster, S.,** Control of sebaceous gland function in the rat by α-melanocyte-stimulating hormone, *J. Endocrinol.,* 64, 503, 1975.
18. **Orias, R. and McCann, S. M.,** Natriuresis induced by alpha and beta melanocyte-stimulating hormone (MSH) in rats, *Endocrinology,* 90, 700, 1972.
19. **Beckwith, B. E. and Sandman, C. A.,** Behavioral influences of the neuropeptides ACTH and MSH: A methodological review, *Neurosci. Biobehav. Rev.,* 2, 311, 1978.
20. **Kastin, A. J., Miller, L. H., Gonzalez-Barcena, D., Hawley, W. D., Dyster-Aas, K., Schally, A. V., De Parra, M. L. V., and Velasco, M.,** Psycho-physiologic correlates of MSH activity in man, *Physiol. Behav.,* 7, 893, 1971.
21. **Ashton, H., Millman, J. E., Telford, R., Thompson, J. W., Davies, T. F., Hall, R., Shuster, S., Thody, A. J., Coy, D. H., and Kastin, A. J.,** Psychopharmacological and endocrinological effects of melanocyte-stimulating hormones in normal man, *Psychopharmacology,* 55, 165, 1977.
22. **Bohus, B., Gispen, W. H., and de Wied, D.,** Effect of lysine vasopressin and $ACTH_{4-10}$ on conditioned avoidance behavior of hypophysectomized rats, *Neuroendocrinology,* 11, 137, 1973.
23. **Garrud, P., Gray, J. A., and de Wied, D.,** Pituitary-adrenal hormones and extinction of reward behavior in the rat, *Physiol. Behav.,* 12, 109, 1974.
24. **Sandman, C. A., George, J. M., Nolan, J. D., van Riezen, H., and Kastin, A. J.,** Enhancement of attention in man with ACTH/MSH 4-10, *Physiol. Behav.,* 15, 427, 1975.
25. **Witter, A., Greven, H. M., and de Wied, D.,** Correlation between structure, behavioral activity and rate of biotransformation of some $ACTH_{4-9}$ analogues, *J. Pharmacol. Exp. Ther.,* 193, 853, 1975.
26. **de Wied, D., Bohus, B., van Ree, J. M., and Urban, I.,** Behavioral and electrophysiological effects of peptides related to lipotropin (β-LPH), *J. Pharmacol. Exp. Ther.,* 204, 570, 1978.
27. **Ferrari, W., Gessa, G. L., and Vargiu, L.,** Behavioral effects induced by intracisternally injected ACTH and MSH, *Ann. N.Y. Acad. Sci.,* 104, 330, 1963.
28. **Lipton, J. M. and Glyn, J. R.,** Central administration of peptides alters thermoregulation in the rabbit, *Peptides,* 1, 15, 1980.
29. **Hedner, P.,** On the effect of long-acting corticotrophin preparations, *Acta Endocrinol.,* 43, 499, 1963.
30. **de Wied, D.,** Inhibitory effect of ACTH and related peptides on extinction of conditioned avoidance behaviour in rats, *Proc. Soc. Exp. Biol. Med.,* 122, 28, 1966.
31. **Futaguchi, S., Odaguchi, K., Tanaka, A., and Hirata, M.,** Stabilization of a sustained-release-type injection vehicle for a synthetic corticotrophin analogue, *J. Pharm. Pharmacol.,* 34, 343, 1982.
32. **Ley, K. F. and Corson, J. A.,** Effects of ACTH and zinc phosphate vehicle on shuttlebox CAR, *Psychon. Sci.,* 20, 307, 1970.
33. **Hirata, M., Futaguchi, S., Tamura, T., Odaguchi, K., and Tanaka, A.,** Rectal absorption of Gly^1-$α^{1-18}$ adrenocorticotropin amide, *Chem. Pharmacol. Bull.,* 26, 1061, 1978.
34. **Ritschel, W. A.,** *Handbook of Basic Pharmacokinetics,* Drug Intelligence Publications, Hamilton, Canada, 1980.
35. **McIntosh, J. E. A. and McIntosh, R. P.,** *Mathematical Modelling and Computers in Endocrinology,* Springer–Verlag, Berlin, 1980.
36. **Lowry, P. J. and McMartin, C.,** Metabolism of two adrenocorticotrophin analogues in the intestine of the rat, *Biochem. J.,* 138, 87, 1974.
37. **Vaucher, Y., Tenore, A., Grimes, J. A., Krulich, L., and Koldovsky, O.,** Absorption of TSH and ACTH in biologically active forms from the gastrointestinal tract of suckling rats, *Endocrinol. Exp.,* 17, 327, 1983.
38. **Verhoff, J. and Witter, A.,** In vivo fate of a behaviorally active ACTH 4-9 analog in rats after systemic administration, *Pharmacol. Biochem. Behav.,* 4, 583, 1976.
39. **Born, J., Fehm–Wolfsdorf, G., Voigt, K. H., and Fehm, H. L.,** Influences of ACTH 4 – 10 or event-related potentials reflecting attention in man, *Physiol. Behav.,* 39, 83, 1987.
40. **Adelmann, H., Graef, V., and Schatz, H.,** Pernasal application of a potent synthetic ACTH 1-17 analogue: dynamics of cortisol and aldosterone levels in serum before and after 7 days of administration, *Horm. Metab. Res.,* 16, 55, 1984.
41. **Wright, A. and Wilson, J. F.,** Absorption of α-MSH from subcutaneous and intraperitoneal sites in the rat, *Peptides,* 4, 5, 1983.
42. **Verhoef, J. C. and van den Wildenberg, H. M.,** Des-enkephalin-γ-endorphin: bioavailability in rats following the subcutaneous and intramuscular route of administration, *Reg. Peptides,* 14, 113, 1986.

43. **Vérine, A., Trouvé-Blanqui, E., Oliver, C., and Boyer, J.**, Influence of corticotrophin on plasma testosterone in normal women, *Acta Endocrinol.*, 94, 201, 1980.
44. **Rudman, D., Hollins, B. M., Kutner, M. H., Moffitt, S. D., and Lynn, M. J.**, Three types of α-melanocyte-stimulating hormone: bioactivities and half-lives, *Am. J. Physiol.*, 245, E47, 1983.
45. **Wilson, J. F. and Harry, F. M.**, Release, distribution and half-life of α-melanotrophin in the rat, *J. Endocrinol.*, 86, 61, 1980.
46. **Wilson, J. F., Anderson, S., Snook, G., and Llewellyn, K. D.**, Quantification of the permeability of the blood-CSF barrier to α-MSH in the rat, *Peptides*, 5, 681, 1984.
47. **Gilham, I. D. and Baker, B. I.**, Unpublished data presented in Figure 1.
48. **Redding, T. W., Kastin, A. J., Nikolics, K., Schally, A. V., and Coy, D. H.**, Disappearance and excretion of labeled α-MSH in man, *Pharmacol. Biochem. Behav.*, 9, 207, 1978.
49. **Kastin, A. J., Nissen, C., Nikolics, K., Medzihradszky, K., Coy, D. H., Teplan, I., and Schally, A. V.**, Distribution of ^{3}H-α-MSH in rat brain, *Brain Res. Bull.*, 1, 19, 1976.
50. **Goos, H. J. Th., de Gran, P., and van Oordt, P. G. W. J.**, Half-life of melanocyte-stimulating hormone and hormone content of the pars intermedia in relation to background adaptation time in Xenopus laevis, *J. Endocrinol.*, 72, 26P, 1977.
51. **Hirata, M., Futaguchi, S., Odaguchi, K., Inouye, K., and Tanaka, K.**, Distribution study of a synthetic ACTH analogue in rat tissue, *Acta Endocrinol.*, 96, 464, 1981.
52. **McMartin, C. and Peters, J.**, Levels of corticotrophin analogues in the blood after infusion into rats, *J. Endocrinol.*, 67, 41, 1975.
53. **Allen, J. P., Kendall, J. W., McGilvra, R., and Vancura, C.**, Immunoreactive ACTH in cerebrospinal fluid, *J. Clin. Endocrinol. Metab.*, 38, 586, 1974.
54. **Jones, C. T., Johnson, P., Kendall, J. Z., Ritchie, J. W. K., and Thorburn, G. D.**, Induction of premature parturition in sheep: adrenocorticotrophin and corticosteroid changes during infusion of synacthen into the foetus, *Acta Endocrinol.*, 87, 192, 1978.
55. **Jones, C. T., Kendall, J. Z., Ritchie, J. W. K., Robinson, J. S., and Thorburn, G. D.**, Adrenocorticotrophin and corticosteroid changes during dexamethasone infusion to intact and synacthen infusion to hypophysectomized foetuses, *Acta Endocrinol.*, 87, 203, 1978.
56. **Nathanielsz, P. W., Jack, P. M. B., Krane, E. J., Thomas, A. L., Ratter, S., and Rees, L. H.**, The role and regulation of corticotropin in the fetal sheep, in *The Fetus and Birth*, Ciba Foundation Symposium 47, O'Connor, M. and Knight, J., Eds., Elsevier, Amsterdam, 1977, 73.
57. **Wood, C. E., Shinsako, J., Keil, L. C., and Dallman, M. F.**, Adrenal sensitivity to adrenocorticotropin in normovolemic and hypovolemic conscious dogs, *Endocrinology*, 110, 1422, 1982.
58. **Franco-Saenz, R., van Gemert, M., and Saffran, M.**, In vivo distribution of radioiodinated $ACTH_{1-24}$ in the rat, *Horm. Res.*, 15, 44, 1981.
59. **Pezalla, P. D., Lis, M., Seidah, N. G., and Chrétien, M.**, Lipotropin, melanotropin and endorphin: In vivo catabolism and entry into cerebrospinal fluid, *J. Can. Sci. Neurol.*, 5, 183, 1978.
60. **Aronin, N., Wiesen, M., Liotta, A. S., Schussler, G. C., and Krieger, D. T.**, Comparative metabolic clearance rates of β-endorphin and β-lipotropin in humans, *Life Sci.*, 29, 1265, 1981.
61. **Liotta, A. S., Li, C. H., Schussler, G. C., and Krieger, D. T.**, Comparative metabolic clearance rate, volume of distribution and plasma half-life of human β-lipotropin and ACTH, *Life Sci.*, 23, 2323, 1978.
62. **Chang, W., Rao, A. J., and Li, C. H.**, Rate of disappearance of human β-lipotropin and β-endorphin in adult male rats as estimated by radioimmunoassay, *Peptide Protein Res.*, 11, 93, 1978.
63. **Lu, C. L., Chan, J. S. D., Léan, A. D., Chen, A., Seidah, N. G., and Chrétien, M.**, Metabolic clearance rate and half-time disappearance rate of human N-terminal and adrenocorticotropin of pro-opiomelanocortin in the rat: a comparative study, *Life Sci.*, 33, 2599, 1983.
64. **Shizume, K. and Irie, M.**, The role of the liver in the inactivation of the melanocyte-stimulating hormone in the body, *Endocrinology*, 61, 506, 1957.
65. **Dupont, A., Kastin, A. J., Labrie, F., Pelletier, G., Puviani, R., and Schally, A. V.**, Distribution of radioactivity in the organs of the rat and mouse after injection of [^{125}I]α-melanocyte-stimulating hormone, *J. Endocrinol.*, 64, 237, 1975.
66. **Golder, M. P. and Boyns, A. R.**, Distribution of [^{131}I]α$^{1-24}$-adrenocorticotrophin in the intact guinea-pig, *J. Endocrinol.*, 49, 649, 1971.
67. **Bennett, H. P. J. and McMartin, C.**, Distribution and degradation of two tritium-labelled corticotrophin analogues in the rat, *J. Endocrinol.*, 82, 33, 1979.
68. **Baker, J. R. J., Bennett, H. P. J., Hudson, A. M., McMartin, C., and Purdon, G. E.**, On the metabolism of two adrenocorticotrophin analogues, *Clin. Endocrinol.*, 5, Suppl., 61S, 1976.
69. **Hudson, A. M. and McMartin, C.**, Mechanisms of catabolism of corticotrophin-(1-24)-tetracosapeptide in the rat in vivo, *J. Endocrinol.*, 85, 93, 1980.
70. **Baker, J. R. J., Bennett, H. P. J., Christian, R. A., and McMartin, C.**, Renal uptake and metabolism of adrenocorticotrophin analogues in the rat: an autoradiographic study, *J. Endocrinol.*, 74, 23, 1977.

71. **Bennett, H. P. J. and McMartin, C.,** Peptide hormones and their analogues: distribution, clearance from the circulation, and inactivation in vivo, *Pharmacol. Rev.,* 30, 247, 1979.
72. **Stouffer, J. E. and Hsu, J. S.,** Structural relationships in the interaction of adrenocorticotropin with plasma proteins, *Biochemistry,* 5, 1195, 1966.
73. **Medzihradszky, K.,** Protein-binding of α-melanotropin. Studies on the relationship between binding, structure and biological activity, in *Peptides 1976, Proceedings of 14th European Peptide Symposium,* Loffet, A., Ed., Editions de l'université de Bruxelles, Belgium, 1976, 401.
74. **Smith, A. C. and Wilson, J. F.,** unpublished data, 1986.
75. **Wilson, J. F. and Morgan, M. A.,** A radioimmunoassay for alpha-melanotropin in rat plasma, *J. Pharmacol. Methods,* 2, 97, 1979.
76. **Muramoto, K. and Ramachandran, J.,** Identification of the corticotropin binding domain of bovine serum albumin by photoaffinity labeling, *Biochemistry,* 20, 3380, 1981.
77. **van Houten, M., Khan, M. N., Khan, R. J., and Posner, B. I.,** Blood-borne adrenocorticotropin binds specifically to the median eminence-arcuate region of the rat hypothalamus, *Endocrinology,* 108, 2385, 1981.
78. **van Houten, M., Khan, M. N., Walsh, R. J., Baquiran, G. B., Renaud, L. P., Bourque, C., Sgro, S., Gauthier, S., Chrétien, M., and Posner, B. I.,** NH_2-terminal specificity and axonal localization of adrenocorticotropin binding sites in rat median eminence, *Proc. Natl. Acad. Sci. U.S.A.,* 82, 1271, 1985.
79. **Varga, J. M., Dipasquale, A., Pawelek, J., McGuire, J. S., and Lerner, A. B.,** Regulation of melanocyte-stimulating hormone action at the receptor level: discontinuous binding of hormone to synchronized mouse melanoma cells during the cell cycle, *Proc. Natl. Acad. Sci. U.S.A.,* 71, 1590, 1974.
80. **Lerner, A. B., Shizume, K., and Bunding, I.,** The mechanism of endocrine control of melanin pigmentation, *J. Clin. Endocrinol. Metab.,* 14, 1463, 1954.
81. **Smith, A. G., Shuster, S., Thody, A. J., Alvarez-ude, F., and Kerr, D. N. S.,** Role of the kidney in regulating plasma immunoreactive beta-melanocyte-stimulating hormone, *Br. Med. J.,* 1, 874, 1976.
82. **Gaspar, L., Chan, J. S. D., Seidah, N. G., and Chrétien, M.,** Urinary molecular forms of human N-terminal of pro-opiomelanocortin: possible deglycosylation and degradation by the kidney, *J. Clin. Endocrinol. Metab.,* 59, 614, 1984.
83. **Cowan, J. S., Davis, A. E., and Layberry, R. A.,** Constancy and linearity of the metabolic clearance of adrenocorticotropin in dogs, *Can. J. Physiol. Pharmacol.,* 52, 8, 1974.
84. **Kem, D. C., Gomez-Sanchez, C., Kramer, N. J., Holland, O. B., and Higgins, J. R.,** Plasma aldosterone and renin activity response to ACTH infusion in dexamethasone-suppressed normal and sodium-depleted man, *J. Clin. Endocrinol. Metab.,* 40, 116, 1975.
85. **Lalonde, J. and Normand, M.,** Metabolic clearance rate of adrenocorticotropin in the rat, *Can. J. Physiol. Pharmacol.,* 55, 1079, 1977.
86. **Imura, H., Matsuyama, H., Matsukura, S., Miyake, T., and Fukase, M.,** Stability of ACTH preparations in human plasma incubated *in vitro, Endocrinology,* 80, 599, 1967.
87. **Trochard, M. C., Vaudry, H., Leboulenger, F., Usategui, R., and Vaillant, R.,** Etude de la dégradation de l'α MSH radioiodée, *in vitro:* influence de quelques inhibiteurs des enzymes protéolytiques, *Compt. Rend. Soc. Biol.,* 170, 1103, 1976.
88. **Gilkes, J. J. H., Bloomfield, G. A., and Rees, L. H.,** Studies on the release and degradation of the human melanocyte-stimulating hormone, *Proc. Roy. Soc. Med.,* 67, 40, 1974.
89. **Marks, N., Stern, F., and Kastin, A. J.,** Biodegradation of α-MSH and derived peptides by rat brain extracts, and by rat and human serum, *Brain Res. Bull.,* 1, 591, 1976.
90. **Reith, M. E. A. and Neidle, A.,** Breakdown and fate of ACTH and MSH, *Pharmacol. Ther.,* 12, 449, 1981.
91. **Burbach, J. P. H.,** Action of proteolytic enzymes on lipotropins and endorphins: biosynthesis, biotransformation and fate, *Pharmacol. Ther.,* 24, 321, 1984.
92. **Bennett, H. P. J., Bullock, G., Lowry, P. J., McMartin, C., and Peters, J.,** Fate of corticotrophins in an isolated adrenal-cell bioassay and decrease of peptide breakdown by cell purification, *Biochem. J.,* 138, 185, 1974.
93. **Rudman, D., Malkin, M. F., Brown, S. J., Garcia, L. A., and Abell, L. L.,** Inactivation of adrenocorticotropin, α- and β- melanocyte-stimulating hormones, vasopressin, and pituitary fraction H by adipose tissue, *J. Lipid Res.,* 5, 38, 1964.
94. **Lambert, T. R. and Wilson, J. F.,** Disappearance of immunoreactive α-melanotrophin from rat plasma following occlusion of different regions of the vascular bed, *J. Endocrinol.,* 88, 437, 1981.
95. **Aronin, N., Liotta, A. S., Shickmanter, B., Schussler, G. C., and Krieger, D. T.,** Impaired clearance of β-lipotropin in uremia, *J. Clin. Endocrinol. Metab.,* 53, 797, 1981.
96. **La Du, B. N., Mandel, H. G., and Way, E. L.,** *Fundamentals of Drug Metabolism and Drug Disposition,* Williams & Wilkins, Baltimore, 1971.
97. **Pardridge, W. M.,** Transport of nutrients and hormones through the blood-brain barrier, *Diabetologia,* 20, 246, 1981.

98. **Rapoport, S. I., Klee, W. A., Pettigrew, K. D., and Ohno, K.,** Entry of opioid peptides into the central nervous system, *Science,* 207, 84, 1980.
99. **Meisenberg, G. and Simmons, W. H.,** Peptides and the blood-brain barrier, *Life Sci.,* 32, 2611, 1983.
100. **Dupouy, J. P., Chatelain, A., and Allaume, P.,** Absence of transplacental passage of ACTH in the rat: direct experimental proof, *Biol. Neonate,* 37, 96, 1980.
101. **Kittinger, G. W., Beamer, N. B., Hagemenas, F., Hill, J. D., Baughman, W. L., and Ochsner, A. J.,** Evidence for autonomous pituitary-adrenal function in the near-term fetal rhesus *(Macaca mulatta), Endocrinology,* 91, 1037, 1972.
102. **Allen, J. P., Cook, D. M., Kendall, J. W., and McGilvra, R.,** Maternal-fetal ACTH relationship in man, *J. Clin. Endocrinol. Metab.,* 37, 230, 1973.
103. **Kastin, A. J., Olson, R. D., Schally, A. V., and Coy, D. H.,** CNS effects of peripherally administered brain peptides, *Life Sci.,* 25, 401, 1979.
104. **Fenstermacher, J. D., Blasberg, R. G., and Patlak, C. S.,** Methods for quantifying the transport of drugs across brain barrier systems, *Pharmacol. Ther.,* 14, 217, 1981.
105. **Pelletier, G., Labrie, F., Kastin, A. J., and Schally, A. V.,** Radioautographic localization of radioactivity in rat brain after intracarotid injection of ^{125}I-α-melanocyte-stimulating hormone, *Pharmacol. Biochem. Behav.,* 3, 671, 1975.
106. **de Rotte, A. A., Bouman, H. J., and van Wimersma Greidanus, Tj. B.,** Relationships between α-MSH levels in blood and in cerebrospinal fluid, *Brain Res. Bull.,* 5, 375, 1980.
107. **Mens, W. B. J., Witter, A., and van Wimersma Greidanus, Tj. B.,** Penetration of neurohypophyseal hormones from plasma into cerebrospinal fluid (CSF): half-times of disappearance of these neuropeptides from CSF, *Brain Res.,* 262, 143, 1983.
108. **Mezey, E., Palkovits, M., de Kloet, E. R., Verhoef, J., and de Wied, D.,** Evidence for pituitary-brain transport of a behaviorally potent ACTH analog, *Life Sci.,* 22, 831, 1978.
109. **Greenberg, R., Whalley, C. E., Jourdikian, F., Mendelson, I. S., Walter, R., Nikolics, K., Coy, D. H., Schally, A. V., and Kastin, A. J.,** Peptides readily penetrate the blood-brain barrier: uptake of peptides by synaptosomes is passive, *Pharmacol. Biochem. Behav.,* 5, Suppl. 1, 151, 1976.
110. **Crandall, M. W. and Drabkin, D. L.,** Cytochrome C in regenerating rat liver and its relation to other pigments, *J. Biol. Chem.,* 166, 653, 1946.
111. **Kastin, A. J., Nissen, C., Schally, A. V., and Coy, D. H.,** Blood brain barrier half-time disappearance, and brain distribution of labeled enkephalin and potent analog, *Brain Res. Bull.,* 1, 583, 1976.
112. **Cornford, E. M., Braun, L. D., Crane, P. D., and Oldendorf, W. H.,** Blood-brain barrier restriction of peptides and the low uptake of enkephalins, *Endocrinology,* 103, 1297, 1978.
113. **Sharma, R. R. and Vimal, R. L. P.,** Theoretical interpretation of extraction (in brain) of peptides including concentration variations, *Brain Res.,* 308, 201, 1984.
114. **Pardridge, W. M. and Mietus, L. J.,** Enkephalin and blood-brain barrier: studies of binding and degradation in isolated brain microvessels, *Endocrinology,* 109, 1138, 1981.
115. **Oliver, C., Mical, R. S., and Porter, J. C.,** Hypothalamic-pituitary vasculature: evidence for retrograde blood flow in the pituitary stalk, *Endocrinology,* 101, 598, 1977.
116. **Lichtensteiger, W. and Lienhart, R.,** Response of mesencephalic and hypothalamic dopamine neurones to α-MSH: mediated by area postrema? *Nature (London),* 266, 635, 1977.
117. **Rudman, D. and Kutner, M. H.,** Melanotropic peptides increase permeability of plasma/cerebrospinal fluid barrier, *Am. J. Physiol.,* 234, E327, 1978.
118. **Sankar, R., Domer, F. R., and Kastin, A. J.,** Selective effects of α-MSH and MIF-1 on the blood-brain barrier, *Peptides,* 2, 345, 1981.
119. **Kastin, A. J. and Fabre, L. A.,** Limitations to effect of α-MSH on permeability of blood-brain barrier to iv ^{99m}Tc-pertechnetate, *Pharmacol. Biochem. Behav.,* 17, 1199, 1982.
120. **Dyster-Aas, K. and Krakau, C. E. T.,** Increased permeability of the blood-aqueous humor barrier in the rabbit's eye provoked by melanocyte-stimulating peptides, *Endocrinology,* 74, 255, 1964.
121. **Goldman, H. and Murphy, S.,** An analog of ACTH/MSH$_{4-9}$, ORG-2766, reduces permeability of the blood-brain barrier, *Pharmacol. Biochem. Behav.,* 14, 845, 1981.
122. **Goldman, H., Krasnewich, D., Murphy, S., and Schneider, D.,** An analog of ACTH/MSH (4-9), ORG-2766, reduces cerebral uptake of morphine, *Peptides,* 3, 649, 1982.
123. **Ohlsson, A. E., Fu, T. C., Jones, D., Martin, B. R., and Dewey, W. L.,** Distribution of radioactivity in the spinal cord after intracerebroventricular and intravenous injection of radiolabeled opioid peptides in mice, *J. Pharmacol. Exp. Ther.,* 221, 362, 1982.
124. **Rudman, D., Scott, J. W., Del Rio, A. E., Houser, D. H., and Sheen, S.,** Melanotropic activity in regions of rodent brain, *Am. J. Physiol.,* 226, 682, 1974.
125. **Verhoef, J., Palkovits, M., and Witter, A.,** Distribution of a behaviorally highly potent ACTH$_{4-9}$ analog in rat brain after intraventricular administration, *Brain Res.,* 126, 89, 1977.
126. **Verhoef, J., Witter, A., and de Wied, D.,** Specific uptake of a behaviorally potent [^{3}H]ACTH$_{4-9}$ analog in the septal area after intraventricular injection in rats, *Brain Res.,* 131, 117, 1977.

127. **Rees, H. D., Verhoef, J., Witter, A., Gispen, W. H., and de Wied, D.,** Autoradiographic studies with a behaviorally potent ^{3}H-ACTH$_{4-9}$ analog in the brain after intraventricular injection in rats, *Brain Res. Bull.*, 5, 509, 1980.
128. **Welch, K. and Pollay, M.,** Perfusion of particles through arachnoid villi of the monkey, *Am. J. Physiol.*, 201, 651, 1961.
129. **Lee, V. C., Burns, R. S., Dubois, M., and Cohen, M. R.,** Clearance from cerebrospinal fluid of intrathecally administered β-endorphin in monkeys, *Anesth. Analg.*, 63, 511, 1984.
130. **Yamada, K. and Furukawa, T.,** The yawning elicited by α-melanocyte-stimulating hormone involves serotonergic-dopaminergic-cholinergic neuron link in rats, *Arch. Pharmacol.*, 316, 155, 1981.
131. **Fekete, M. and de Wied, D.,** Potency and duration of action of the ACTH 4-9 analog (ORG 2766) as compared to ACTH 4-10 and [D-Phe^7] ACTH 4-10 on active and passive avoidance behavior of rats, *Pharmacol. Biochem. Behav.*, 16, 387, 1982.

INDEX

A

Acetic acid, 185
Acetylation, 111
 endorphin, 119
 enzymes, 110
 intracellular, 114
 MSH, 112—114, 119
 N-terminal, 105, 109, 112—113
 of POMC, 115
 of POMC-derived peptides, 96—98
Acetyl-CoA, 32
Acetylcholine, 140
Acetyltransferase, 32—33, 73
N-Acetyltransferase, 32—33
ACTH, 1, 48, 57, 68, 73, 87, 96, 106, 109—111, 140, 142, 150—151
 α-MSH and, 40
 anterior lobe, 95
 big, 69
 bioassay, 43
 biosynthetic intermediate, 28, 30
 CLIP and, 40
 C-terminal amidation and, 42
 melanosome dispersion and, 45
 N-terminal 13 of, 26
 POMC gene and, 27
 -positive axons, 62
ACTH 1, 42
N,O-diacetyl-ACTH(1-13)amide, 27—28, 30—33
Adaptation, 1, 8—9, 40
Adenohypophysial secretory cells, 6
Adenohypophysis, 6, 7, 11, 14, 17
Adenylate cyclase, 90, 92, 117, 133
Adrenal, 28—30, 76, 94
Adrenalin, 104, 134
α-Adrenergic blocking agents, 47
Adrenocorticotropin, see ACTH
Adsorption chromatography, 46
Agonists
 β-adrenergic, 135
 dopaminergic, 69, 92, 109, 131—132, 145
 GABA receptor, 117
Agouti mice, 131
Alanine, 187
Albumin, 201
Alkaline phosphatase, 14
Alanine, 42, 44, 74, 113
Aldosterone, 86, 94
Alpha receptors, 133
Aluminum oxide column, 46
Alzet osmotic pump, 10
Amidation, 74, 97
Amine Precursor Uptake and Decarboxylation (APUD), 6
Amines, 19
 control of pars intermedia by biogenic, 128—131

 conversion of amino acids to, 6
Amino acids, 40, 69, 73—75, 110—111, see also individual amino acids
 ACTH N-terminal, 26
 aromatic, 180
 conversion of to amines, 6
 C-terminal amidation of α-MSH and, 31
 homology of, 75
 oxytocin, 11
 radioactive, 107
γ-Aminobutyric acid, see GABA
γ-Aminoisobutyric acid, 201
Aminopeptidase, 42, 195—196
Aminopeptidase-B, 96
Amphibians, see individual species
Anguilla anguilla, 189
Animal habitats, 7
Amygdala, 93
Anolis carolinensis, 19
 β-endorphin and, 33
 MSH and, 113
 α-MSH and, 43
 pars intermedia, 113
 serotonin and, 139
 skin darkening and, 130, 174—175, 177—178
Anolis spp., 7, 10—11, 46
Ansa lenticularis, 57
Antagonists
 α-adrenoreceptor, 47, 49, 133
 β-adrenoreceptor, 47, 134
 D-2 receptor, 93, 134
 dopaminergic, 134, 149
 $GABA_A$ receptor, 136, 149
Anterior lobe, pituitary
 ACTH and, 94
 β-lipotropin processing in, 96
 POMC genes and, 69, 93, 111
Antibodies
 catecholamine, 129
 dopamine, 130
 β-endorphin, 57
 α-MSH, 57, 59
 γ-MSH, 47
 salmon MSH, 47
 monoclonal, 17
Anticlockwise hysteresis, 203
Antipyrine, 201
Antisera, 2
 glutamic acid decarboxylase, 140
 GFAP, 17
 MCH, 152
 thyrotropin-releasing hormone, 142
Anurans, see individual species
APO, see Apomorphine
Apomorphine (APO), 131—133
APUD, see Amine Precursor Uptake and Decarboxylation

Arcuate nucleus, 56—58
Arginine, 1, 26—29, 31, 43—45, 49—50, 69, 73—74, 96, 172, 186, 195, 197
[^{3}H]-Arginine, 86—90, 94, 97
Arginine-vasopressin, 49, 95, 151
Arginine-vasotocin, 10, 49
Asparagine, 44, 73, 77, 110
Astrocytes, 197
ATPase, 17
Atrial natriuretic factor, 151—153
AtT 20 tumor cells, 29
Autoradiography
　corticotrophic peptides and, 192
　nerve fibers and, 130
　serotonin and, 138
Avian pancreatic peptide, 148
Axonal tract tracing, 56
Axon(s), 11
　ACTH-positive, 61
　α-MSH, 61
　POMC-containing, 60—61
　terminals, 134

B

Background adaptation
　coordinate vs. noncoordinate regulation of peptide release and, 118
　evolution of mechanisms surrounding, 119
　MSH acetylation and, 115
　α-MSH mediated, 97
　as neuroendocrine reflex, 104—105
　neurohypophysial peptides and, 152
　pigmentation change as, 1, 8—9, 40
　POMC
　　biosynthesis during, 107—109
　　-derived peptide secretion and, 106, 115—116
　　gene and, 78
　　mRNA levels and, 75—76, 86—90
　　thyrotropin-releasing hormone and, 142
Baclofen, 136—137
Bacterial, 70, 72, 79
Bicuculline, 136, 149
Bioassays, see also Immunoassays and Radioimmunoassays
　colorimetric, 140
　corticotropic peptide, 192
　fish-scale, 47, 50
　frog skin, 49
　lizard skin, 49
　MCH, 46
　melanocyte, 173, 184
　melanotropin, 43
　melanophore, 73
　MSH pharmokinetics and, 184
　nuclear transcription run-on, 109
Birds, 6
Blanching
　GABA and, 136
　MAO inhibitors and, 131
　pituitary removal and, 104
Blebs, 15

Blood, 139, 195
　-brain barrier, 197—201
　peripheral, 94
　proteins, 17
Blot hybridization, 70
Bony fishes, see individual species
Bovine, 40—41
　β-MSH, 42
　γ-MSH, 43
　MSH, 40, 41
　pituitary, 26, 28, 73
　POMC, 71, 74, 86
　serum albumin, 193
Brain, see also specific parts of brain
　APUD cells, 6
　homogenates, 178
　paste standards of known radioactivity, 90
　uptake index, 198, 200
　ventricle, third, 6
Bufo arenarum
　aminergic innervation and, 130
　dopamine and, 131
　hyperpigmentation and, 128
　intermediate lobe secretion and, 129
　pars intermedia, 129
Bufo spp., 7

C

Calcium, 10, 139
　channel blockers, 144
　intravesicular, 17
　ion, 144
Camels, 40—42, see also individual species
cAMP, 90—92, 98, 118
Capillaries, 7
Capping sites, 70
Carassius auratus, 33, 47
Carbohydrates, 29, 113
Carboxypeptidase(s), 42, 197
　B, 96
　neurons, 129
　perikarya, 129
　Y, 49
Cartilaginous fishes, see individual species
Catecholamine(s), 47, 49, 95, 104, 117, 129, 131, 133, 140, 146, see also specific substances
Catfish, 45—46, see also individual species
Cathepsin D, 42
Cats, see also individual species
　corticotrophs in, 10
　GABAergic neurons and, 135
　serotonin and, 138
Caudal zones of continuity, 8
CCAAT box, 71
CCK, see Cholecystokinin
Cell lines, see specific cell lines
Cells, see specific cell types
Central nervous system (CNS), 202—203
　cells, 6
　glial cells, 17
　heterogeneity of neural tissue, 106

melanotropin penetration into, 198
MSH in, 56—62
neuroendocrine reflex and, 105
tissue binding within, 201
Cerebrospinal fluid (CSF), 197—199, 201, 203—204
Cetaceans, see individual species
Cholecystokinin (CCK), 151
Cholinesterase, 140
Chromaffin cells, 96
Chromatin, 79
Chromatography
  corticotrophic peptides and, 192
  melanotropin metabolism and, 196
  MSH pharmokinetics and, 184
  neuropeptide Y and, 146
  POMC-derived peptide secretion and, 115
  Sephadex® gel, 146
Chromatophores, 104
Chromosome studies, 76
Chronic stress paradigms, 29
Chrysiptera cyanea, 47, 48
Chrysiptera hollisi, 47, 48
Chymotryptic mapping techniques, 109
Chymotrypsin, 172
α-Chymotrypsin, 176
Cleft
  hypophysial, 6—8, 15
  pars intermedia, 14
Cloning, 1
CM cellulose, 46
Codons, 73, 77
Co-injection studies, 79
Colchicine, 57
Collagen, 18
Colloid vesicles, 8
Colorimetric assay, 140
Concomitant release, 40, 115
Corticosteroids, 185
Corticotrophs
  ACTH-related peptides in, 30
  amidating enzymes and, 32
  anterior pituitary, 151
  N-acetyltransferase and, 32
  pars distalis, 1, 40
  POMC, 26, 28
  pars intermedia, 1, 10
Corticotropin(s), 1, 28, 40—41, 112, 184, 186—190, 193—198, 201, 204, see also ACTH and α-MSH
  -containing neurons, 150
  control of pars intermedia, 150—151
  family, 26
  -like intermediate lobe peptide (CLIP), 28, 31, 40—41, 73—74, 87, 109—111
  opiate, 69
  releasing factor, 150
CNS, see Central nervous system
Cryoultramicrotomy, 149
CSF, see Cerebrospinal fluid
Cushing's disease, 142
Cyclostomes, see individual species
CsTFA-EDTA gradient, 94

1,2-Cyclohexadion, 49
Cyprinus carpio, 47
Cysteine, 45, 49, 75

D

D-1 receptors, 91, 133
D-2 receptors, 91, 131, 133—134
Dantrolene, 144
DEAE cellulose, 46
Dendrites, 61
Depolarization, 17
Desmosomes, 13, 14
Diastereomers, 173
Dibasic cleavage sites, 111
Dibenamine, 49
Dideoxy chain termination method, 71
Diencephalon, 138
Digestive tract, 6
Distal lobe, pituitary, 130
Diurnal rhythms, 56
DNA
  adrenal synthesis of in rats, 43
  carrier, 73
    cloned intermediate lobe, 68
    clones, 68, 69, 73, 75—76, 110
    libraries, 70, 72—73
    MSH and, 40
    POMC and, 43, 60
  pBR332 vector, 72
  phage, 70
  probes, 60, 72
  recombinant techniques, 68, 76
    mRNA level quantification and, 108
    POMC gene and, 70, 72
  transcription, 79
Dogfish, see also individual species
  ACTH and, 42
  MSH and, 41
  α-MSH and, 30
  γ-MSH and, 43
Dogs, 191, see also individual species
Domperidone, 133, 145
Doormouse, 7, see also individual species
Dopamine, 11, 19, 59, 104, 108, 115, 119, 136, 153
  agonists, 69, 109, 131—132, 145
  antagonists, 149
  antibodies, 130
  control of pars intermedia, 128—133
  fibers, 11, 90, 130
  β-hydroxylase, 129
  neurointermediate lobe and, 145
  receptors, 90—92, 97—98
Dorsal infundibular nucleus, 130
Dumbbell-shaped organ, 7
Dot-blot analysis, 78
Dynorphin, 59, 151

E

Ectoderm, 6
Edman degradation, 49

Eels, 139, see also individual species
Elasmobranches, see individual species
Electron microprobe techniques, 10
Electron microscopy, 10, 17, 107, 115, 129, 131, 141
Electrophoresis, 46, 76, 107, 109
Electrothermic lesions, 131
Encephalo-neurohypophysial portal system, 7
End-feet, 12—14
Endocytosis, 15
Endopeptidase, 172, 186, 195
Endoplasmic reticulum, 8—12, 111, 139
Endorphin, 41, 87, 109, 111—113, 119
N-acetyl-α-Endorphin, 27
β-Endorphin, 1, 29—30, 56, 68, 71, 74—75, 86, 93, 97—98, 106, 110, 112—113, 116, 143, 150, 188
  acetylation, 96
  binding sites, 94
  immunoreactive, 118
  nonacetylated, 28, 33
  N-terminal acetylation, 32—33
N-acetyl-β-Endorphin, 27
N-acetyl-γ-Endorphin, 27
Endothelial cells, 197
Enkephalin, 59, 71, 75, 151, 198
Enzymatic-isotopic method, 140
Enzymes
  acetylation, 96—97, 109, 113—114
  amine conversion to amino acids by, 6
  granule-associated amidating, 32
  proteolytic proceesing, 29—32, 34, 73, 75, 112, 176—178, 181, 186
Ependymal cells, 15
Epithelium, 6
2-Br-α-(CB154)-Ergocryptine, 131
Estradiol, 17, 153
Estrogen, 17
Ethidium bromide, 94
Eukaryotes, 79
Evolution
  α-MSH amidation and, 32
  POMC gene, 69, 71—72, 78
  Xenopus laevis POMC mRNAs and proteins, 74—75
Exocytosis, 10, 114
Exon(s), 26, see also Gene(s)
  boundaries, 70
  -intron splice junctions, 70
  POMC gene, 69—72, 76
  post-transcriptional reshuffling of, 33
Exopeptidase, 42
Extracellular signalling, 86

F

Fast blue tract tracing, 61—62
Ferrets, 7
Fetal, 7
Fishes, see individual species
Fish-scale bioassay, 47, 50
Fluorescein, 58

Formaldehyde-induced fluorescence, 138
Fornix, 57
Follicles, 6—7, 14
Folliculo-stellate cells, 11—19
  extracellular channel system, 17—19
  pars intermedia, 11—14
Forskolin, 92
Freeze-etch techniques, 14
Frogs, see individual species
Frog-skin assay, 49

G

GABA, 11, 140, 146, 153
  control of pars intermedia, 134—138
  radiolabeled, 135—136
  transaminase, 137
$GABA_A$ receptors, 136—137
GABAergic neurons, 134—135, 146
GAD, see Glutamic acid decarboxylase
Ganglionectomy, 134
Gene(s), see also Pseudogenes and Transfer genes
  coding for enzymes and, 112
  duplication, 78
  eukaryotic, 70, 72
  expression in neuron subpopulations, 62
  families, 78
  glucocorticoid-controlled, 72
  homology between, 78
  melanotropic peptide, 68
  mutation, 78—79
  over-expression of, 79
  POMC, 26, 32—34, 98
    A, 76, 78, 93
    B, 76, 78, 93
    evolution of, 78
    human, 71
    organization, 26—29
    Xenopus laevis, 68—78
  recombination, 79
  regulation, 79
  transcription inhibition, 109
Gerbils, 7, see also individual species
GH_3 cells, 144
Glial, 17
Glucocorticoid(s)
  adrenal cortex synthesis of, 28, 30, 94
  -controlled genes, 72
  POMC gene transcription and, 69
$[^3H]$-Glucosamine, 73
Glutamic acid decarboxylase (GAD), 134—135, 137, 140
Glutamine, 1, 26, 42—44, 49, 141
Glycine, 1, 26, 29, 31—33, 43, 45, 73—74, 151
1-Glycine-corticotropin-1,18-amide, 185
Glycosylations, 77, 110, 112
Glyoxylate, 32
Goldberg/Hogness box, 71
Goldfish, 134, 135, see also individual species
Golgi apparatus, 8, 12, 111
  proteolytic processing of POMC in, 30

saccules, 10
secretory granules, 8
serotonin and, 139
Gonadotrophs, 15
Gonadotropin, 16, 151
Granular cells, 8—10
Guinea pig, 17, see also individual species

H

Haloperidol, 132—133, 149
Hamsters, 15, see also individual species
Hares, 10, see also Rabbits and individual species
Heat-alkali treatment, 178
Heptadecapeptides, 40, 46, see also specific compounds
Heptapeptides, 1, 45, see also specific compounds
Hexagrammos otakii, 48
High performance liquid chromatography (HPLC)
dopamine and, 130
MCH and, 47
MSH and, 47, 113
neuropeptide Y and, 146—147
POMC and, 107, 109
Histamine, 140
Histidine, 1, 26, 29, 32—33, 42—45, 69, 73—74, 141, 172, 195
Histidylhistidine, 185
Histofluorescence technique, 129
Holoestian fishes, see individual species
Homologous gene assay transcription system, 79
Homopolymeric tailing, 73
Horses, 40—41, see also individual species
Horseradish peroxidase (HRP), 14—15, 62
HPLC, see High performance liquid chromatography
HRP, see Horseradish peroxidase
Human
β-MSH, 42
POMC, 71—72, 74, 86—87
Hybridization
in situ, 56, 59, 90—91
probes, 70, 75
Hydrogen peroxide, 49
Hydrophilia, 75, 185
Hydrophobia, 75, 185
6-Hydroxydopamine (6-OHDA), 11, 129
5-Hydroxytryptophan, 139
Hyla regilla, 130
Hyperpigmentation, 128, 139
Hypertonic saline, 94
Hypertrophy, 17, 128
Hypophysial, 6—8, 15
Hypothalamus
catecholamines in, 129
corticotropin releasing factor and, 150
electrothermic lesions of, 131
inhibitory influence of, 104
MCH and, 40, 46—47, 50
MSH and, 105—106, 116
α-MSH and, 56—57, 61—62
pars intermedia regulation by, 115, 119, 128, 131
POMC and, 56, 60—61, 93
thyrotropin-releasing hormone and, 141
Hypothermia, 185
Hypsosyomus spp., 47
Hysteresis, 203

I

Immunoassays, 184, see also Bioassays and Radioimmunoassays
Immunocytochemical, 6, 10
Immunogold technique, 146—147
Immunohistochemistry, 58
Incerto-hypothalamic system, 129
Inhibitory paracrine factor, 17
Innervation, 10—11
Intermediate lobe, pituitary
cloned cDNA, 68
histamine and, 140
immunoreactive fibers in, 130
innervation of, 128
melanotrophs, 75, 107, 109, 112—116, 118—119
MSH and, 40
POMC, 69, 86—98
serotonin and, 138—139
somatostatin and, 146
Intermedin, 40, 104
Intra-adenohypophysial portal system, 8
Intraglandular cleft, see Hypophysial
Intron(s), 26, 70, 72, see also Gene(s)
Inulin, 199, 201—202
Ions, 10
Ca^{++}, 17, 144
Na^+, 94
Isoproterenol, 134—135
Isotocin, 49

J

Jirds, 7, see also individual species

K

Kidneys, 94
Killifish, 46, see also individual species
Klauberina riversiana, 7

L

Lampetra fluviatilis, 47
Lampreys, 7
Lead hematoxylin (PbH), 10
Lebistes reticulatus, 47
Leporidae, 129, see also individual species
Leucine, 42, 44, 151
Leu-enkephalin, 143, 200
Ligands, 91
Lipids, 16—17, 197
β-Lipotropin (β-LPH), 1, 27—30, 41, 68, 86, 96, 106, 110, 118, 191, 195—197
γ-Lipotropin (γ-LPH), 28—29, 87

Lipotropins, 26, 28
Lizards, see individual species
Lizard-skin assay, 49
β-LPH, see β-Lipotropin
γ-LPH, see γ-Lipotropin
LSD, see Lysergic acid diethylamide
Lungfish, 6
LY-171555, 92, 133
Lysergic acid diethylamide (LSD), 129
Lysyl endopeptidase, 47
Lysine, 27—29, 42, 45, 49, 73—74, 96, 187—188, 190, 196, 198
D-Lysine, 185—186, 188, 195, 197, 199, 201—202, 204
Lysosomes, 12, 15—16

M

Macacus spp., 40—42
Macrophages, 16
Maculae adhaerentes, see Desmosomes
Maculae occludentes, 14
Magnocellular paraventricular neurons, 61
Mammals, see individual species
Mammilothalamic tract, 57
Mammotropes, 16—17, 137, 152
Mannitol, 201
MAO, see Monoamine oxidase
Marginal cells, 15, 17
Mast cells, 11, 139—140
Median eminence, 7, 130
Melanin
  -concentrating hormone, 40, 45—46, 48—50, 131, 152—153
    active site, 49
    antisera, 152
    bioassay, 46
    isolation and structure of, 45—48
    melanophorotropic activity of, 48—49
    melanotrophic activity of, 47
    molecular weight of, 46
    phylogenetic distribution of, 47—48
    prohormone, 47
    site of origin, 46—47
    structure-activity relationships, 47—49
  dispersion, 1, 40, 68
  granules, 40, 50
Melanocyte(s)
  bioassay, 173, 184
  hair follicle, 181
  punctuate, 87
Melanogenesis, 181
Melanophore(s), 1, 47, 50
  dermal, 68, 104—105, 109
  index, 43, 86, 88, 104, 107, 131
  inhibiting factor (MIF), 11
  integumentary, 40
Melanophore stimulating hormone, see MSH entries
Melanosomes, 43, 45, 49, 104
Melanotrophs, 109, 111, 137, 152
  amphibian, 108

colloid vesicles and, 8
intermediate lobe, 75, 106—107, 109—110, 112—114, 119
multiple factors regarding secretion from, 116—117
pars intermedia and, 1, 9, 40, 68, 78—79, 119
secretion in, 10, 17
stereological analysis of, 11
Melanotropic
  cells, 26, 42
  peptides
    genes coding for, 68
    incubation of in purified proteolytic enzymes, 176—177
    relative stability of in brain homogenates, 178
    structure and chemistry, 40, 42—50
Melanotropin(s)
  absorption and bioavailability, 186—189
  active site, 42
  bioassay, 43
  core sequence, 26
  cyclic, 173
  distribution, 187—189
  dosages, 184—186
  enzymology, 172—181
  formulations, 186
  metabolism and clearance, 194—197
  multihormonal control of secretion, 128—153
  pars intermedia structure and function, 6—19
  pharmacodynamics, 203—204
  pharmacokinetics, 184—204
    central, 201—203
    peripheral, 186—187
    quantitation of concentrations, 184
    specialized barriers and secretions, 197—201
    tissue and plasma protein binding, 189—194
Meriones spp., 7
Mesencephalon, 7, 129, 138, 150
Mesocortical system, 129
Mesotocin, 10, 141, 151
Metencephalon, 150
Met-enkephalin, 74, 143, 200
Metalloprotease, 96
Methionine, 1, 26, 29, 41—45, 49, 71, 74, 185—186, 188, 190, 195, 199, 201—202, 204
4-Methionine sulfoxide, 195
Mice, see individual species
Mitochondria, 12—13
Modulators, see specific modulators
Monkeys, see individual species
Monoamine oxidase (MAO)
  α, 10
  β, 19
  inhibitors, 129
Monoamines, 200
Monosodium glutamate (MSG), 200
Morphine, 201
MSG, see Monosodium glutamate
MSH, 8, 10, 17, 19, 29—30, 40, 84, 71, 105—106
  active site, 43—45
  acetylation, 113—114, 119

Cys^4-substituted analogues of, 173—180
Cys^{10}-substituted analogues of, 175—180
dopamine as release inhibiting factor of, 129
half-life of iodinated, 172
Met^4-substituted analogues of, 179, 181
Nle^4-substituted analogues of, 172—173, 175—176
D-Phe^7-substituted analogues of, 173, 175—176, 179—181
racemization of amino acids in, 45
-related peptides, 56—57
secretion
  induction, 141—145, 151
  inhibition, 104, 109, 116, 133, 135—139,
stereisomers, 50
synthetic, 179
α-MSH, 1, 26, 28—33, 74—75, 79, 86—87, 90, 92, 95—98, 104, 106, 110—114, 118, 172—176, 179, 184, 192—193, 196, 200—203
  antibodies, 57, 59
  antisera, 47, 62
  axons, 61
  bioassay, 43, 73
  in CNS, 56—63
  dopamine and, 131—132
  hypothalamic regulation of, 116
  iodine labeled, 187
  isolation and structure of, 41—43
  -like peptide, 69
  melanosome dispersion and, 45
  monoacetylated, 27
  multiple factors regulating secretin of, 116—117
  neurons, 58, 62
  N-terminal acetylation of, 109, 112
  precursor of, 69
  radioimmunoassayed peptide distribution, 56
  release inhibiting factors, 115, 148—149
  structure-activity relationship of, 43—45
  synthetic, 189
  tritiated, 202
des-N-acetyl-α-MSH, 42, 45, 96—97, 109—114, 145, 190
N,O-diacetyl-α-MSH, 73, 112
β-MSH, 1, 26, 29—30, 40—42, 45, 49, 69—70, 74—76, 87—88, 97, 107, 111, 172—174, 176, 184, 191, 194—195, 197, 199, 201
γ-MSH, 1, 26, 28—31, 40, 43, 49, 70, 74—76, 88, 107—108, 111—113
  18K product, 118
  antibodies, 57
  bioassay, 43
γ$_3$-MSH, 27, 29—30
α-MSH1, 33, 42
α-MSH2, 32—33, 42
MSH2, 42
Muscimol, 136—137

N

Natriuresis, 185
Nelson's syndrome, 142
Nerve(s), 19, 129
Neural, 6
Neuroactive substances, 59, see also specific substances
Neuroectodermal markers, see specific markers
Neuroendocrine
  model systems, 76, 104—119
    biosynthesis and processing of amphibian POMC, 107—115
    secretion of POMC-derived peptides, 115—119
  transducer cells, 69
Neurohypophysis, 6, 7, 50
Neurointermediate lobe, pituitary
  atrial natriuretic factor and, 152
  coordinate vs. noncoordinate regulation of peptide release and, 118
  dopamine and, 131—132, 145, 149
  factors regulating secretion from, 117
  GABA control of, 136—137
  heat-alkali treatment, 178
  histamine and, 141
  MCH, 40
  MSH, 41
  neuropeptide Y and, 148—149
  noradrenaline and, 134
  serotonin and, 138—140
  thyroid, 141, 143
  Xenopus laevis, 73—74, 76
Neurons
  α-MSH, 56, 58, 62
  aminergic, 104
  arcuate nucleus, 56
  arcuate-periarcuate, 56
  catecholaminergic, 129
  corticotropin releasing factor containing, 150
  GABAergic, 134—146
  magnocellular paraventricular, 61
  multiple neurotransmitter, 56
  neuropeptide Y containing, 58, 146
  neurophysin containing, 150
  NLT/pars lateralis, 47
  noradrenergic, 133
  oxytocinergic, 151
  POMC, 56—57, 59, 61
  serotoninergic, 138
  somatostatinergic, 146
Neuron-specific enolase (NSE), 6
Neuropeptide-Y (NPY), 117, 147—148, 153
  control of pars intermedia, 148—150
  C-PON peptide of, 147—149
  neurons, 58, 146
Neurophysin, 150
Neuropil, 61
Neurotransmitters, 19, 86, see also specific neurotransmitters
  adrenergic, 48
  aminergic, 129
  cholinergic, 129
  control of pars intermedia, 140—141
  multiple, 56
  pars intermedia regulation by, 105

Nifedipine, 144
Nigro-striatial system, 129
o-Nitrophenylsulfenyl chloride, 49
Nongranular cells
   pars intermedia, 8, 11—12, 14, 15
   suggested roles for, 15—17
Noradrenaline, 130, 133—134
Northern
   blot analysis, 70, 73, 75, 78, 93—94
   gel analysis, 90, 95
Norvaline, 49
NPY, see Neuropeptide-Y
NSE, see Neuron-specific enolase
NTP, 27, 40
Nuclear transcription run-on assays, 78, 109
Nuclei, see specific nuclei
Nucleotide(s), 70, 72, 78
Nucleus accumbens, 57

O

n-Octanol, 198
Oldendorf technique, 200
Onchorynchus keta
   MCH and, 46
   MSH and, 41
   α-MSH and, 33, 42
   β-MSH and, 42
   POMC mRNA and, 28
Oocytes, 79
Opiate, see also specific opiates
   corticotropins, 69
   peptides, 1, 69, 74
   proteins, 75
   receptors, 28, 74
Optic microscopy, 129
Oral, 6
Osmoregulation, 86, 94
Ovine
   MSH, 40—41
   β-MSH, 42
   POMC gene, 72
Oxytocin, 10—11, 49, 151
immunoreactive fibers, 10—11

P

Pallor, see Blanching
PAM, see Peptidyl-glycine α-amidation monooxygenase
Parathyroid hormone (PTH), 118
Paraventricular
   nucleus
      corticotropin-releasing factor and, 150
      α-MSH and, 57
      thyrotropin-releasing factor and, 141
   organ
      catecholamines and, 130
      dopamine and, 134
Parenchyma, 8, 15, 18, 19
Parenchymal cells, 11, 19, 129—130

Pars distalis, 15, see also specific substructures
   corticotrophs, 1, 26, 28, 40
   extracts, 30
   ß-LPH and, 28
   nongranular cells, 16
   POMC, 26, 29—30, 77
Pars intermedia, see also specific substructures
   biogenic amine control of, 128—130
   corticotropin releasing factor control of, 150—151
   development, 6
   dopamine control of, 128—133
   evolution of, 15
   fibers
      Type I, 10
      Type II, 11
   function, 6—19
   GABA control of, 134—138
   granular cells, 8—11
   gross anatomy of, 6—19
   hypothalamic regulation of, 115, 119, 128, 131
   innervation of, 10—11
   β-LPH, 29
   melanotrophs, 1, 68, 78—79, 119
   morphology, 6—19
   MSH, 104, 115
   α-MSH, 33, 104
   as neuroendocrine transducer cell, 69
   neurohypophysial peptide control of, 151—152
   neuropeptide Y control of, 148—150
   non-granular cells, 11—19
   noradrenaline control of, 133—134
   N-terminal acetylation in, 32
   parenchymal cells, 129—130
   POMC gene and, 26, 28—30, 32, 105—106, 128
   prohormone synthesis in, 40
   pulse-chase analyses and, 30
   regulatory peptide control of, 141—152
   serotonin control of, 138—140
   size of correlated with animal habitat, 7
   structure, 6—19
   thyroid-releasing hormone control of, 141—146, 149
   vascular supply of, 7—8
   Xenopus laevis, 75—78
Pars lateralis, 47
Pars nervosa, 12, 117
Pars tuberalis, 7
PAS, see Periodic acid Schiff
PbH, see Lead hematoxylin
Peptidergic, 10, 129
Peptides, see also specific peptides
   13K ACTH-like, 73
   acetylated, 56
   α-amidated bioactive, 74
   chain elongation of, 45
   cleavage of, 172
   concentration time graphs, 186
   coordinate vs. noncoordinate release of, 117—119
   corticotropic, 192
   CSF, 203—204
   desacetyl, 56

lipotropin-related, 28
melanotropic, 1—2, 172—178
α-MSH like, 69
neurohypophysial, 151—152
opiate, 1, 69, 74
POMC-derived, 56, 76, 106—107, 152, 188
  acetylation of, 98
  co-localization of, 57, 59
  pools, 118
  radiolabeled, 107, 115
  secretion of, 115—119
  signal, 78
  regulatory, 141—152
Peptidyl-glycine α-amidation monooxygenase (PAM), 32, 34, 97
Perfusion technique, 136
Pericapillary space, 14—15, 18
Perikarya
  catecholaminergic, 129
  dopaminergic, 129
  dorsolateral hypothalamic, 63
  POMC-positive, 62
  serotoninergic, 129, 138
Periodic acid Schiff (PAS), 10
Perivascular space, 17, 18
Periventricular nucleus, 56
Phagocytosis, 15—16
Phentolamine, 47
Phenylalanine, 1, 26, 29, 42—45, 69, 73—74, 172, 180, 186, 188, 195, 197, 199, 201—202, 204
D-Phenylalanine, 173, 178—181
L-Phenylalanine, 173, 180
Phosphorylation, 93
Photoaffinity labelling, 193
Photometric reflection meter, 43
Phylogeny, 33, 47—48
Picrotoxin, 136—137
Pigmentation change, 1, 8—9
Pimephales promelas, 47
Pinocytosis, 15, see also Pinocytotic
Pinocytotic, see also Pinocytosis
  cells, 12
  vesicles, 14—15, 17
Piriform cortex, 57
Pituitary, see also specific structures
  anterior, 88, 137, 151
  implantation of into pale tadpoles, 1
  melanotropin biosynthesis, 26—34
  skin coloration control by, 1, 40
Pituicytes, 17, 19
Placenta, 197
Plaque hybridization, 70
Plasma
  heparinized, 192
  membrane, 13, 14, 17
  proteins, 194
Plexus intermedius, 7, 11, 14, 18
Poecilia spp., 46
Poly(A) tail, 70
Polyadenylation signals, 71—73, 77
POMC, see Pro-opiomelanocortin

Posterior hypothalamic nucleus, 57
Porcine
  MSH, 40—41
  pars intermedia cells, 136
  POMC, 74
Post-translational processing, 34
Preoptic recess organ-hypophysial tract, 129
Pro-ACTH/β-LPH, 68
Prodynorphin, 75
Prohormones, 40, 50, 56, 68—69, 73, 75, 78—79, 86, 97, 106—112
Proenkephalin-A, 75
Prokaryotes, 79
Prolactin, 46, 49, 118, 136—137, 141—143, 145, 152—153
Proline, 31, 42, 44—45, 49, 141, 151
Pro-opiocortin, see Pro-opiomelanocortin
Pro-opiomelanocortin (POMC), 8, 10—11, 40—41, 56, 58—59
  16K glycopeptide, 87, 96
  31K precursor, 87
  -A, 76, 93, 110
  absorption and bioavailability of, 188
  acetylation of, 115
  amino acid sequence of, 44, 74, 79, 86
  anterior pituitary, 111
  atrial natriuretic factor and, 152
  -B, 76, 93, 110
  biosynthesis and processing of, 106—115
    intermediate lobe pituitary, 86—99
    post-translational, 95—98, 105
    proteolytic, 28—31
    salt loading and, 86, 93—95
    transcriptional and translational level, 86—87
  cDNA, 43, 61
  co-localization of peptides and, 57
  -containing axons, 61—62
  -derived peptides
    acetylation of, 96—98
    amidation of, 97
    radiolabeled, 107
    secretion of, 115—119
  distribution of, 187
  enzymology, 95—96
  fiber distribution maps, 61
  genes, 32, 76, 78
    evolution of, 69, 78
    expression of, 34, 98
    organization of, 26—29
  -like precursor, 69
  mRNA, 26—28, 33, 59—61, 72—78, 87—88, 90, 93, 95, 97—98, 108
  neurons, 56—57, 59, 61
    pars distalis, 26, 76
    pars intermedia, 26, 104—119
    -positive dendrites, 61
    -positive perikarya, 6
  structure of, 68
Prohormones, 8, 10, see also specific hormones
Propargylglycine, 49
Prostoglandins, 144—145

Proteases, 172
Protein(s), 68
  brain, 46
  blood, 17
  evolution of *Xenopus laevis*, 74—75
  granular cells and, 8
  kinases, 93
  morphological characters of production, 8
  nuclear, 79
  opioid, 75
  phosphorylation, 93
  plasma, 189—193
  POMC, 72—73, 76
  precursor, 74
  secretory, 75
  tryptic mapping characteristics of, 106
Proteolytic
  cleavage, 26, 28, 75, 95—96, 105, 110—111
  processing enzymes, 31—32, 34, 73, 75, 178, 180—181
Pseudogenes, 78, 93, see also Gene(s)
Ptergoplichys spp., 47
PTH, see Parathyroid hormone
Pulse-chase analyses, 1, 30, 109, 111, 115
Purkinje cells, 135
Pyroantimonate, 10

R

Rabbits, see also Hares and individual species
  adenylate cyclase system and, 133
  GABAergic neurons and, 135
  pars intermedia, 17, 128
Racemization, 45
Radioimmunoassays, 109, see also Bioassays and Immunoassays
  MCH, 47
  MSH, 113, 184
  neuropeptide Y and, 146
  POMC-derived peptide secretion and, 116
  thyrotropin-releasing hormone and, 141
Rana berlandieri, 140
Rana catesbiana, 139, 141, 146
Rana esculenta, 130, 134, 151
Rana nigromaculata, 130
Rana pipiens
  acetylcholine and, 140
  adrenaline and, 134
  aminergic innervation and, 130
  neuropeptide Y and, 148
  noradrenaline and, 134
  thyrotropin-releasing hormone and, 141—142
Rana spp., 10, 17
Rana ridibunda
  ACTH and, 151
  ACTH(1-13)amide N-acetylation and, 33
  aminergic innervation and, 130
  corticotropin releasing factor and, 150
  dopamine and, 108, 117
  GABA and, 136
  histamine and, 140
  intermediate lobe, 108, 112, 115—117
  MSH acetylation and, 113—114
  α-MSH secretion inhibition and, 131
  N,O-diacetyl-α-MSH and, 75
  neurohypophysial peptides and, 151
  neuropeptide Y and, 146, 148
  pars distalis, 15
  serotonin and, 139
  thyrotropin-releasing hormone and, 141, 144—145
Rana temporaria
  adrenaline and, 134
  dopamine and, 134
  MCH and, 47
  perikarya, 129
Rathke's pouch, 6, 7, 11
Rat(s), see also individual species
  adrenal DNA synthesis in, 43
  amygdala, 93
  cerebral cortex, 93
  CNS, 60
  corticotrophs in, 10
  dopamine and, 109
  estradiol and, 17
  GABAergic neurons and, 135
  hypothalamus, 47, 50, 93
  mammotrophs, 137
  MCH, 47
  melanotrophs, 118
  γ_3-MSH, 29
  natriuresis and, 185
  pars distalis, 14, 30
  pars intermedia, 7, 10—11, 136
  POMC, 29, 71—72, 74, 76, 86
  serotonin and, 138—139
  sympathectomized, 134
Receptors, 11, 86
  α-, 133
  β-adrenergic, 134
  catecholamine, 117
  D-1, 91, 133
  D-2, 91, 131, 133—134
  dopamine, 90—92, 97—98
  $GABA_A$, 117, 137, 149
  $GABA_B$, 117
  genes coding for, 78
  opiate, 28, 74
  thyrotropin-releasing hormone, 142
Reptiles, see also individual species
  melanin granules in, 50
  α-MSH amidation and, 32
  pars intermedia in, 11
  POMC in, 30
Reserpine, 131
Restriction enzyme mapping, 70
Retina, 104
Rhodamine, 58
Rhomboid nucleus, 56—57
RNA
  brain, 70, 72—73
  messenger
    ACTH/LPH precursors and, 1

evolution of, 74—75
initiation site, 70
in situ hybridization histochemistry, 59—60
POMC, 26—29, 33, 60, 70, 72—78, 86—87, 90, 93—95, 97, 108
nuclear, 79
Rostral zones of continuity, 8

S

^{35}S, 90—91
S1 mapping, 78
Salamanders, 15, see also individual species
Saline, 185
Saliva, 197
Salmo gairdneri, 47
Salmon, see also individual species
  ACTH, 42
  hypothalamus, 47
  MCH, 40, 46—48, 50
  MSH, 41
  α-MSH, 30, 33
  β-MSH, 42
  γ-MSH, 43
  pituitary POMC systems, 31
  POMC, 31, 43, 68
  prolactin, 49
Salt-loading, 86, 93—96
Sauvagine, 150
SCH-23390, 92, 133
Scyliorhinus canicula, 41, 42
Sebastes schlegeli, 47
Sebum, 185
Second messengers, 108, see also specific substances
Secretagogues, 107, 117, 133, see also specific substances
Secretory
  cells, 8, 11, 15—16, 19
  granules, 8—10, 17, 30, 107, 111, 113—114
Sequencing, 1
Serine, 29, 32, 43—44, 72, 85, 96, 112, 186, 188
D-Serine, 190, 196
Serotonin, 11, 138—140
Serotoninergic
  fibers, 11
  neurons, 138
  perikarya, 129, 138
Shaking and stretching response, 203
Signal peptide sequence, 77
SKF-38393, 92, 133
Skin melanophore index, 86, 88
Sodium, 94, 185
Solid phase methods, 184
Solitary tract nucleus, 59
Solution hybridization, 78
Somatic cells, 79
Somatostatin, 49, 143—144, 146, 151, 153
Somatostatinergic neurons, 146
Spacer units, 75
Spinal cord, 129, 138
Spontaneous release, 115—116

Squalus acanthias, 33, 41, 42
Stereoisomers, 50
Stereological analysis, 8, 11
Steric hindrance, 31
Steroids, 107
Stomodeum, 6
Stretching and yawning syndrome, 185
Substantia nigra, 129
Sucrose, 201
Sulpiride, 133—134, 149
Sunlight, 7
Superagonists, 45
Superfusion techniques, 116, 137
Sympathomimetic drugs, see specific drugs
Synacthen Depot®, 185

T

Tadpoles, see also individual frog species
  melanophores, 49
  hypophysectomized, 40
  pale, 1
TATA box, 71, 77
TCA, see Trichloroacetic acid
Telencephalon
  catecholameinergic perikarya and, 129
  corticotropin releasing factor-containing neurons and, 150
  serotoninergic neurons and, 138
Teleosts, see Fishes and individual fish species
Termination codons, 71, 77
Tetradecamers, 73
Tetranitromethane, 49
Tetrapeptides, 42, 45, 69
Tetrapods, 6, see also individual species
Thalamus, 56—57
Thioamide, 142
Threonine, 29, 42, 44, 49
Thyrotropin-releasing hormone (TRH), 15, 117
  antisera, 142
  control of pars intermedia, 141—146, 149
  receptors, 142
Thyroid-stimulating hormone (TSH), 143
Thyroxin (T_4), 143
Tilapia spp., 152
Tissue-specific differential post-translational, 26
Toads, 8, 72, 74, see also individual species
Trans factor, 93
Transfer genes, 78, see also Gene(s)
Translational initiation, 71, 77
TRH, see Thyrotropin-releasing hormone
Trichloroacetic acid (TCA), 88—89
Tridecapeptides, 40
Tripeptides, 117
Tritium, 189
Triturus cristatus, 148
Trout, 152, see also individual species
Tryptic mapping techniques, 109
Tryptophan, 1, 26, 29, 42—45, 49, 69, 73—74, 139, 172, 176, 178, 197
TSH, see Thyroid-stimulating hormone

Tuberoinfundibular system, 129
Tumors
  ATt20 pituitary, 1, 29, 105, 112, 142
  pars distalis, 6
  pars intermedia, 6
Tyrode-phentolamine, 47
Tyrosine, 29, 43, 49—50, 73—74, 93, 112, 172, 199
  hydroxylase, 129—130
  neuropeptide, 146
  tritiated, 199

U

Ultracentrifugation studies, 30
Ultrastructural studies, 8, 106, 107
Urine, 194
Urotensin I, 150

V

Vacuoles, 10
Valine, 29, 31, 44—45, 197, 201
Vasoactive intestinal peptide, 151
Vasopressin, 151
Vectors, 72
Verapamil, 144

W

Wallabies, 15

X

Xenopus laevis
  ACTH(1-13)amide *N*-acetylation and, 33
  adrenalin and, 134
  background adaptation and, 104, 107, 109
  cholinergic innervation and, 140
  dopamine and, 109, 129, 133
  ß-endorphin and, 33
  evolution of, 70
  GABA and, 136
  intermediate lobe, 109, 115—117
  λ-genomic library, 72
  MCH and, 47
  melanotrophs, 114
  MSH and, 40, 113, 172
  desAc-N-α-MSH and, 98
  neurointermediate lobe, 74, 76, 117
  neuropeptide Y and, 146, 148
  noradrenaline and, 134
  oocytes, 80
  6-OHDA and, 11
  pars intermedia, 68, 75—78, 129
  POMC, 86, 106
    -A, 74
    -derived peptides, 118
    gene structural organization, 68—78
    mRNA and protein structures, 28, 72—75
  serotonin and, 139
  sulpiride and, 149
  thyrotropin-releasing hormone and, 144
Xenopus tropicalis, 76
Xiphophorus helleri, 47

Y

Yellow mice, 131
YM-09151-2, 92, 133

Z

Zinc phosphate, 195
Zona incerta, 57, 62